Undergraduate Lecture Notes in Physics

Undergraduate Lecture Notes in Physics (ULNP) publishes authoritative texts covering topics throughout pure and applied physics. Each title in the series is suitable as a basis for undergraduate instruction, typically containing practice problems, worked examples, chapter summaries, and suggestions for further reading.

ULNP titles must provide at least one of the following:

- An exceptionally clear and concise treatment of a standard undergraduate subject.
- A solid undergraduate-level introduction to a graduate, advanced, or non-standard subject.
- A novel perspective or an unusual approach to teaching a subject.

ULNP especially encourages new, original, and idiosyncratic approaches to physics teaching at the undergraduate level.

The purpose of ULNP is to provide intriguing, absorbing books that will continue to be the reader's preferred reference throughout their academic career.

Series editors

Neil Ashby
Professor Emeritus, University of Colorado, Boulder, CO, USA

William Brantley
Professor, Furman University, Greenville, SC, USA

Matthew Deady
Professor, Bard College Physics Program, Annandale-on-Hudson, NY, USA

Michael Fowler
Professor, University of Virginia, Charlottesville, VA, USA

Morten Hjorth-Jensen
Professor, University of Oslo, Oslo, Norway

Michael Inglis
Professor, SUNY Suffolk County Community College, Long Island, NY, USA

Heinz Klose
Professor Emeritus, Humboldt University Berlin, Germany

Helmy Sherif
Professor, University of Alberta, Edmonton, AB, Canada

More information about this series at http://www.springer.com/series/8917

Carlo Maria Becchi · Massimo D'Elia

Introduction to the Basic Concepts of Modern Physics

Special Relativity, Quantum and Statistical Physics

Third Edition

 Springer

Carlo Maria Becchi
Department of Physics
University of Genoa
Genova
Italy

Massimo D'Elia
Department of Physics
University of Pisa
Pisa
Italy

ISSN 2192-4791 ISSN 2192-4805 (electronic)
Undergraduate Lecture Notes in Physics
ISBN 978-3-319-20629-5 ISBN 978-3-319-20630-1 (eBook)
DOI 10.1007/978-3-319-20630-1

Library of Congress Control Number: 2015946993

Springer Cham Heidelberg New York Dordrecht London
1st + 2nd edition: © Springer-Verlag Italia 2007, 2010
3rd edition: © Springer International Publishing Switzerland 2016

Printed on acid-free paper

Springer International Publishing AG Switzerland is part of Springer Science+Business Media (www.springer.com)

Preface

During the last years of the nineteenth century, the development of new techniques and the refinement of measuring apparatuses provided an abundance of new data whose interpretation implied deep changes in the formulation of physical laws and in the development of new phenomenology.

Several experimental results lead to the birth of the *new physics*. A brief list of the most important experiments must contain those performed by H. Hertz about the photoelectric effect, the measurement of the distribution in frequency of the radiation emitted by an ideal oven (the so-called *black body* radiation), the measurement of specific heats at low temperatures, which showed violations of the Dulong–Petit law and contradicted the general applicability of the equipartition of energy. Furthermore, we have to mention the discovery of the electron by J.J. Thomson in 1897, A. Michelson and E. Morley's experiments in 1887, showing that the speed of light is independent of the reference frame, and the detection of line spectra in atomic radiation.

From a theoretical point of view, one of the main themes pushing for new physics was the failure in identifying the ether, i.e., the medium propagating electromagnetic waves, and the consequent Einstein–Lorentz interpretation of the Galilean *relativity principle*, which states the equivalence among all reference frames having a linear uniform motion with respect to fixed stars.

In the light of the electromagnetic interpretation of radiation, of the discovery of the electron and of Rutherford's studies about atomic structure, the anomaly in black body radiation and the particular line structure of atomic spectra lead to the formulation of *quantum theory*, to the birth of *atomic physics* and, strictly related to that, to the quantum formulation of the *statistical theory of matter*.

Modern Physics, which is the subject of these notes, is well distinct from *Classical Physics*, developed during the XIX century, and from *Contemporary Physics*, which was started during the Thirties (of XX century) and deals with the nature of *Fundamental Interactions* and with the physics of matter under extreme conditions. The aim of this introduction to Modern Physics is that of presenting a quantitative, even if necessarily also concise and schematic, account of the main features of *Special Relativity*, of *Quantum Physics* and of its application to the *Statistical Theory of Matter*. In usual textbooks these three subjects are presented together only at an introductory and descriptive level, while analytic presentations can be found in distinct volumes, also in view of examining quite complex technical aspects. This state of things can be problematic from the educational point of view.

Indeed, while the need for presenting the three topics together clearly follows from their strict interrelations (think for instance of the role played by special relativity in the hypothesis of de Broglie's waves or of that of statistical physics in the hypothesis of energy quantization), it is also clear that this unitary presentation must necessarily be supplied with enough analytic tools so as to allow a full understanding of the contents and of the consequences of the new theories.

On the other hand, since the present text is aimed to be introductory, the obvious constraints on its length and on its prerequisites must be properly taken into account: it is not possible to write an introductory encyclopedia. That imposes a selection of the topics which are most qualified from the point of view of the physical content/mathematical formalism ratio.

In the context of special relativity, after recalling the classical analysis of the ether hypothesis, we introduce Lorentz's transformations and their action on Minkowski space-time, discussing the main consequences of the new interpretation of space and time. Then we introduce the idea of covariant formulation of the laws of nature, considering in particular the new formulation of energy-momentum conservation. Finally, we discuss the covariant formulation of electrodynamics and its consequences on field transformation laws and Doppler effect.

Regarding Schrödinger quantum mechanics, after presenting with some care the origin of the wave equation and the nature of the wave function together with its main implications, like *Heisenberg's Uncertainty Principle*, we have emphasized its qualitative consequences on energy levels. The main analysis begins with one-dimensional problems, where we have examined the origin of discrete energy levels and of band spectra as well as the tunnel effect. Extensions to more than one dimension have been limited to very simple examples in which the Schrödinger equation is easily separable, like the case of central forces. Among the simplest separable cases we discuss the three-dimensional harmonic oscillator and the cubic well with completely reflecting walls, which are however among the most useful systems for their applications to statistical physics. In a further section we have discussed a general solution to the three-dimensional motion in a central potential based on the harmonic homogeneous polynomials in the Cartesian particle coordinates. This method, which simplifies the standard approach based on the analysis of the Schrödinger equation in spherical coordinates, is shown to be perfectly equivalent to the standard one. It is applied in particular to the study of bound states in spherical wells, of the hydrogen atom spectrum, of that of the isotropic harmonic oscillator and, finally, of elastic scattering.

Going to the last subject, which we have discussed, as usual, on the basis of Gibbs construction of the statistical ensemble and of the related distribution, we have chosen to consider those cases which are more meaningful from the point of view of quantum effects, like degenerate gases, focusing in particular on distribution laws and on the equation of state. In order to put into evidence the strict connection between the statistical results and thermodynamics, we have extended Gibbs construction to the classical gases, considering also the real gas case. We have then presented the statistical meaning of entropy and of the thermodynamic potentials, concluding the chapter with the discussion of the phase transition in the van der Waals real gas model.

In order to accomplish the aim of writing a text which is introductory and analytic at the same time, the inclusion of significant collections of problems associated with each chapter has been essential. We have possibly tried to avoid mixing problems with text complements; however, moving some relevant applications to the exercise section has the obvious advantage of streamlining the general presentation. Therefore in a few cases we have chosen to insert relatively long exercises, taking the risk of dissuading the average student from trying to give an answer before looking at the suggested solution scheme. On the other hand, we have tried to limit the number of those (however necessary) exercises involving a mere analysis of the order of magnitudes of the physical effects under consideration. The resulting picture, regarding problems, should consist of a sufficiently wide series of applications of the theory, being simple but technically nontrivial at the same time: we hope that the reader will feel that this result has been achieved.

Going to the chapter organization, the one about *Special Relativity* is divided into five sections, dealing respectively with Lorentz transformations, with the covariant form of Maxwell's equations, and with relativistic kinematics. The chapter on *Wave Mechanics* is made up of nine sections, going from an analysis of the photoelectric effect to the Schrödinger equation and from the potential barrier to the analysis of band spectra and to the Schrödinger equation in central potentials. Finally, the chapter on the *Statistical Theory of Matter* is divided into seven sections, going from Gibbs distribution, to the equation of state, to perfect quantum gases and to the classical real ones. The statistical results are interpreted in thermodynamical terms, introducing the thermodynamic potentials and giving a simple example of a phase transition.

Genova   Carlo Maria Becchi
Pisa   Massimo D'Elia
July 2015

Suggestion for Introductory Reading

- K. Krane: *Modern Physics*, 2nd edn (John Wiley, New York 1996)

Physical Constants

- Speed of light in vacuum: $c = 2.998 \times 10^8$ m/s
- Planck's constant: $h = 6.626 \times 10^{-34}$ J s $= 4.136 \times 10^{-15}$ eV s
- $\hbar \equiv h/2\pi = 1.055 \times 10^{-34}$ J s $= 6.582 \times 10^{-16}$ eV s
- Boltzmann's constant: $k = 1.381 \times 10^{-23}$ J/°K $= 8.617 \times 10^{-5}$ eV/°K
- Electron charge: $e = 1.602 \times 10^{-19}$ C
- Electron mass: $m_e = 9.109 \times 10^{-31}$ Kg $= 0.5110$ MeV/c^2
- Proton mass: $m_p = 1.673 \times 10^{-27}$ Kg $= 0.9383$ GeV/c^2
- Electric permittivity of free space: $\epsilon_0 = 8.854 \times 10^{-12}$ F/m
- Magnetic permeability of free space: $\mu_0 = 4\pi \times 10^{-7}$ N/A^2

Contents

Chapter 1
Introduction to Special Relativity

The relativity principle, first formulated by Galileo,[1] states that the laws of Nature are the same in all inertial frames. Those are identified with the class of reference frames in which any body not subject to external forces stays at rest or moves at constant speed. As we will discuss later, that implies that all inertial frames move at constant velocity with respect to each other. The actual realization and verification of the principle requires to state the transformation laws according to which an observer in one inertial frame reports the results of experiments performed in another frame: while this does not coincide with the principle itself, it is the unavoidable step to implement it in our mathematical description of the laws of Nature.

For a long time, the correct transformation laws were believed to be those named after Galileo himself. Suppose that two inertial frames, O and O', are such that they coordinate axes are parallel to each other and O' moves at constant velocity V, directed along the x direction, with respect to O. Then it is always possible to set clocks in both frames such that

$$x' = x - Vt; \quad y' = y; \quad z' = z; \quad t' = t. \tag{1.1}$$

Equation (1.1) contains various additional assumptions, beyond the relativity principle itself, the most relevant being the possibility of setting an absolute time flow which is equal for all observers. Galilean transformations predict that accelerations of moving bodies be the same in different inertial frames: that provides a consistent framework for Classical Mechanics.

Such a framework underwent a serious crisis in the 19th century, after J.C. Maxwell had completed the formulation of the laws of electromagnetism, usually known as Maxwell equations. There are various ways to realize that such equations are not compatible with Galileo's transformation laws.

---

[1]Remember, for instance, of the famous discussion about experiments in a moving ship, which is reported in his Dialogue.

© Springer International Publishing Switzerland 2016
C.M. Becchi and M. D'Elia, *Introduction to the Basic Concepts of Modern Physics*, Undergraduate Lecture Notes in Physics, DOI 10.1007/978-3-319-20630-1_1

Let us start by considering the force exchanged between two charged particles at rest with respect to each other. From the point of view of an observer at rest with the particles, the force is given by Coulomb law, which is repulsive if the charges have equal sign. An observer in a moving reference frame, instead, must also consider the magnetic field produced by each particle, which acts on moving electric charges according to the Lorentz force law. If the velocity of the particles is orthogonal to their relative distance, it can be easily checked that the Lorentz force is opposite to the Coulomb one and reduces the electrostatic force by a factor $(1 - v^2/c^2)$, where $c \equiv 1/\sqrt{\mu_0 \epsilon_0}$ and ϵ_0 and μ_0 are respectively the electric permittivity and the magnetic permeability in vacuum. The difference leads to different accelerations in the two reference frames, in contrast with Eq. (1.1).

The core in the discrepancy between Galilean transformations and Maxwell equations can be associated with the fact that the latter contain a new fundamental constant which has the dimensions of a velocity, $c \simeq 3 \times 10^8$ m/s. Indeed, it is well known that Maxwell equations describe the propagation of electromagnetic waves with speed c, according to the d'Alambert equation:

$$\frac{1}{c^2} \frac{\partial^2}{\partial t^2} u(\boldsymbol{x}, t) - \nabla^2 u(\boldsymbol{x}, t) \equiv \Box u(\boldsymbol{x}, t) = 0 \qquad (1.2)$$

where $u(\boldsymbol{x}, t)$ stands for any component of the electromagnetic fields. According to Galilean transformations velocities get added when going from one inertial reference frame to the other: the vector corresponding to the velocity of a luminous signal in one inertial reference frame O must be added to the velocity of O with respect to a new inertial frame O' to obtain the velocity of the luminous signal as measured in O'. Hence, for a generic value of the relative velocity, the speed of the signal in O' will be different, implying that, if Maxwell equations are valid in O, they are not valid in a generic inertial reference frame O'. Moreover, one can check directly[2] that, apart from the change in the speed of propagation, the form of Eq. (1.2) itself is not invariant under the transformations in (1.1).

Various possibilities were opened at that time to solve the inconsistency. The first one, claiming that Maxwell equations were wrong, was soon discarded, based on the overwhelming and accurate experimental evidence in favour of them. The second one, claiming that the relativity principle was wrong and that an absolute reference frame existed where Maxwell equations were true, seemed the most natural for a while.

Indeed, at the beginning of 19th century, Young's experiments on interference had proven that the wave theory of light was correct. Since then, people had been questioning about which medium could permit the propagation of light from distant

---

[2]From (1.1) one obtains

$$\partial/\partial x = \partial/\partial x' \; ; \quad \partial/\partial y = \partial/\partial y' \; ; \quad \partial/\partial z = \partial/\partial z'$$

$$\partial/\partial t = \partial/\partial t' - v_x \, \partial/\partial x' - v_y \, \partial/\partial y' - v_z \, \partial/\partial z' = \partial/\partial t' - \boldsymbol{v} \cdot \boldsymbol{\nabla}'$$

from which the new wave equation in the reference frame O' is obtained by combining derivatives.

stars to us. All wave phenomena known at that time had a corresponding oscillating substance, so that wave propagation in true vacuum was inconceivable. The most natural solution seemed that based on the assumption that electromagnetic waves correspond to deformations of an extremely rigid and rarefied medium, which was named *ether*. Therefore, such an absolute reference system, where Maxwell equations were true, was already provided: it was the one at rest with ether.

Since ether had no other apparent property apart from propagating light, it just remained, in order to validate its existence, to prove that one could find out one's own velocity with respect to ether by means of appropriate experiments. Due to the large value of c and to the fact that most effects related to the motion with respect to ether are quadratic in v/c, that was not completely trivial. Looking at the change in the force between two electric charges was certainly not a good idea, since in this case violations from Coulomb force are not easily detectable: for instance, in the case of two electrons accelerated through a potential gap equal to 10^4 V, one would need a precision of the order of $v^2/c^2 \simeq 4 \times 10^{-4}$ in order to reveal the effect, and force measurements can hardly reach such levels of accuracy.

It was much more convenient to try measuring the motion of Earth with respect to the ether, by studying effects related to variations in the speed of light. Taking into account that Earth rotates along its orbit with a velocity v_T such that $v_T/c \sim 10^{-4}$, an experiment able to reveal the possible change of velocity of the Earth with respect to the ether in two different periods of the year would require a precision of at least one part over ten thousand. We will show how A. Michelson and E. Morley were able to reach that precision by using interference: however they (and other experiments later) did not found any evidence of motion with respect to ether.

There was a third possibility, actually, which was waiting for A. Einstein, who first proposed it. He assumed that both the relativity principle and Maxwell equations were true, but that the transformation laws in (1.1) were wrong. H. Lorentz had already found, before Einstein, a new class of linear transformations which preserved Maxwell equations, without however gaining further physical insight. Einstein had both the genius and the courage to take the relativity principle as a founding principle for the laws of Nature, bringing againg beauty and simplicity in our description of them, with a number of new associated phenomena that has revolutionized our world since then.

1.1 From Ether Theory to the Postulates of Relativity

The properties of ether as the medium underlying light propagation have been a theoretical and experimental puzzle till the theory of Special Relativity has blown ether away. One of the main conceptual problems was related to the propagation of light in materials. At the beginning of the 19th century, it was already well known that the speed of light depends on the material and is always lower than in vacuum. This variation was attributed to an increase in the density of ether, caused by the presence of the material itself, with the density being proportional to n^2, where n is the refraction index of the medium.

If matter is able to influence the density of ether, it is natural to ask whether a moving piece of material will drag ether, hence causing a shift in the speed of light. The first successful experiment trying to answer this question was carried out in 1851 by Fizeau, who managed to measure the speed of light propagating in moving fluids. The experiment was based on interference of light and is described schematically in Problem 1.5. The outcome of Fizeau's experiment seemed to be in agreement with the hypothesis of partial drag proposed by Fresnel, according to which only the excess part of ether, caused by the presence of the medium, is dragged by it. Indeed, Fizeau measured a non-zero velocity shift for water ($n \simeq 1.3333$) and no drag effect at all for air ($n \simeq 1.0003$). The analysis reported in Problem 1.5 will show how such a result is actually in agreement with the relativistic law for the addition of velocities, rather than with Fresnel's hypothesis.

The fact that air has no dragging effect on ether justifies an experiment carried out on Earth (hence in the presence of air) and trying to measure the relative motion of Earth with respect to ether: that was the missing piece to prove the existence of the absolute reference frame. The experimental analysis was done by Michelson and Morley (starting in 1887), who made use of a two-arm interferometer similar to what reported in Fig. 1.1. The light source L generates a beam which is split into two parts by a half-silvered mirror S. The two beams travel up to the end of the arms 1 and 2 of the interferometer, where they are reflected back to S: there they recombine and interfere along the tract connecting to the observer in O. The observer detects the phase shift, which can be easily shown to be proportional to the difference ΔT between the times needed by the two beams to go along their paths. Indeed, the total phase Φ gained by each beam can be constructed by considering that a phase 2π is taken for each travelled wavelength, so that

Fig. 1.1 A sketch of Michelson-Morley interferometer

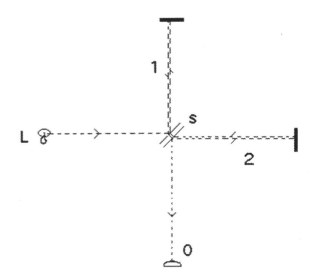

$$\Phi = 2\pi \sum_i \frac{L_i}{\lambda_i} = 2\pi \sum_i \nu \frac{L_i}{v_i} = 2\pi\nu \sum_i t_i = 2\pi\nu T \qquad (1.3)$$

where the sum is performed over the various pieces of the path, of length L_i, for which the phase velocity is constant and equal to v_i, and we have used the relation $v_i = \nu \lambda_i$ linking v_i to the wavelength λ_i, where ν is the light frequency, which is obviously the same along the whole path of both beams. The quantities t_i are the travelling times along the various pieces, whose sum gives the total time T. If the two arms have the same length l and light moves with the same velocity c along the two directions, then $\Delta T = 0$ and constructive interference is observed in O.

If however the interferometer is moving with respect to ether with a velocity v, which we assume for simplicity to be parallel to the second arm, then the path of the first beam will be seen from the reference frame of the ether as reported in the figure below and the time T needed to make the path will be given by Pythagoras' theorem:

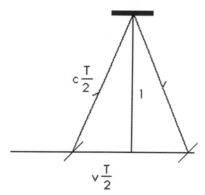

$$c^2 T^2 = v^2 T^2 + 4 l^2 \qquad (1.4)$$

from which we infer

$$T = \frac{2\,l/c}{\sqrt{1 - v^2/c^2}}. \qquad (1.5)$$

If we instead consider the second beam, we have a time t_1 needed to make half-path and a time t_2 to go back, which are given respectively by

$$t_1 = \frac{l}{c - v}, \qquad t_2 = \frac{l}{c + v} \qquad (1.6)$$

so that the total time needed by the second beam is

$$T' = t_1 + t_2 = \frac{2\,l/c}{1 - v^2/c^2} = \frac{T}{\sqrt{1 - v^2/c^2}}. \qquad (1.7)$$

We can finally write the phase difference $\Delta\Phi$, which for small values of v, $v/c \ll 1$, is

$$\Delta\Phi = 2\pi\nu(T - T') \simeq 2\pi\nu\frac{T}{2}\frac{v^2}{c^2} \simeq 2\pi\nu\frac{l\,v^2}{c^3} = 2\pi\frac{l}{\lambda}\frac{v^2}{c^2}. \tag{1.8}$$

This result shows that the experimental apparatus is in principle able to reveal the motion of the laboratory with respect to ether. Indeed, assuming that one is able to reveal phase differences as small as $2\pi/20$, we obtain, setting $\lambda \simeq 0.5 \times 10^{-6}$ m (a tipical wavelength for visible light) and $l \simeq 2$ m, that the experiment should be able to reveal values of v/c as small as 10^{-4}, which roughly corresponds to the orbital speed of Earth. Hence, if we compare the outcome of two such experiments separated by an interval of 6 months, we should be able to reveal the motion of Earth with respect to ether. The experiment, repeated by Michelson and Morley in several different times of the year, each time rotating the apparatus along different directions, always gave negative results, i.e. phase differences compatible with zero. At that time it was not obvious at all that the correct interpretation of Michelson-Morley experiment was that ether does not exist. Indeed, many physicists tried to find various possible patches to the ether theory, in order to make it compatible with the experimental evidence. The most notable is probably that due to H. Lorentz and G. FitzGerald, who proposed that objects moving with respect to ether undergo a contraction in the direction of motion (Lorentz-FitzGerald contraction), due to some modification in the electromagnetic interactions inside materials. In particular, one would need a contraction factor $\sqrt{1 - v^2/c^2}$ in order to cancel the difference in the travel times along the two orthogonal paths, as it is clear from Eq. (1.7). It is interesting to notice that such a contraction factor actually appears in the theory of Relativity (we will discuss it later in this Chapter) and is indeed usually named after Lorentz and FitzGerald, even if the physical interpretation of its origin is completely different.

All the different variations on the ether theory and the very need for an absolute reference system became useless once the Theory of Relativity, formulated by Albert Einstein in 1905, blew ether away. Einstein was guided towards the formulation of the theory more by symmetry principles (e.g., by the structure of Maxwell equations and by their symmetry under exchange of motion between the sources and the observer) than by experiments on ether. The theory is based on the two following postulates:

1. the laws of Physics are the same in all inertial reference frames;
2. the velocity of light in vacuum is the same in all inertial reference frames.

Actually, the second postulate could be included in the first, once one considers the speed of light to be among the fundamental laws of Physics. Based on such postulates, in the next Section we will derive the most general transformation connecting coordinates in two different inertial frames.

1.2 Lorentz Transformations and Their Main Consequences

We are looking for the most general coordinate transformation between inertial systems which is compatible with the Relativity principle. There are many constraints that we can impose on the transformation, and which are satisfied by Galilean transformations as well, before we introduce the invariance of the speed of light. Let us call, from now on, O and O' the two inertial systems under consideration. We suppose that in each of them an orthonormal set of coordinate axes has been defined, xyz for O and $x'y'z'$ for O', and that a set of synchronized clocks at rest with respect to each other has been prepared, separately for each system, so that both O and O' can assign to something happening somewhen and somewhere, i.e. to what we call an *event*, a unique set of space-time coordinates, (t, x, y, z) for O and (t', x', y', z') for O'. We want to find the relation between (t', x', y', z') and (t, x, y, z). The general requirements are the following: any motion at constant velocity in one system must be so also in the other system; in addition, we require homogeneity and isotropy of space to hold in both systems.[3]

A motion at constant velocity in O is a succession of events linked by the following linear relations:

$$x = x_0 + w_x\, t; \quad y = y_0 + w_y\, t; \quad z = z_0 + w_z\, t \qquad (1.9)$$

where w is the velocity. It is easy to realize that, if we want that such relations get transformed into a similar set of relations (i.e. into a uniform linear motion) for any choice of the constant parameters, the transformation, which is of course assumed to be continuous and invertible, must involve only linear or constant terms in the coordinates. If we set coordinates in both systems such that the event $x = y = z = t = 0$ coincides with the event $x' = y' = z' = t' = 0$ (this is always possible by a translation of the origins) then we are dealing with a linear transformation between two vector spaces of dimension four, which is specified by a 4×4 matrix Λ:

$$
\begin{aligned}
t' &= \Lambda_{tt}\, t + \Lambda_{tx}\, x + \Lambda_{ty}\, y + \Lambda_{tz}\, z \\
x' &= \Lambda_{xt}\, t + \Lambda_{xx}\, x + \Lambda_{xy}\, y + \Lambda_{xz}\, z \\
y' &= \Lambda_{yt}\, t + \Lambda_{yx}\, x + \Lambda_{yy}\, y + \Lambda_{yz}\, z \\
z' &= \Lambda_{zt}\, t + \Lambda_{zx}\, x + \Lambda_{zy}\, y + \Lambda_{zz}\, z.
\end{aligned}
\qquad (1.10)
$$

It is easy to prove that, given any point at rest in O, it will have a velocity with respect to O' which is uniquely obtained in terms of the coefficients of Λ, hence it is the same for all points. Therefore one can define the velocity of O with respect to O', $v_{O'O}$, as well as that of O' with respect to O, $v_{OO'}$.

---

[3]It is easy to realize that the lack of either isotropy of homogeneity would violate the relativity principle. Phenomenological observations, hence physical laws, would change by just rotating or translating the experimental apparatus.

From now on, in order to get rapidly to the main results, we consider a particular case in which the coordinate axes in O and O' are parallel to each other and $\boldsymbol{v}_{OO'}$ is directed along the x axis, i.e. we set $\boldsymbol{v}_{OO'} = (v, 0, 0)$. We call this transformation a *pure boost* (i.e. with no rotation) along the x axis. We are thus looking for a particular subset of transformations characterized by a single continuous parameter v; the generalization of this case will be discussed later on.

As a first constraint linked to this particular choice, we must neglect all terms mixing different spatial coordinates, otherwise the coordinate axes would not be parallel to each other:

$$\Lambda_{xy} = \Lambda_{xz} = \Lambda_{yx} = \Lambda_{yz} = \Lambda_{zx} = \Lambda_{zy} = 0.$$

Next we obtain a set of constraints based on the isotropy of space. Λ cannot depend on the particular orientation of the y, z axes, hence any combined rotation around the x and x' axes for both systems (i.e. leaving the y, y', z, z' still parallel to each other) must leave Λ unchanged. If we consider, in particular, a rotation by π, which changes the signs of y, y', z, and z' and leaves x, x', t and t' unchanged, we deduce that no terms can exist mixing the two different sets of coordinates, i.e.

$$\Lambda_{ty} = \Lambda_{tz} = \Lambda_{yt} = \Lambda_{zt} = 0.$$

We have already reduced the free parameters of Λ from 16 to 6:

$$
\begin{aligned}
t' &= \Lambda_{tt}(v)\, t + \Lambda_{tx}(v)\, x \\
x' &= \Lambda_{xt}(v)\, t + \Lambda_{xx}(v)\, x \\
y' &= \Lambda_{yy}(v)\, y \\
z' &= \Lambda_{zz}(v)\, z
\end{aligned}
\tag{1.11}
$$

where we have explicitly indicated the dependence of the coefficients on v, the relative speed of O' with respect to O. Isotropy can be used again by considering the inversion of both the x and the x' axis: if at the same time we invert the sign of v, Λ must remain unchanged. From that we deduce that $\Lambda_{tt}(v)$, $\Lambda_{xx}(v)$, $\Lambda_{yy}(v)$ and $\Lambda_{zz}(v)$ are even functions of v, while $\Lambda_{tx}(v)$ and $\Lambda_{xt}(v)$ are odd. A general requirement that we do is that Λ be a continuous function of v and that it tends to the identity transformation for $v \to 0$, so that $\Lambda_{tt}(0) = \Lambda_{xx}(0) = \Lambda_{yy}(0) = \Lambda_{zz}(0) = 1$ and $\Lambda_{tx}(0) = \Lambda_{xt}(0) = 0$.

Let us consider now the relative velocities of O and O' and their relation to the coefficients. The origin of frame O', represented by $x' = y' = z' = 0$, moves in O according to $y = z = 0$; $x = vt$, so that

$$\Lambda_{xt}(v) = -v\Lambda_{xx}(v).\tag{1.12}$$

In a similar way, we can consider the motion of the origin of O, $x = y = z = 0$, as seen from O', where it moves with velocity $\boldsymbol{v}_{O'O}$. What do we know about $\boldsymbol{v}_{O'O}$? Using the same argument of invariance under combined rotations around the x, x' axes one sees that $\boldsymbol{v}_{O'O}$ cannot point to any direction but x' itself, hence we can write

$v_{O'O} = (w, 0, 0)$ and the world line of the origin of O in O' is given by $y' = z' = 0$ and $x' = wt'$, so that

$$\Lambda_{xt}(v) = w\Lambda_{tt}(v). \tag{1.13}$$

We would be tempted to state immediately that $w = -v$, however such a conclusion is not so obvious and would be wrong in a non-isotropic Universe. So, let us follow the precise chain of deductions leading to it. First of all, w must be a well defined and universal function of v, $w = f(v)$, which according to the relativity principle must be independent of the particular starting reference frame O. We can state the following further properties of $f(v)$:

1. if O' applies f to w, the result must be v, i.e. $f(f(v)) = v$, f coincides with its inverse;
2. from Eqs. (1.12) and (1.13) and from the even/odd properties of the transformation coefficients, we deduce $f(-v) = -f(v)$. This can be linked directly to space isotropy, since the even/odd properties of the coefficients are a consequence of that;
3. finally, if we ask that for $v \to 0$ the transformation goes to identity, then $f(0) = 0$.

It is easy to prove that the only two continuous functions satisfying the properties above are $f(v) = \pm v$. On the other hand, if we set $w = f(v) = v$ we obtain, from (1.12) and (1.13), that $\Lambda_{tt}(v) = -\Lambda_{xx}(v)$, which is not compatible with Λ going to the identity for $v \to 0$. Hence the only possibility is the (seemingly trivial) $w = -v$, so that

$$\Lambda_{tt}(v) = \Lambda_{xx}(v). \tag{1.14}$$

Let us now consider the inverse transformation Λ^{-1} which brings back from O' to O: it must be of the same form as Λ itself in Eq. (1.11) and is obtained by substitution of $v \to w = -v$, i.e. $(\Lambda(v))^{-1} = \Lambda(-v)$. If we impose this condition, together with the known even/odd properties of the coefficients, we obtain

$$
\begin{aligned}
t &= \Lambda_{tt}(-v)\, t' + \Lambda_{tx}(-v)\, x' = \Lambda_{tt}(v)\, t' - \Lambda_{tx}(v)\, x' \\
x &= \Lambda_{xt}(-v)\, t' + \Lambda_{xx}(-v)\, x' = -\Lambda_{xt}(v)\, t' + \Lambda_{xx}(v)\, x' \\
y &= \Lambda_{yy}(-v)\, y' = \Lambda_{yy}(v)\, y' \\
z &= \Lambda_{zz}(-v)\, z' = \Lambda_{zz}(v)\, z'
\end{aligned}
\tag{1.15}
$$

which compared to (1.11), (1.12) and (1.14) leads to $\Lambda_{yy}(v) = \Lambda_{zz}(v) = 1$ and, if we set $\gamma(v) \equiv \Lambda_{xx}(v)$, to

$$\Lambda_{tx} = \frac{1 - \gamma^2}{\gamma v}.$$

We are thus left with one only unknown function $\gamma(v)$ defining the most general pure boost transformation between inertial systems:

$$
\begin{pmatrix} t' \\ x' \\ y' \\ z' \end{pmatrix} = \begin{pmatrix} \gamma & -\frac{\gamma^2-1}{\gamma v} & 0 & 0 \\ -\gamma v & \gamma & 0 & 0 \\ 0 & 0 & 1 & 0 \\ 0 & 0 & 0 & 1 \end{pmatrix} \begin{pmatrix} t \\ x \\ y \\ z \end{pmatrix} \tag{1.16}
$$

where we have expressed space-time event coordinates as column vectors and the notation stands for the standard row-by-column product. Notice that up to now the invariance of the speed of light has not been used anywhere, therefore Galilean transformations must be of the same form as in (1.16), which is indeed the case by setting $\gamma = 1$.

We can impose the invariance of c by considering a particular light beam trajectory in O, e.g., $x = ct$ and $y = z = 0$, and by requiring that (1.16) transforms it into a trajectory with the same speed, i.e. $x' = ct'$ and $y' = z' = 0$. After some elementary algebra we obtain $1/\gamma^2 = 1 - v^2/c^2$ which, requiring also $\lim_{v \to 0} \gamma(v) = 1$, has the unique solution

$$
\gamma = \frac{1}{\sqrt{1 - v^2/c^2}}. \tag{1.17}
$$

We can rewrite the transformation in a simpler form by adopting ct in place of t, which permits us to deal with a homogeneous set of coordinates, and by defining $\beta \equiv v/c$:

$$
\begin{pmatrix} ct' \\ x' \\ y' \\ z' \end{pmatrix} = \begin{pmatrix} \gamma & -\gamma\beta & 0 & 0 \\ -\gamma\beta & \gamma & 0 & 0 \\ 0 & 0 & 1 & 0 \\ 0 & 0 & 0 & 1 \end{pmatrix} \begin{pmatrix} ct \\ x \\ y \\ z \end{pmatrix}. \tag{1.18}
$$

Equation (1.18) represents the Lorentz transformation for a pure boost along the x axis, which goes back to (1.1) when $c \to \infty$. We can thus look at Galilean transformations as the correct realization of the Relativity principle as long as we do not know about the existence of a universal, invariant velocity of Nature (i.e. if $c = \infty$); it is electromagnetism which has made us aware of such a velocity for the first time.

We notice that the transformation (1.18) becomes ill-defined or complex for $|\beta| \geq 1$, thus suggesting that relative velocities cannot exceed c. However evidence is not compelling and the question about c as a limiting velocity of Nature requires a more careful discussion.

1.2.1 Transformation Laws for Velocities

One of the main consequences of Lorentz transformations is a different addition law for velocities, which is expected from the invariance of the speed of light. Let us consider a particle which, as seen from reference frame O, is in (x, y, z) at time t and in $(x + \Delta x, y + \Delta y, z + \Delta z)$ at time $t + \Delta t$, thus moving with an average velocity $(V_x = \Delta x/\Delta t, V_y = \Delta y/\Delta t, V_z = \Delta z/\Delta t)$. Since the coordinate transformation is linear, it applies to coordinate differences as well, therefore in reference frame O' we have, using (1.18), $\Delta y' = \Delta y$, $\Delta z' = \Delta z$ and

$$\Delta x' = \gamma(\Delta x - v\Delta t) \ , \qquad \Delta t' = \gamma\left(\Delta t - \frac{v}{c^2}\Delta x\right), \tag{1.19}$$

from which we obtain

$$V'_x \equiv \frac{\Delta x'}{\Delta t'} = \frac{\Delta x - v\Delta t}{\Delta t - \frac{v}{c^2}\Delta x} = \frac{V_x - v}{1 - \frac{vV_x}{c^2}} \ , \qquad V'_{y/z} = \frac{1}{\gamma}\frac{V_{y/z}}{1 - \frac{vV_x}{c^2}} \tag{1.20}$$

instead of $V'_x = V_x - v$ and $V'_{y/z} = V_{y/z}$, as predicted by Galilean laws. It requires only some simple algebra to prove that, according to (1.20), if $|V| = c$ then also $|V'| = c$, as it should, in agreement with the invariance of the speed of light (see Problem 1.7 for more details).

1.2.2 Invariant Quantities and Space-Time Geometry

The parameters γ and $\gamma\beta$ appearing in (1.18) are easily verified to satisfy the relation $\gamma^2 - (\gamma\beta)^2 = 1$. That means that it is always possible to rewrite them in terms of hyperbolic functions of a new parameter χ, with $\chi \in [-\infty, \infty]$, as follows

$$\gamma = \cosh \chi \ ; \quad \gamma\beta = \sinh \chi. \tag{1.21}$$

The parameter χ is usually called rapidity and can be used in place of $\beta = \tanh \chi$ to specify the motion of O' with respect to O. Expressed in terms of χ, Eq. (1.18) takes the form

$$\begin{pmatrix} ct' \\ x' \\ y' \\ z' \end{pmatrix} = \begin{pmatrix} \cosh \chi & -\sinh \chi & 0 & 0 \\ -\sinh \chi & \cosh \chi & 0 & 0 \\ 0 & 0 & 1 & 0 \\ 0 & 0 & 0 & 1 \end{pmatrix} \begin{pmatrix} ct \\ x \\ y \\ z \end{pmatrix} \tag{1.22}$$

which closely resembles a rotation in the x–t plane, with trigonometric functions replaced by hyperbolic ones. The analogy is even closer when we consider the combination of two Lorentz boosts along the same axis. If $\Lambda(\beta_1)$ is the transformation connecting O to O' and $\Lambda(\beta_2)$ that connecting O' to O'', it can be easily verified that the transformation connecting O to O'' is $\Lambda(\beta) = \Lambda(\beta_1)\Lambda(\beta_2)$ with $\beta = (\beta_1 + \beta_2)/(\beta_1\beta_2)$, which in terms of rapidity means $\chi = \tanh^{-1}\beta = \chi_1 + \chi_2$. Therefore, as for usual rotations, two boosts along the same axis combine in such a way that rapidities are added linearly, as for usual angles.

However, while rotations keep invariant the square modulus of a vector, it is easy to verify, using (1.22), that

$$(ct')^2 - x'^2 - y'^2 - z'^2 = (ct)^2 - x^2 - y^2 - z^2. \tag{1.23}$$

Moreover, since any two space-time events (ct_1, x_1, y_1, z_1) and (ct_2, x_2, y_2, z_2) transform according to the same linear transformation, one also has

$$c^2 t'_1 t'_2 - x'_1 x'_2 - y'_1 y'_2 - z'_1 z'_2 = c^2 t_1 t_2 - x_1 x_2 - y_1 y_2 - z_1 z_2. \tag{1.24}$$

To better understand the meaning of these invariant quantities, let us introduce a new class of objects, called *four-vectors*: they are four-component column vectors and are characterized by their linear transformation properties when going from one inertial frame to the other, which for a Lorentz boost is given by (1.22). Space-time events, constructed by adding ct to the standard spatial coordinates (x, y, x), are four-vectors by definition. We can define, in the vector space of four-vectors, a scalar product, which is assigned in terms of a symmetric matrix g. Let x and y be two four-vectors, then

$$x \cdot y \equiv x^T g y \equiv \begin{pmatrix} x_0 & x_1 & x_2 & x_3 \end{pmatrix} \begin{pmatrix} 1 & 0 & 0 & 0 \\ 0 & -1 & 0 & 0 \\ 0 & 0 & -1 & 0 \\ 0 & 0 & 0 & -1 \end{pmatrix} \begin{pmatrix} y_0 \\ y_1 \\ y_2 \\ y_3 \end{pmatrix} \tag{1.25}$$

where by x^T we mean the row vector obtained from x by standard matrix transposition. The 4×4 symmetric matrix g is known as the *metric tensor*, it defines a scalar product which differs from the canonical product in $\mathbb{R}^4$: indeed, it has 1 positive eigenvalue and three negative ones (this is often summarized saying that g has a signature $(1, 3)$) and coincides with the standard scalar product only on the spatial coordinates (apart from a global minus sign). What we have learned is that the scalar product defined by g is left invariant by the Lorentz boosts, Eq. (1.22). As a consequence, also the quantity $x^T g x = x_0^2 - x_1^2 - x_2^2 - x_3^2$, which is known as the *invariant squared length* of x, is conserved by Lorentz boosts. We have now to address two important questions: (i) is the invariance true only for the particular Lorentz boost in (1.22), or does it hold for the most general transformation linking different inertial frames? (ii) where does the invariance come from?

In order to answer the first question, we have to understand what is the form of the most general transformation. We have considered the particular case in which the axes of O and O' are parallel to each other, and the relative velocity is directed along the x axes. The generalization is straightforward: the relative velocity can be directed along any direction, the axes of the two systems can be rotated arbitrarily with respect to each other. Such a generalization can be implemented by performing an arbitrary spatial rotation before doing the boost, and an arbitrary spatial rotation after that; the corresponding transformation matrix is given by the row-by-column multiplication of the three matrices:

$$\Lambda = \begin{pmatrix} 1 & 0 & 0 & 0 \\ 0 & & & \\ 0 & & \mathcal{R}' & \\ 0 & & & \end{pmatrix} \begin{pmatrix} \gamma & -\gamma\beta & 0 & 0 \\ -\gamma\beta & \gamma & 0 & 0 \\ 0 & 0 & 1 & 0 \\ 0 & 0 & 0 & 1 \end{pmatrix} \begin{pmatrix} 1 & 0 & 0 & 0 \\ 0 & & & \\ 0 & & \mathcal{R} & \\ 0 & & & \end{pmatrix} \tag{1.26}$$

where $\mathcal{R}$ and $\mathcal{R}'$ stand for two orthogonal 3×3 matrices representing spatial rotations.[4] Since both $\mathcal{R}$ and $\mathcal{R}'$ leave the standard scalar product on spatial coordinates invariant, it is clear that their introduction does not alter the invariance of the scalar product defined in (1.25), which is then conserved for each of the three steps making up the total transformation in (1.26). Hence we can state that the invariance of this product is a general property of Lorentz transformations.

Let us now question about the origin of the invariance. The requirement about the invariance of c imposes that, whenever $c^2t^2 - x^2 - y^2 - z^2 = 0$, i.e. whenever the event (ct, x, y, z) is on the world line of a light beam emitted from the origin at time zero, then also $c^2t'^2 - x'^2 - y'^2 - z'^2 = 0$, and viceversa, i.e.

$$c^2t^2 - x^2 - y^2 - z^2 = 0 \quad \Longleftrightarrow \quad c^2t'^2 - x'^2 - y'^2 - z'^2 = 0. \tag{1.27}$$

We have found, instead, that the two quantities are equal, whatever value they take; we are going to see why. The quantity $c^2t'^2 - x'^2 - y'^2 - z'^2$, because of the linear relation between the coordinates of O and O', is also a quadratic form in the original coordinates; what we know about this quadratic form is that it must vanish on the 4-dimensional cone defined by the first equation in (1.27): that constrains the quadratic form so strongly[5] that the only possibility is that it is proportional to $c^2t^2 - x^2 - y^2 - z^2$. Therefore, we can write

$$c^2t'^2 - x'^2 - y'^2 - z'^2 = \lambda(v)(c^2t^2 - x^2 - y^2 - z^2), \tag{1.28}$$

where the proportionality coefficient λ is in principle a function of the velocity v of O' with respect to O, which is strongly constrained by isotropy. Indeed, when

---

[4]Based on Eq. (1.26), we can make a statement about the total number of independent parameters characterizing a general Lorentz transformation. Each rotation matrix brings 3 parameters, hence we have 6 new parameters on the whole, however one of them is redundant, since a rotation $\mathcal{R}$ around the x axis can be cancelled by an opposite rotation $\mathcal{R}'$ around the same axis. Therefore, considering the single parameter v characterizing the boost, we have a total of 6 independent parameters.

[5]If the form vanishes in (ct, x, y, z), it must vanish also in $(-ct, x, y, z)$, that forbids the presence of the mixed terms xt, yt and zt. Moreover, for each fixed t, the form must vanish on a spatial spherical surface, that forbids the presence of the mixed terms xy, xz and yz. Finally, the opening angle of the cone fixes the ratio of the diagonal terms.

we consider the inverse transformation, taking into account that it corresponds to a Lorentz transformation with velocity $-\boldsymbol{v}$, we obtain $\lambda(-\boldsymbol{v}) = 1/\lambda(\boldsymbol{v})$. On the other hand isotropy imposes that λ can depend only on $v = |\boldsymbol{v}|$. Hence we have $\lambda(v) = 1/\lambda(v)$ which implies $\lambda = \pm 1$, with the negative solution to be discarded because of the required continuity as $v \to 0$. Therefore we obtain Eq. (1.23), which we now understand to be a consequence of the invariance of c combined with linearity and isotropy.

The vector space of events, provided with the scalar product in (1.25), constitutes what is usually called *Minkowski spacetime*, after H. Minkowski who first studied its geometric structure. The condition of invariance of the scalar product can be taken as the very definition of Lorentz transformations, in the same way as spatial rotations can be associated with orthogonal matrices. Let x and y be any two four-vectors, and let x' and y' be the corresponding four-vectors in O', then we have

$$x'^T g \, y' = (\Lambda x)^T g \, (\Lambda y) = x^T \Lambda^T g \, \Lambda y = x^T g \, y \qquad (1.29)$$

where we have used the property $(\Lambda x)^T = x^T \Lambda^T$ for the transpose of a matrix-by-vector product. Equation (1.29) holds for any pair x and y if and only if Λ satisfies

$$\Lambda^T g \, \Lambda = g. \qquad (1.30)$$

Equation (1.30) defines a group of matrices[6] which is known as the Lorentz group and coincides, apart from a few elements representing discrete transformations, with the set of general transformations represented in (1.26).

Indeed, taking the determinant of both members in (1.30), one sees that all elements of the group satisfy $\det \Lambda = \pm 1$, while those represented in (1.26), if $\mathcal{R}$ and $\mathcal{R}'$ are pure rotations, have $\det \Lambda = 1$: this is also expected, since a matrix of the form (1.26) can be obtained by continuously deforming the identity transformation, hence they must have unitary determinant. On the other hand, when we add two further discrete transformations, which are not represented by Eq. (1.26) and which have $\det \Lambda = -1$, we are able to reproduce the whole Lorentz group, they are

$$T = \begin{pmatrix} -1 & 0 & 0 & 0 \\ 0 & 1 & 0 & 0 \\ 0 & 0 & 1 & 0 \\ 0 & 0 & 0 & 1 \end{pmatrix}; \quad P = \begin{pmatrix} 1 & 0 & 0 & 0 \\ 0 & -1 & 0 & 0 \\ 0 & 0 & -1 & 0 \\ 0 & 0 & 0 & -1 \end{pmatrix} \qquad (1.31)$$

which are known respectively, and for obvious reasons, as time reversal and parity transformation.

A property common to all Lorentz transformations is that they leave the four-dimensional integration measure invariant. Indeed, since the integration measure

---

[6]If Λ_1 and Λ_2 satisfy Eq. (1.30), this is true also for $\Lambda_1 \Lambda_2$. The identity matrix clearly satisfies the relation. Moreover from the condition (1.30) we deduce that $(\det \Lambda)^2 = 1$, so that Λ is surely invertible.

transforms under a change of variables according to the Jacobian determinant of the transformation, we have $d^3x' \, dt' = |\det \Lambda| \, d^3x \, dt = d^3x \, dt$.

In a vector space provided with a metric it is possible to assign the squared distance between every pair of points. However, the Minkowski metric is not positive definite, so that various non-trivial possibilities exist for the associated distance, which we are going to classify. Let us consider two different events in space-time, (ct_1, x_1) and (ct_2, x_2): their difference $(c\Delta t, \Delta x)$ is also a four-vector, whose invariant squared length defines the distance between the two events and can take positive, negative, or null values:

1. If $\Delta x^2 \equiv c^2 \Delta t^2 - |\Delta x|^2 > 0$ we say that the two events are at *time-like* distance. In this case the two events can be thought as lying on the world line of an object moving slower than light, $|\Delta x|/|\Delta t| < c$. The distance will stay positive for all inertial reference frames; moreover, if $c\Delta t = c(t_2 - t_1) > 0$ in one frame, meaning that event 2 happens after event 1, the same time ordering will hold for all other inertial frames. This is clear from the fact that otherwise, by continuity, one could also find a frame where $t_2 = t_1$, but there one would have $\Delta x^2 \leq 0$, thus violating length invariance. It is therefore possible to establish an absolute time ordering for such pairs of events.

2. If $\Delta x^2 = 0$ we say that the two events are at *light-like* distance. In this case the two events can be thought as lying on the world line of an object moving exactly at the speed of light. Also in this case time ordering is well defined, since Δt can vanish in some frame only if Δx vanishes as well, meaning that the events are actually the same event.

3. If $\Delta x^2 < 0$ we say that the two events are at *space-like* distance. In this case the two events can be thought as lying on the world line of an object moving faster than light, $|\Delta x|/|\Delta t| > c$, and an absolute time ordering is not possible. To prove that, suppose we have $\Delta x_0 = c(t_2 - t_1) > 0$ in O, then we need to find at least one frame O' where $\Delta x_0' < 0$. We can consider for simplicity a pure boost along Δx, then we have

$$\Delta x_0' = \gamma \left(\Delta x_0 - \beta |\Delta x| \right) \tag{1.32}$$

so that $\Delta x_0' \leq 0$ if $\beta \geq \Delta x_0/|\Delta x|$, which is possible since $\Delta x_0/|\Delta x| < 1$. Simultaneous events ($\Delta x_0 = 0$) represent a particular case of space-like distance. However we learn that simultaneity is not an absolute concept any more: two events which are simultaneous in one frame, can be time-ordered in either ways in other frames.

Another way to look at previous classification is that, given a particular reference event $\bar{x} = (c\bar{t}, \bar{x})$, we can divide all other events in space-time in the following classes: (i) those which are at light-like distance from $\bar{x}$, thus lying on a conical surface (which is called the *light-cone* and has the vertex in $\bar{x}$), which is further divided into two halves, the past light-cone and the future light-cone; (ii) those which are at time-like distance and have a well defined time ordering with respect to $\bar{x}$, lying either within the past or within the future cone; (iii) those which are at space-like distance and are said to be *contemporary* to $\bar{x}$, since an inertial frame always exist where the two events are simultaneous.

1.2.3 Faster Than Light?

The discussion about the geometric structure of space-time leads us to some further considerations. First of all, we can now clarify why we abhor signals propagating faster than light. Time-ordering is at the basis of our way to classify natural phenomena and deduce the laws of Nature from experimental observations: we need to establish causality relations between different events (e.g., event A causes event B) and for that we need an absolute time-ordering, A cannot be the cause of B if it happens after B in some reference frame. This is usually called the *causality principle*. If we had signals able to propagate a cause-effect relation and travelling faster than light, they could connect events at space-like distance, thus destroying the causality principle. That means that things moving faster than light are allowed, in principle, but they cannot bring any information with them, i.e. anything capable of establishing a cause-effect relation between events.

Think for instance of a very large circular room, of radius R, with a laser apparatus in the center of it which projects a beam on the internal circular wall. If the laser starts rotating, with angular velocity ω, the light spot corresponding to the beam projection on the wall will move with velocity ωR. Nothing prevents ωR from being larger than c, however this is not a problem at all: the light spot carries information from the laser apparatus, but it cannot carry any information from one point of the wall to the other, it is just a projection. There are actually such sorts of fastly rotating beams in our Universe, they are called *pulsars*.

However it is clear that, should it happen that any future experiment demonstrates the existence of signals carrying information (e.g., particles) and travelling faster than light, then one should seriously reconsider the causality principle itself.

The second consideration is that we see the end of the concept of time as an absolute quantity. While this is already clear from the form of Lorentz transformations, we learn that even time ordering may be a relative concept. We are going to discuss more about time in the next subsection.

1.2.4 New Phenomena: Time Dilation and Length Contraction

Lorentz transformations lead to some new phenomena. Let us consider a clock which is placed at rest in the origin of frame O', which is related to frame O by a pure boost along the x axis, as in (1.18). The ends of any time interval $\Delta T'$ measured by that clock will be associated with two different events[7] $(x_1' = 0, t_1')$ and $(x_2' = 0, t_2')$: they may correspond to two different beats of the clock, with $t_2' - t_1' = \Delta T'$. The above events have different coordinates in frame O, which by (1.18) are $(x_1 = \gamma v t_1', \ t_1 = \gamma t_1')$ and $(x_2 = \gamma v t_2', \ t_2 = \gamma t_2')$: therefore they are separated by a different time interval $\Delta T = \gamma \Delta T'$, which is in general dilated ($\gamma > 1$) with respect to the original one. This result, which can be summarized by saying that a moving clock seems to slow down, is usually known as *time dilation*, and is experimentally confirmed by observing subatomic particles which spontaneously disintegrate with very well known mean life times: the mean life of moving particles increases with respect to that of particles at rest with the same law predicted for moving clocks (see also Problem 1.17). Notice that, reasoning along the same lines, we conclude that also a clock at rest in the origin of O seems to slow down according to an observer in O'. There is no paradox in having two different clocks each slowing down with respect to the other: as long as both reference frames are inertial, the two clocks come close to each other and can be directly compared (synchronized) only once. The same is not true if at least one of the two clocks is accelerated: this case will be discussed in the next subsection.

Notice that time dilation is also in agreement with what observed regarding the travel time of beam 1 in Michelson's interferometer, which is $2l/c$ when observed at rest and $2l/(c\sqrt{1 - v^2/c^2})$ when in motion. Instead, in order that the travel time of beam 2 be the same, we need the length l of the arm parallel to the direction of motion to be reduced by a factor $\sqrt{1 - v^2/c^2}$, i.e. that an arm moving parallel to its length appears contracted. This is the famous contraction first supposed by Lorentz and FitzGerald, which we now show to be a general consequence of Lorentz transformations. Indeed, let us consider a segment of length L', at rest in reference frame O', where it is identified by the trajectories $x_1'(t_1') = 0$ and $x_2'(t_2') = L'$ of its two endpoints. For an observer in O the two trajectories appear as

$$x_1 = \gamma v t_1', \quad t_1 = \gamma t_1'; \quad x_2 = \gamma \left(L' + v t_2' \right), \quad t_2 = \gamma \left(t_2' + \frac{vL'}{c^2} \right). \quad (1.33)$$

The length of the moving segment is measured in O as the distance between its two endpoints, located at the same time $t_2 = t_1$, i.e. $L = x_2 - x_1$ for $t_2' = t_1' - vL'/c^2$, so that

$$L = \gamma \left(L' + v(t_2' - t_1') \right) = L'\sqrt{1 - \frac{v^2}{c^2}}, \quad (1.34)$$

thus confirming that, in general, any body appears contracted along the direction of its velocity (*length contraction*).

---

[7]We do not consider here the y, z coordinates, which do not play any role.

1.2.5 On the Concept of Proper Time

If a clock moves along a closed trajectory, it will come back to the original point after a while, so that possible effects related to time dilation can be proved experimentally by comparing it twice to a reference clock at rest. However in this case the moving clock is accelerated and it is not possible to find an inertial frame at rest with it, so that the issue must be treated more carefully.

Let $x(t)$ be the world line followed by the clock, as expressed in a particular inertial frame O (the function $x(t)$ of course depends on the chosen frame). The clock velocity $v(t) = dx(t)/dt$ in general depends on t, however it is always possible, for each t, to find an inertial frame $O_I(t)$ which moves at speed $v(t)$ with respect to O and is therefore at rest with respect to the clock. $O_I(t)$ is the inertial frame istantaneously at rest with the clock, which changes from time to time.

Let us consider the space-time interval between two very close points on the trajectory: $(cdt, dx) = (cdt, x(t + dt) - x(t))$. Such an interval is a four-vector, its particular form depends on the inertial frame, however its invariant squared length

$$ds^2 = c^2dt^2 - |dx|^2$$

is frame independent. In particular, in frame $O_I(t)$, which is istantaneously at rest with the clock ($dx' = 0$) the interval has the simple form $(d\tau, 0)$, where $d\tau$ is the infinitesimal time elapsed for the clock along its trajectory, which is therefore proportional to the invariant interval

$$d\tau^2 = ds^2/c^2$$

and is known as the infinitesimal *proper time*. All inertial frames will agree in saying that $d\tau$ is the time elapsed for an observer istantaneously at rest with the clock, hence $d\tau$ is an invariant quantity by construction.

Proper time can be expressed in terms of the time dt elapsed in the original frame O as follows:

$$d\tau = \sqrt{c^2dt^2 - |dx|^2}/c = dt\sqrt{1 - |dx/dt|^2/c^2} = dt/\gamma(t) \qquad (1.35)$$

where $\gamma(t) \equiv 1/\sqrt{1 - v^2(t)/c^2}$. Equation (1.35) can be integrated over a finite piece of trajectory, to obtain a finite proper time interval:

$$\Delta\tau_{AB} = \int_A^B dt\sqrt{1 - v^2(t)/c^2} \qquad (1.36)$$

where A and B stands for two generic points on the world line. Notice that we always have $\Delta\tau_{AB} \leq \int_A^B dt = \Delta t_{AB}$, so that a moving clock will be always slower than one at rest. Therefore, the famous twin, leaving for a long space trip and coming back after making a closed trajectory, will be effectively younger, at the end of the trip, than the twin staying at rest at home. In this case, of course, the travelling twin cannot follow the same line of reasoning and expect that it is instead the twin at home to be younger: only observers in inertial frames are allowed to make use of Eq. (1.36) in order to determine the proper time elapsed for a moving object.

It is interesting to consider the following question. Suppose we have to make a trip, leaving point x_A at time t_A and reaching point x_B at time t_B, having complete freedom on the choice of the path, apart from the two endpoints which are fixed in space and time and are clearly supposed to be at a time-like distance. Is there an optimal choice in order to arrive in B as younger or as older as possible? We are looking for trajectories which minimize or maximize the proper time for going from A to B, i.e. for extrema of the following functional of the world line $x(t)$:

$$\tau_{AB} = \int_{t_A}^{t_B} dt \left(1 - \frac{|v|^2(t)}{c^2} \right)^{1/2}. \tag{1.37}$$

It is quite easy to prove that the maximum possible value is attained for a rectilinear uniform motion from A to B. Indeed, since τ_{AB} is a Lorentz invariant quantity, we can compute it in any inertial frame, in particular in the frame O' whose origin passes through both A and B. In such a frame we have $x_A' = x_B' = 0$ and

$$\tau_{AB} = \int_{t_A'}^{t_B'} dt' \left(1 - \frac{|v'|^2(t')}{c^2} \right)^{1/2}. \tag{1.38}$$

The trajectory $x'(t') = 0$ $(v'(t') = 0)$ $\forall\, t'$ satisfies the given boundary conditions and maximizes the integrand at all times, so that it represents an absolute maximum for the proper time. As for the mininum value, the faster we move the younger we keep, and if we manage to go as fast as light we would need zero proper time: unfortunately, as we are going to see later on, that would also require an infinite amount of energy.

Various experiments have been performed in order to verify Eq. (1.36) for the computation of the proper time, starting with the famous experiment by Hafele and Keating in 1971. They put synchronized atomic clocks (with a precision of the order of 10^{-9} s) on a series of intercontinental flights, making different closed trips around the world, and compared the times of the clocks after the trip among themselves and with a similar clock at rest on Earth.

1.3 Covariant Formulation of Relativity

The analysis about the geometric structure of space-time needs some further mathematical developments which, even if not strictly necessary for an elementary treatment of Special Relativity, are well worth doing in order to reach a new perspective on the way we formulate the laws of Nature.

Let us start with some general considerations on vector spaces and linear transformations. Given a generic N-dimensional vector space on the real field, each element x is given a unique representation in terms of coordinates x_i once we fix a basis e_i, $i = 1, 2, \ldots N$

$$\mathbf{x} = \sum_{i=1}^{N} x_i \mathbf{e}_i = x_i \mathbf{e}_i;$$

here and from now on we assume the convention that, unless otherwise stated, an index appearing twice in the same monomial implies a sum over it (this is the so-called *Einstein notation*, which saves a lot of summation symbols).

Each vector is assigned once for all, however its coordinate representation depends on the choice of the basis. Any change of basis implies a linear transformation of coordinates, which can be generically represented by an invertible $N \times N$ matrix L and is implemented, adopting an index representation of the row-by-column multiplication, as

$$x'_i = L_{ij} x_j. \tag{1.39}$$

Going back to our language, space-time events are vectors and a change of basis means switching to a new reference frame.

The change of basis can be easily inferred from the transformation law of coordinates, Eq. (1.39). Indeed, since $x_i e_i$ must stay unchanged, the only possibility is

$$\mathbf{e}_i = (L^{-1})_{ij}^{T} \, \mathbf{e}_j. \tag{1.40}$$

It is easy to understand where the "transpose of inverse" in (1.40) comes from, by discussing a simple one-dimensional example. Suppose we are measuring the length of a given object (the vector), if we double the unit of measure (the basis vector), the numerical value of the measure (the coordinate) will be halved.

Suppose now we are given a generic scalar function over the vector space, i.e. a function $f(x) = f(x_1, x_2, \ldots, x_N)$ whose value on a given vector is independent of the frame choice. The N-tuple of its first derivatives

$$\nabla_i f \equiv \frac{\partial}{\partial x_i} f,$$

which is usually called the *gradient*, has instead a dependence on the coordinate choice, which we would like to specify. In order to do that, we can apply the standard chain rule for a change of variables, obtaining:

$$\nabla'_i f = \frac{\partial}{\partial x'_i} f = \frac{\partial f}{\partial x_j} \frac{\partial x_j}{\partial x'_i} = \frac{\partial f}{\partial x_j} (L^{-1})_{ji} = (L^{-1})_{ij}^{T} \nabla_j f \tag{1.41}$$

showing that the components of the gradient transform as the basis vectors do. That might sound strange at first, since we are used to think of the gradient as a vector-like object. We might have expected its components to transform as vector coordinates, however this is true only for orthogonal transformations, like rotations, for which $(L^{-1})^T$ and L coincide, since $L^T L = \text{Id}$ by definition.

Therefore, when dealing with non-orthogonal transformations, we need to distinguish two different kinds of vector-like entities. Objects transforming like vector coordinates are called *contravariant* vectors, while objects transforming like gradient components are called *covariant*, since they transform as the basis vectors do. When working in the index notation, it is necessary to adopt a proper convention to distinguish them: from now on, an upper index will stand for a contravariant vector, and a lower index for a covariant one; moreover, greek letters will be used for indices referring to space-time quantities. Therefore, space-time events will be represented as contravariant vectors x^μ ($x^\mu = ct$ for $\mu = 0$ and $x^\mu = x_i$ for $\mu = 1, 2, 3$) and the gradient as $\partial_\mu f \equiv \partial f / \partial x^\mu$.

One can also define entities carrying more that one index, which are called *tensors*. For instance, the matrix of second derivatives,

$$\partial_\mu \partial_\nu f = \frac{\partial}{\partial x^\mu} \frac{\partial}{\partial x^\nu} f$$

is a (symmetric) tensor with two covariant indices, which transforms by a row-by-column multiplication of two $(L^{-1})^T$ matrices, one for each of the two indices. The product of coordinate differentials, $dx^\mu dx^\nu$, is a (symmetric) tensor with two contravariant indices. The set of quantities $dx^\mu \partial_\nu f$ instead reprents a tensor with one covariant and one contravariant index. In this last case we can define an operation, called *trace* or *contraction*, by setting $\mu = \nu$ and summing over the common index, this is easily verified to be invariant, since one index trasforms with L and the other with $(L^{-1})^T$; for the particular given example such an invariant coincides with the total differential

$$df = dx^\mu \partial_\mu f = dx^\mu \frac{\partial f}{\partial x^\mu}.$$

For tensors with more than two indices, the contraction, instead of leading to an invariant quantity, will just reduce the number of indices by two.

A new aspect comes into play when we consider the particular class of linear transformations which leave a given scalar product invariant. This is the case of Lorentz transformations Λ, which satisfy (1.30), i.e. $\Lambda^T g \Lambda = g$, where g is the metric tensor. That can be conveniently rewritten as

$$(\Lambda^T)^{-1} = g \Lambda g^{-1}; \quad \Lambda = g^{-1} (\Lambda^t)^{-1} g, \tag{1.42}$$

from which it follows that if v is a contravariant vector, $v \to \Lambda v$, then gv is covariant, indeed $gv \to g\Lambda v = g\Lambda g^{-1} gv = (\Lambda^t)^{-1}(gv)$. Similarly, if w is covariant, then $g^{-1}w$ is contravariant. Therefore the metric tensor can be used to transform contravariant into covariant vectors and viceversa.[8]

A convenient way to consider that is to represent g itself as a tensor with two covariant indices, $g_{\mu\nu}$, and its inverse g^{-1} as a tensor with two contravariant indices, $g^{\mu\nu}$, so that the correspondence between different kinds of vectors becomes a simple operation of lowering/raising indices by contraction with the metric tensor:

$$x_\mu = g_{\mu\nu}x^\nu; \quad \partial^\mu f = g^{\mu\nu}\partial_\nu f. \tag{1.43}$$

In the case of Minkowski metric, g has a particularly simple form, $g = \mathrm{diag}(1, -1, -1, -1)$ and coincides with its inverse, so that such a lowering/raising corresponds to just a sign inversion for spatial coordinates, e.g., $(ct, \boldsymbol{x}) \rightarrow (ct, -\boldsymbol{x})$. However the formalism applies to general metric tensors as well, with the caveat that the contravariant form $g^{\mu\nu}$ actually corresponds to g^{-1}, so that we can write

$$\delta^\mu{}_\nu = g^{\mu\rho}g_{\rho\nu} \tag{1.44}$$

where $\delta^\mu{}_\nu$ (Kronecher delta) is an appropriate representation of the identity matrix in tensor notation.

Given such a correspondence, the scalar product itself can be represented in terms of a contraction between a covariant and a contravariant vector:

$$x \cdot y = g_{\mu\nu}x^\mu y^\nu = x^\mu y_\mu, \tag{1.45}$$

while the transformation matrices, Λ and $(\Lambda^{-1})^T$, can be seen as tensors with one covariant and one contravariant index:

$$(\Lambda x)^\mu = \Lambda^\mu{}_\nu x^\nu; \quad ((\Lambda^{-1})^T \partial f)_\mu = \Lambda_\mu{}^\nu \partial_\nu f$$

where $(\Lambda^T)^{-1} = g\Lambda g^{-1}$, hence $\Lambda_\mu{}^\nu \equiv g_{\mu\rho}g^{\nu\sigma}\Lambda^\rho{}_\sigma$. Finally, the transformation properties of multi-index tensors can be deduced accordingly for each upper-lower index, for instance:

$$dx^\rho \partial_\mu f \quad \rightarrow \quad \Lambda^\rho{}_\sigma \Lambda_\mu{}^\nu dx^\sigma \partial_\nu f. \tag{1.46}$$

---

[8]The unambiguous correspondence between covariant and contravariant vectors, which is established by a non-degenerate scalar product, can be well understood in terms of dual space. This is the space of all linear scalar functions defined on the vector space, which can be proven to be a linear space having the same dimension of the starting one and whose elements $\boldsymbol{w}^*$ can be expressed, by a canonical correspondence, by covariant vector coordinates $w_i^*, f_{\boldsymbol{w}^*}(\boldsymbol{v}) = w_j^* v^j$. The scalar product, $\langle \boldsymbol{v}_1, \boldsymbol{v}_2 \rangle$, associates a scalar linear function of $\boldsymbol{v}_1$, hence an element of the dual space, to each vector $\boldsymbol{v}_2$.

1.3.1 Covariant Formulation of the Laws of Nature

Most physical laws are expressed as particular relations between mathematical quantities. The Relativity Principle states that such laws must be the same in all inertial frames: the formalism we have developed gives us a simple and elegant way to implement directly such a requirement in the allowed structure that physical laws can take.

Let us start by discussing standard rotational invariance. Equations relating scalar quantities are obviously invariant under rotations. However also Newton's second law, $F = ma$, respects it, even if it relates vectors, which are not invariant quantities. The point is that both members of the equation change in the same way, i.e. they are *covariant* under rotations, so that if the equation holds in one particular frame, it will hold in all other rotated frames.

The same is true when we consider transformations between inertial frames: a physical law must be a relation between quantities which have the same transformation properties under Lorentz transformations. The transformation properties of a tensor are established by the number of its upper and lower indices, see for instance (1.46), hence equations which are covariant under Lorentz transformations must relate tensors which have the same index structure. If f is a scalar function, equations like $f = x^\mu$ or $\partial_\mu f = f x^\mu$ would be obviously wrong, while an equation like $\partial_\mu f = x^\nu \partial_\nu \partial_\mu f$, which is covariant, could be correct, at least from the point of view of frame independence.

We are going to see two implementations of this principle in the following, first to understand what is the correct generalization of the law of momentum conservation, then to rewrite Maxwell equations in a new elegant form which is manifestly Lorentz covariant.

1.4 Relativistic Kinematics

The conservation of the total energy and of the total momentum of an isolated system are among the fundamental laws of Classical Mechanics. They can be associated with the symmetries of the system under temporal and spatial translations and can be expressed by stating that their total time derivative must vanish, or equivalently by stating that their values before and after some internal process of the system, like the scattering of particles, are the same:

$$E_{in}^{tot} = E_{fin}^{tot}; \qquad P_{in}^{tot} = P_{fin}^{tot}. \tag{1.47}$$

Such equations must hold in all inertial frames. From the point of view of rotations, the statement is trivial: the first equation relates scalar quantities, the second relates vectors, hence both equations are covariant under rotations. The situations is a bit trickier when we consider Galilean transformations.

Let us take a system which is initially composed of N free particles, of mass m_i and velocities v_i. Due to some unspecified interaction, the system evolves towards a final state composed of N' free particles of masses μ_j and velocities w_j. Then, Eq. (1.47) can be rewritten, in non-relativistic mechanics, as

$$\sum_i \frac{1}{2} m_i |v_i|^2 = \sum_j \frac{1}{2} \mu_j |w_j|^2; \quad \sum_i m_i v_i = \sum_j \mu_j w_j. \tag{1.48}$$

Let us now check the invariance properties of such equations under Galilean transformations, $v_i \rightarrow v_i{}' = v_i - V$ and $w_i \rightarrow w_i{}' = w_i - V$. We consider just momentum conservation, the reader can easily check that the same conclusions applies to energy conservation. The second equation in (1.48), when rewritten in terms of the new velocities, reads

$$\sum_i m_i v_i{}' + V \sum_i m_i = \sum_j \mu_j w_j{}' + V \sum_j \mu_j \tag{1.49}$$

which coincides momentum conservation in the new frame if and only if

$$\sum_i m_i = \sum_j \mu_j \tag{1.50}$$

meaning that the total mass of the system must stay unchanged. We conclude that momentum conservation is not a covariant law of Newtonian mechanics, unless it is accompanied by an additional law: *mass conservation*. A fancier way to state the same concept is to say that, in Newtownian mechanics, mass is condemned to stay mass forever.

What about Lorentz transformations? A few explicit examples show that the classical definitions of kinetic energy and momentum, like those assumed in (1.48), do not work. Let us consider, for instance, the scattering of two particles of equal mass m moving in the $x - y$ plane, with initial velocities (v, v) and $(0, -v)$, and final velocities $(v, -v)$ and $(0, v)$: standard momentum conservation holds. However it fails, if we adopt the relativistic transformation laws for velocities, Eq. (1.20), when we go to a new reference frame O' which moves with velocity $(v, 0)$ with respect to the first. The main reason is that velocities now transform non-linearly: different linear combinations of velocities will transform in different ways, and even mass conservation will not suffice any more.

We are thus left with the problem of finding a new expression for energy and momentum in Special Relativity, in order to make their conservation a Lorentz invariant statement. What we learned in last section will be our lighthouse: the new conservation laws must be equations relating quantities with well defined transformation properties. Since classical momentum is a vector under rotations, and since rotations are a subgroup of Lorentz transformations, it is most natural to look for an equation involving four-vectors: that will suffice to get an answer. Then, based on the well known relation between symmetries and conservation laws, we will convince ourselves that we got the right answer.

1.4.1 Four-Velocity and Four-Momentum

We are looking for a generalization of momentum as the spatial component of a four-vector: the search must necessarily start with the case of a single free particle, since the total momentum of a larger system will be the sum of the single particle ones and, as such, it will share the same linear transformation properties as the free particle momentum.

Let us consider a particle moving at constant speed v, how many independent four-vectors can be associated with it? Each single infinitesimal piece of trajectory, corresponding to a time interval dt, is characterized by the four-position x^μ and by the four-displacement dx^μ. In absence of external fields, no more independent four-vectors exist. The position x^μ is not invariant under translations and changes with time: it has nothing to do with momentum. The displacement $dx^\mu = (cdt, dx)$ is directly related to velocity, $v = dx/dt$, which however does not transform as the spatial part of a four-vector. Nevertheless, we can build a four-vector by dividing dx^μ by $d\tau$, i.e. the differential of the particle proper time, which is a scalar quantity: that defines the four-velocity

$$u^\mu \equiv \frac{dx^\mu}{d\tau}; \quad d\tau = dt/\gamma(v); \quad \gamma(v) = (1 - |v|^2/c^2)^{-1/2}. \quad (1.51)$$

We can compute its components and its invariant square modulus explicitly:

$$u^0 = \gamma c; \quad u = \gamma v; \quad u_\mu u^\mu = (u^0)^2 - |u|^2 = c^2 \quad (1.52)$$

hence u^μ is a time-like four-vector of fixed squared length. The four-velocity u^μ is the only constant four-vector which can be associated with the motion of a free particle: we can build a new four-vector, which has the dimensions of momentum, by multiplying it by the particle mass:

$$p^\mu = m\,u^\mu. \quad (1.53)$$

Of course p^μ, which is called four-momentum, is a four-vector only if m is a Lorentz invariant quantity: we can take this condition as the very definition of m, by stating that the inertial mass of a particle is determined by measuring its acceleration in a system in which the particle is subject to a known force and is initially at rest. Later we will discuss again about the relation of mass to four-momentum.

The spatial part of p^μ is given by

$$p = m\,\gamma(v)v \quad (1.54)$$

which in the non-relativistic limit tends, at the leading order in v/c, to the standard classical momentum mv. The meaning of its temporal component is slightly less trivial, its non-relativistic expansion in powers of v/c

$$p^0 = m\gamma(v)c = \frac{mc}{\sqrt{1 - v^2/c^2}} \simeq \frac{1}{c}\left(mc^2 + \frac{1}{2}mv^2 + O\left(\frac{v^4}{c^4}\right)\right) \quad (1.55)$$

suggests that it might be related to the energy of the particle. We can then proceed further along these lines and make the hypothesis that, for a system of N free particles, the quantity

$$P_{tot}^{\mu} = \sum_{a=1}^{N} p_a^{\mu} \quad (1.56)$$

be associated with the total spatial momentum and with the total energy (temporal component) of the system, so that the conservation laws for energy and momentum would be written in one single, covariant equation:

$$P_{tot,\ in}^{\mu} = P_{tot,\ fin}^{\mu}. \quad (1.57)$$

However, even if the choice made for the four-momentum of a free particle, equation (1.53), had no possible alternatives, we need a more rigorous treatment of the question, in order to be sure that the temporal and spatial components of P_{tot}^{μ} are indeed associated with energy and momentum. Classical mechanics teaches us that energy and momentum are the conserved quantities associated with the invariance under temporal and spatial translations; Noether theorem teaches us how to derive such conserved quantities once the Lagrangian function of the system is known: therefore, in order to finalize our discussion, we need to derive the Lagrangian for a free relativistic particle.

Before doing that, let us stress that, for a general non-uniform motion, more four-vectors can be associated with a given world line. For instance, the derivative of the four-velocity with respect to the proper time defines the so-called four-acceleration:

$$a^{\mu} \equiv \frac{du^{\mu}}{d\tau}, \quad (1.58)$$

which is easily proved to be a space-like vector. Indeed, if we derive the equation $u_{\mu}u^{\mu} = c^2$ with respect to τ, we obtain

$$0 = \frac{d}{d\tau}(u_{\mu}u^{\mu}) = 2u_{\mu}\frac{du^{\mu}}{d\tau} = 2u_{\mu}a^{\mu}$$

meaning that a^{μ} and u^{μ} are orthogonal vectors in Minkowski space. Since u^{μ} is time-like, we can always find a reference frame where its spatial components vanish. In such a frame, which is the one instantaneously at rest with the particle, orthogonality implies $a^0 = 0$: a^μ is a purely spatial vector and coincides with usual acceleration, the fact that a^μ is space-like then trivially follows, the condition $a_\mu a^\mu < 0$ being frame independent.

1.4.2 The Lagrangian of a Free Relativistic Particle

Classical Mechanics is governed by the *minimum action* principle. A Lagrangian $\mathcal{L}(t, q_i, \dot{q}_i)$ is usually associated with a mechanical system: it has the dimensions of an energy and is a function of time, of the coordinates q_i and of their time derivatives $\dot{q}_i$. Given a time evolution law for the coordinates $q_i(t)$ in the time interval $t_1 \leq t \leq t_2$, we define the action:

$$A = \int_{t_1}^{t_2} dt\, \mathcal{L}(t, q_i(t), \dot{q}_i(t)) \tag{1.59}$$

which is then a functional of $q_i(t)$. The minimum action principle states that the equations of motion are equivalent to finding a trajectory for which the action is minimum (or at least stationary) in the given time interval, with the constraint of having the initial and final coordinates of the system, $q_i(t_1)$ and $q_i(t_2)$, fixed. That makes it clear that $\mathcal{L}$ is defined but for the addition of any function like $dF(t, q_i)/dt = \sum_i \dot{q}_i\, \partial F(t, q_i)/\partial q_i + \partial F(t, q_i)/\partial t$, i.e. a total time derivative, since that modifies the action by a quantity which depends only on the initial and final coordinates, leaving the solution unchanged. For a non-relativistic particle in one dimension, a possible choice is $\mathcal{L}_{n.r.} = (1/2)m\dot{x}^2 + \text{const}$, and it is obvious that, among all possible time evolutions, the uniform linear motion is the one which minimizes the action.

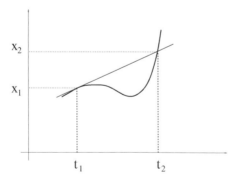

For a system of particles with positions r_i, $i = 1, \ldots n$, and velocities v_i, a deformation of the time evolution law: $r_i \rightarrow r_i + \delta r_i$ ($\delta r_i(t_1) = \delta r_i(t_2) = 0$) corresponds to a variation of the action:

$$\delta A = \int_{t_1}^{t_2} dt \sum_{i=1}^{n} \left[\frac{\partial \mathcal{L}}{\partial r_i} \cdot \delta r_i(t) + \frac{\partial \mathcal{L}}{\partial v_i} \cdot \delta v_i(t) \right] = \int_{t_1}^{t_2} dt \sum_{i=1}^{n} \left[\frac{\partial \mathcal{L}}{\partial r_i} - \frac{d}{dt} \frac{\partial \mathcal{L}}{\partial v_i} \right] \cdot \delta r_i(t)$$

so that the requirement that A be stationary for arbitrary variations $\delta r_i(t)$ is equivalent to the system of Lagrangian equations

$$\frac{\partial \mathcal{L}}{\partial r_i} - \frac{d}{dt}\frac{\partial \mathcal{L}}{\partial v_i} = 0 \,. \tag{1.60}$$

Let us now discuss how the principle of relativity is implemented in such a formulation of mechanics. The obvious requirement is that, if a particular world line minimizes the action in one inertial frame, it does so in all other inertial frames. A simple, even if not strictly necessary, way to achieve that is to choose a Lagrangian which makes the action associated with a particular world line invariant under frame trasformations. This is the standard choice made in Newtonian mechanics to ensure invariance under spatial rotations: both the action and the Lagrangian are scalar quantities; instead, regarding Galilean transformations, one is simply satisfied with the fact that the Lagrangian changes just by a total time derivative, leaving the equations of motion unchanged.

When we consider Special Relativity and Lorentz transformations, the most natural approach, in analogy with that for spatial rotations, is to require that the action be an invariant quantity, i.e. the action must depend on the trajectory $q_i(t)$ in such a way that it does not change when changing the reference frame. If we are considering the case of a single, point-like particle, we are left with the question of finding the most general Lorentz invariant quantity we can associate with its world line $x(t)$. We already know that, in the absence of any external field (free particle), each infinitesimal piece of the trajectory is characterized by its location x^μ and by the infinitesimal displacement dx^μ: discarding the first four-vector, since it is not invariant under translations, one is left with $ds^2 = dx^\mu dx_\mu = c^2 d\tau^2$, i.e. the infinitesimal proper time, as the only invariant quantity.[9]

Adding the considerations above to the requirement that the action be an additive function of the path, i.e. that the action corresponding to the union of two paths with a common endpoint be the sum of the actions corresponding to the single paths, we see that the only possible choice for the action of a free relativistic particle is

$$A = k \int_{t_1}^{t_2} d\tau = k \int_{t_1}^{t_2} dt \sqrt{1 - \frac{v^2(t)}{c^2}}, \tag{1.61}$$

which fixes the Lagrangian $\mathcal{L}_{free} = k\sqrt{1 - (v^2/c^2)}$ apart from an overall constant k. We already know that, among all world lines with fixed endpoints, the total proper time is maximized by the one corresponding to a constant velocity, hence the action in (1.61) leads to the correct equations of motion for the free particle. In order to

---

[9]The situation changes in the presence of external forces acting on the particle, which in general can be represented in terms of four-vector or tensor fields. This is the case of electromagnetic interactions, for which one can associate the value of the four-potential A^μ (to be defined later in this Chapter) to each point of the trajectory and construct a new invariant quantity, $A^\mu\, dx_\mu$ (see in particular Eq. (1.103)).

fix the constant k, we can compare to the non-relativistic case: for velocities much smaller than c, we can use a Taylor expansion

$$\mathcal{L}_{free} = k\left(1 - \frac{1}{2}\frac{v^2}{c^2} - \frac{1}{8}\frac{v^4}{c^4} + \cdots\right) \tag{1.62}$$

which, when compared to $\mathcal{L}_{n.r.} = mv^2/2$, gives $k = -mc^2$.

Let us now consider a scattering process involving n particles. The Lagrangian of the system at the beginning and at the end of the process, i.e. far away from when the interactions among the particles are not negligible, must be equal to the sum of the Lagrangians of the single particles, i.e.

$$\mathcal{L}(t)|_{|t|\to\infty} \to \sum_{i=1}^{n} \mathcal{L}_{free,i} = -\sum_{i=1}^{n} m_i c^2 \sqrt{1 - \frac{v_i^2}{c^2}}, \tag{1.63}$$

where $\mathcal{L}(t)$ is in general the complete Lagrangian describing also the interaction process and $\mathcal{L}_{free,i}$ is the Lagrangian for the i-th free particle.

If no external forces are acting on the particles, $\mathcal{L}$ is invariant under translations, i.e. it does not change if the positions of all particles are translated by the same vector $\boldsymbol{a}: \boldsymbol{r}_i \to \boldsymbol{r}_i + \boldsymbol{a}$. This *invariance requirement* can be written as:

$$\frac{\partial \mathcal{L}}{\partial \boldsymbol{a}} = \sum_{i=1}^{n} \frac{\partial \mathcal{L}}{\partial \boldsymbol{r}_i} = 0. \tag{1.64}$$

Combining this with the Lagrangian equations (1.60), we obtain the *conservation law*:

$$\frac{d}{dt}\sum_{i=1}^{n} \frac{\partial \mathcal{L}}{\partial \boldsymbol{v}_i} = 0, \tag{1.65}$$

which is a particular case of Noether's theorem, which associates a conserved quantity to every symmetry of the system which can be parametrized in terms of a continuous parameter. Equation (1.65) means that the sum of the vector quantities $\partial \mathcal{L}/\partial \boldsymbol{v}_i$ does not change with time, hence in particular:

$$\sum_{i=1}^{n} \frac{\partial \mathcal{L}_{free,i}}{\partial \boldsymbol{v}_i}\Big|_{t\to-\infty} = \sum_{i=1}^{n} \frac{\partial \mathcal{L}_{free,i}}{\partial \boldsymbol{v}_i}\Big|_{t\to\infty}. \tag{1.66}$$

In the case of relativistic particles, taking into account the following identity:

$$\frac{\partial}{\partial \boldsymbol{v}}\sqrt{1 - \frac{v^2}{c^2}} = -\frac{\boldsymbol{v}}{c^2\sqrt{1 - v^2/c^2}} \tag{1.67}$$

and setting $v_i|_{t\to-\infty} = v_{i,I}$ and $v_i|_{t\to\infty} = v_{i,F}$, Eq. (1.65) reads

$$\sum_{i=1}^{n} \frac{m_i v_{i,I}}{\sqrt{1 - v_{i,I}^2/c^2}} = \sum_{i=1}^{n} \frac{m_i v_{i,F}}{\sqrt{1 - v_{i,F}^2/c^2}}. \qquad (1.68)$$

Invariance under translations is always related to the conservation of the total momentum of the system, hence we infer that $mv/\sqrt{1 - v^2/c^2}$, i.e. the spatial part of what we have defined as the four-momentum p^μ, is indeed the correct generalization of *momentum* for a relativistic particle.

Similarly, if the Lagrangian does not depend explicitly on time, it is possible, using again (1.60), to derive

$$\frac{d}{dt}\mathcal{L} = \sum_i \left(\dot{v}_i \cdot \frac{\partial \mathcal{L}}{\partial v_i} + v_i \cdot \frac{\partial \mathcal{L}}{\partial r_i} \right)$$

$$= \sum_i \left(\dot{v}_i \cdot \frac{\partial \mathcal{L}}{\partial v_i} + v_i \cdot \frac{d}{dt}\frac{\partial \mathcal{L}}{\partial v_i} \right) = \frac{d}{dt} \sum_i v_i \cdot \frac{\partial \mathcal{L}}{\partial v_i} \qquad (1.69)$$

which is equivalent to the conservation law

$$\frac{d}{dt}\left[\sum_i v_i \cdot \frac{\partial \mathcal{L}}{\partial v_i} - \mathcal{L} \right] \equiv \frac{d}{dt} H = 0. \qquad (1.70)$$

The conserved quantity H, associated with the invariance under time translations, is known as the *Hamiltonian function* of the system and is usually identified with its total energy. In the case of free relativistic particles, H reads

$$\sum_i \left(v_i \cdot \frac{m_i v_i}{\sqrt{1 - v_i^2/c^2}} + m_i c^2 \sqrt{1 - \frac{v_i^2}{c^2}} \right) = \sum_i \frac{m_i c^2}{\sqrt{1 - v_i^2/c^2}} \qquad (1.71)$$

so that our identification of the temporal component of four-momentum, $p^0 = m\gamma(v)c^2$, with the energy of a relativistic particle stands now on a more solid basis. The quantity mc^2 is the energy possessed by the particle while it is at rest, so that $mc^2(\gamma(v) - 1)$ is the relativistic generalization of the kinetic energy.

Till now we have considered the case in which the final particles coincide with the initial ones. However, particles can in general change their nature during the scattering process, melting together or splitting, losing or gaining mass, so that the final set of particles is different from the initial one. It is possible, for instance, that the collision of two particles leads to the production of new particles, or that a single particle spontaneously decays into two or more different particles. We are not interested at all, in the present context, in the specific dynamic laws regulating the interaction process; however we can say that, in absence of external forces, the invariance of the Lagrangian under spatial and time translations is valid anyway, together with the associated conservation laws. If we refer in particular to the initial and final states, in which the system can be described as composed by non-interacting free particles, such conservation laws can be written as

$$\sum_{i=1}^{n_I} \frac{m_i^{(I)} c^2}{\sqrt{1 - v_{i,I}^2/c^2}} = \sum_{j=1}^{n_F} \frac{m_j^{(F)} c^2}{\sqrt{1 - v_{j,F}^2/c^2}}, \tag{1.72}$$

$$\sum_{i=1}^{n_I} \frac{m_i^{(I)} v_{i,I}}{\sqrt{1 - v_{i,I}^2/c^2}} = \sum_{j=1}^{n_F} \frac{m_j^{(F)} v_{j,F}}{\sqrt{1 - v_{j,F}^2/c^2}} \tag{1.73}$$

and represent the four components of the simple, covariant equation reported in (1.57), i.e. $P^\mu_{tot,I} = P^\mu_{tot,F}$. As a consequence, they will hold in every inertial frame by construction, with no need to put further constraints on the mass of the particles which are present before and after the process, as it happens instead in Newtonian mechanics, where the sum of the particle masses must be conserved. This is a fundamental aspect of relativistic mechanics, which opens the way to a completely new phenomenology, in which mass can be converted into kinetic energy and viceversa. Everybody knows the relevance that this has acquired in the recent past.

Equations (1.54) and (1.55) define the momentum p and the energy E of a free particle as a function of its mass m and of its velocity v. It is very useful to invert such relations, taking four-momentum itself as the fundamental quantity. Velocity and mass can be easily obtained as a function of $(E/c, p)$:

$$v = \frac{pc^2}{E}; \quad m^2 c^2 = \frac{E^2}{c^2} - p^2 = p^\mu p_\mu, \tag{1.74}$$

where $p \equiv |p|$, showing also that mass[10] is related to the invariant squared length of the time-like four-vector p^μ, which fixes the relation between energy and momentum, also known as *dispersion relation*: $E = \sqrt{m^2 c^4 + p^2 c^2}$. That also permits to smoothly define the limit of zero-mass particles, like photons, which are those for which $E = |p|c$ and $|v| = c$, i.e. having a light-like four-momentum[11] For massive particles, one can consider a non-relativistic approximation, in which $p \ll mc$, $v \sim p/m$ and $E \simeq mc^2 + p^2/(2m)$, and a ultra-relativistic approximation, in which $p \gg mc$, $v \sim c$ and $E \sim pc$, like for massless particles.

---

[10] In some texts alternative definitions of mass can be found, which make a distinction between rest mass m and relativistic mass γm. We retain that a single definition of mass, related to the invariant squared length of the particle four-momentum, is more useful in order to make the treatment simpler and to avoid confusion.

[11] One can also consider a new class of particles having space-like four-momentum, i.e. $|p| > E/c$. Those particles would have a negative squared mass and velocities always greater than c and are called *tachyons*. Their existence, of course, would put serious questions to the principle of causality.

1.4.3 Energy-Momentum Conservation in Relativistic Processes

We are going to analyze some examples which exemplify the new phenomena which become possible, at least from a kinematic point of view, because of the new relativistic definition of energy and momentum. More examples are discussed in the various problems reported at the end of this Chapter. Before doing that, let us define some quantities related to the total four-momentum of a system of particles, which will be useful in the following.

$P_{tot}^{\mu} = \sum_a p_a^{\mu}$, being the sum of four-vectors, transforms as a four-vector as well.[12] That means that we can associate a Lorentz invariant quantity to its Minkowski squared length, $P_{tot}^{\mu} P_{\mu\, tot}$, which is called the invariant mass of the system

$$M_{inv}^2 \, c^2 \equiv E_{tot}^2/c^2 - |\boldsymbol{P}_{tot}|^2, \tag{1.75}$$

since, as for the single particle case, it provides the relation linking energy and momentum. It is often convenient to consider the so-called *center-of-mass* (CM) frame, which, in analogy with the non-relativistic case, is defined as the reference frame where the total spatial momentum vanishes. If we are sitting in a frame where $\boldsymbol{P}_{tot} \neq 0$, it is easy to find the boost needed to move to the CM frame, i.e. the CM velocity $\boldsymbol{v}_{cm}$; indeed, a simple application of Lorentz transformations to P_{tot}^{μ} shows that

$$\boldsymbol{v}_{cm} = \frac{\boldsymbol{P}_{tot}}{E_{tot}} c^2, \tag{1.76}$$

from which it also follows that

$$\boldsymbol{v}_{cm} = \frac{\sum_a \boldsymbol{p}_a c^2}{\sum_a E_a} = \frac{\sum_a E_a \boldsymbol{v}_a}{\sum_a E_a} = \frac{d}{dt}\left(\frac{\sum_a E_a \boldsymbol{r}_a}{\sum_a E_a}\right). \tag{1.77}$$

Hence, despite its name, which is borrowed from non-relativistic mechanics, the CM position is more like an energy-weighted average of the particle positions. Notice that, according to (1.75) and (1.76), the CM frame is well defined only if P_{tot}^{μ} is a time-like vector, i.e. if the invariant mass of the system is different from zero. We can shed new light on the meaning of M_{inv} by noticing that, in the CM frame, it is proportional to the total energy

$$E_{tot}^{(CM)} = M_{inv} c^2. \tag{1.78}$$

Last relation between the invariant mass of a system and its total energy, measured in its rest frame, is probably the clearest expression of Einstein's famous relation between energy and mass.

---

[12] P_{tot}^{μ} is time-like, since it is the sum of time and light-like four-vectors with positive time components.

M_{inv} is the most useful definition of the mass of a system as a whole, since it is the quantity connecting E_{tot}, $\boldsymbol{P}_{tot}$ and v_{cm} like in the single particle case. Equation (1.78) gives the possibility to define the connection between M_{inv} and the standard non-relativistic definition of the *total mass* as the sum of the masses of the single components. Indeed, we can write

$$M_{inv}c^2 = E_{tot}^{(CM)} = \sum_a \sqrt{m_a^2 c^4 + |\boldsymbol{p}_a^{(CM)}|^2 c^2} \geq \sum_a m_a c^2 \qquad (1.79)$$

meaning that the invariant mass coincides with sum of the masses only if the particles are at rest in the CM frame; any kinetic energy possessed by the particles in such a frame goes into a contribution to the mass of the system as a whole.

Our considerations on conservation and transformation properties of energy and momentum permit to fix in a quite simple way the kinematic constraints related to interaction processes: let us illustrate this point with a first example. In relativistic diffusion processes it is possible to produce new particles starting from particles which are commonly found in Nature. The collision of two hydrogen nuclei (protons), which have a mass $m \simeq 1.67 \times 10^{-27}$ kg, can generate the particle π, which has a mass $\mu \simeq 2.4 \times 10^{-28}$ kg. Technically, some protons are accelerated in the reference frame of the *laboratory*, till one obtains a beam of momentum P, which is then directed against hydrogen at rest. That leads to proton–proton collisions from which, apart from the already existing protons, also the π particles emerge (it is possible to describe schematically the reaction as $p + p \rightarrow p + p + \pi$).

A natural question regards the minimum momentum or energy of the beam particles needed to produce the reaction: in order to get an answer, it is convenient to consider this problem as seen from the center of mass frame, in which the two particles have opposite momenta, which we suppose to be parallel, or anti-parallel, to the x axis: $P_1 = -P_2$, and equal energies $E_1 = \sqrt{P_1^2 c^2 + m^2 c^4} = E_2$. In this reference frame the total momentum vanishes and the total energy is $E = 2E_1$. Conservation of momentum and energy constrains the sum of the three final particle momenta to vanish, and the sum of their energy to be equal to E. The required energy is minimal if all final particles are produced at rest, the kinematic constraint on the total momentum being automatically satisfied in this case (this is the advantage of doing computations in the CM frame). We can then conclude that the minimum value of E in the center of mass is $E_{min} = (2m + \mu)c^2$. However that is not exactly the answer to our question: we have to find the value of the energy of the beam protons corresponding to a total energy in the center of mass equal to E_{min}. That can be done by noticing that, in the center of mass, both colliding protons have energy $E_{min}/2$, so that we can compute the relative velocity βc between the CM frame and the laboratory as that corresponding to a Lorentz transformation leading from a proton at rest to a proton with energy $E_{min}/2$, i.e. solving the equation

$$\frac{1}{\sqrt{1 - \beta^2}} = \frac{E_{min}}{2mc^2} = \frac{2m + \mu}{2m}.$$

While, as said above, the total momentum of the system vanishes in the center of mass frame and the total energy equals E_{min}, in the laboratory the total momentum is obtained by a Lorentz transformation as

$$P_L = \frac{\beta}{\sqrt{1-\beta^2}}\frac{E_{min}}{c} = \sqrt{\frac{1}{1-\beta^2}-1}\frac{E_{min}}{c} = (2m+\mu)c\sqrt{\frac{(2m+\mu)^2}{4m^2}-1}$$

$$= \frac{2m+\mu}{2m}c\sqrt{4m\mu+\mu^2}.$$

This is also the answer to our question, since in the laboratory all momentum is carried by the proton of the beam.

An alternative way to obtain the same result, without making explicit use of Lorentz transformations, is by noticing that, if E_L is the total energy in the laboratory, $P_L^2 - E_L^2/c^2$ is an invariant quantity, which is therefore equal to the same quantity computed in the CM frame, i.e.

$$\frac{E_L^2}{c^2} - P_L^2 = (2m+\mu)^2c^2.$$

Writing E_L as the sum of the energy of the proton in the beam, which is $\sqrt{P_L^2c^2+m^2c^4}$, and of that of the proton at rest, which is mc^2, we have the following equation for P_L:

$$\frac{1}{c^2}\left[\sqrt{P_L^2c^2+m^2c^4}+mc^2\right]^2 - P_L^2 = 2m^2c^2+2m\sqrt{P_L^2c^2+m^2c^4} = (2m+\mu)^2c^2$$

which finally leads to the same result obtained above.

In the example above, part of the kinetic energy of the initial particles is converted into rest energy for the new particle emerging after the collision: this is the usual strategy adopted by high energy experiments since many years, in order to explore the existence of new particles and understand the nature of fundamental interactions. An experiment like the one we have just discussed, in which one of the colliding particles is at rest, is usually called a *fixed target experiment* and is the easiest to be performed: it was indeed by this kind of experiment that pions were first produced in a laboratory. Last equation shows that in this case, for very high values of the incident beam energy, $P_L \gg mc$, the energy in the CM frame, which is the one really available for new particle production, grows asymptotically like the square root of P_Lc. A much more convenient setup is obtained by the collision of two equal energy beams proceeding in opposite directions: in this case the laboratory coincides with the CM frame. An easy computation shows that, in this case, $E_{tot}^{(CM)}$ grows like P_Lc, thus permitting to proceed much faster across the energy frontier: this is the reason why this strategy has been adopted by most recent experiments, starting from AdA (Frascati) in 1960 and ending with the Large Hadron Collider (LHC) at CERN.

Let us now discuss a different example, in which mass in converted into kinetic energy of the final products, a process which is at the basis of energy production via nuclear reactions. Consider a particle of mass M, which then decays into two equal particles of mass m; to simplify computations, let us suppose the particle of mass M is initially at rest, so that $E_{tot} = Mc^2$ and $P_{tot} = 0$. The final state depends on 6 parameters, namely the 6 components of the momenta q_1 e q_2 of the two final particles: 4 degrees of freedom are fixed by four-momentum conservation, while two of them are free, i.e. they cannot be fixed by simple kinematic considerations and depend on the specific dynamics of the interaction.[13] Momentum and energy conservation implies

$$q_1 + q_2 = 0 \quad \to \quad q \equiv q_1 = -q_2 \tag{1.80}$$

$$Mc = 2\sqrt{q^2 c^2 + m^2 c^4} \quad \to \quad E_1 = E_2 = \frac{Mc^2}{2}; \quad q^2 = \frac{M^2 c^2}{4} - m^2 c^2.$$

Last equation implies $M > 2m$, meaning that the sum of the final masses cannot exceed M, however it can be lower: the mass defect is then converted into kinetic energy of the final particles.

Let us close this Section by some considerations regarding the very definition of energy. On the basis of what we have deduced about the transformation properties of energy and momentum, it is important to notice that we can identify them as the components of a four-vector only if the energy of a particle is defined as $mc^2/\sqrt{1 - v^2/c^2}$, thus fixing the arbitrary constant usually appearing in the definition of energy. The energy balance of various experiments, in which an identical initial state can give rise to different numbers and species of final particles, proves that this is indeed the correct choice. In order to understand why an additional arbitrary constant to the relativistic definition of energy in not allowed, one should look at elementary particles in the context of Quantum Field Theory, where they are viewed as excitations over a vacuum, ground state of the theory. An arbitrary constant can then be associated only with the vacuum energy itself, while what we call "particle energies" are actually energy differences with respect to the vacuum state: contrary to energy itself, such differences are not allowed to be redefined by arbitrary additive constants.

1.5 Covariant Formulation of Electromagnetism

It is time to go back to the starting point of our discussion. We have seen that the implementation of the relativity principle by means of Galilean transformations revealed to have serious flaws after the theory of electromagnetism had been formulated in terms

---

[13]In the CM frame, the two unknown parameters are the two angles specifying the direction in which the final particles are emitted.

of Maxwell equations. Now we are going to see how such equations can be rewritten in a form which makes them explicitly covariant under Lorentz transformations. Let us start by briefly recalling the theory of electromagnetic fields in vacuum space; we will first adopt the standard vector notation, making use of S.I. units.

Maxwell equations establish a set of relations between the electric and magnetic fields, $E(x, t)$ and $B(x, t)$, and the electromagnetic sources, which can be written in terms of an electric charge density $\rho(x, t)$ and an electric current density $J(x, t)$. We can consider for simplicity ρ as a continuous distribution, such that $\rho(x, t)d^3x$ gives the total electric charge contained in a given volume d^3x around x at a given time t, while the current density can be conveniently rewritten as $J(x, t) = \rho(x, t)v(x, t)$, where $v(x, t)$ is a vector field representing the velocity at which charges located in x at time t are moving. One can formulate the theory either in terms of integral equations or in terms of partial differential equations, we adopt the second strategy. Maxwell equations then read:

$$\nabla \cdot E = \frac{\rho}{\epsilon_0} \tag{1.81}$$

$$\nabla \wedge E = -\frac{\partial B}{\partial t} \tag{1.82}$$

$$\nabla \cdot B = 0 \tag{1.83}$$

$$\nabla \wedge B = \mu_0 \left(J + \epsilon_0 \frac{\partial E}{\partial t} \right). \tag{1.84}$$

One additional equation is needed to express charge conservation, i.e. the fact that electric charges can move from one point to the other but cannot disappear; that is expressed in differential form in terms of a continuity equation:

$$\frac{\partial \rho}{\partial t} + \nabla \cdot J = 0 \tag{1.85}$$

which when integrated tells us that the total charge disappearing from a given volume is equal to the flux of current going out of that volume (hence, the charge goes somewhere else and is not destroyed).

The set of Maxwell equations, given suitable boundary conditions, can be partially integrated by introducing auxiliary fields. Indeed, from (1.83) one sees that it is always possible to find a vector field $A(x, t)$ such that

$$B = \nabla \wedge A, \tag{1.86}$$

then from (1.82) it follows that one can find an additional scalar field $\phi(x, t)$ such that

$$E = -\nabla \phi - \frac{\partial A}{\partial t}. \tag{1.87}$$

There is some freedom in the choice of the auxiliary fields, indeed it is easy to verify that, given any scalar function $\Lambda(\boldsymbol{x}, t)$, the new auxiliary fields

$$\phi' = \phi + \frac{\partial \Lambda}{\partial t}; \qquad \boldsymbol{A}' = \boldsymbol{A} - \boldsymbol{\nabla}\Lambda \tag{1.88}$$

lead to the same electromagnetic fields. This is known as *gauge invariance* and the transformation in (1.88) is known as *gauge transformation*. Gauge freedom can be constrained by imposing particular conditions on the auxiliary field, corresponding to particular gauge choices, in the following we will consider what is usually known as Lorenz (or Landau) gauge

$$\boldsymbol{\nabla} \cdot \boldsymbol{A} + \frac{1}{c^2}\frac{\partial \phi}{\partial t} = 0. \tag{1.89}$$

While Eqs. (1.82) and (1.83) are automatically satisfied by the introduction of the auxiliary fields, (1.81) and (1.84) can be re-expressed as differential equations for $\boldsymbol{A}$ and ϕ. Making use of the identity $\boldsymbol{\nabla}\wedge(\boldsymbol{\nabla}\wedge\boldsymbol{A}) = \boldsymbol{\nabla}(\boldsymbol{\nabla}\cdot\boldsymbol{A}) - \nabla^2\boldsymbol{A}$, of the condition (1.89) and of the relation $\epsilon_0\mu_0 \equiv 1/c^2$, it is easy to verify that, in the Lorenz gauge, such equations read:

$$\frac{1}{c^2}\frac{\partial^2\phi}{\partial t^2} - \nabla^2\phi = \Box\phi = \frac{\rho}{\epsilon_0} \tag{1.90}$$

$$\frac{1}{c^2}\frac{\partial^2\boldsymbol{A}}{\partial t^2} - \nabla^2\boldsymbol{A} = \Box\boldsymbol{A} = \mu_0\boldsymbol{J}. \tag{1.91}$$

In absence of external sources, $\boldsymbol{J} = 0$ and $\rho = 0$, such equations describe the propagation of electromagnetic waves.

We can now proceed and try to reformulate the theory above in terms of covariant equations. In order to do that, we need to know the transformation properties under Lorentz transformations of the various quantities illustrated above, i.e. we need to understand if they can be rewritten in terms of well defined tensorial structures. Since we do not know how to start, we need some suggestion from Nature, which comes in this form: *the electric charge is a Lorentz invariant quantity*. There is plenty of experimental evidence for that, consider for instance the various systems which are made up of charged particles and whose total charge results independent of the internal motion of their components.

From the invariance properties of the electric charge we can derive the transformation properties for ρ and $\boldsymbol{J}$. The velocity field $\boldsymbol{v}$ defined above describes a set of trajectories, which can be parametrized in terms of either standard time t or proper time τ (computed along the given trajectory). Then we can define a four-velocity field $u^\mu(\boldsymbol{x}, t)$ associated with $\boldsymbol{v}(\boldsymbol{x}, t)$, whose components are $(u^0, \boldsymbol{u}) = \gamma(v)(c, \boldsymbol{v})$. Since $\rho\, d^3x$, i.e. the total charge contained in the infinitesimal volume d^3x, is Lorentz invariant, also the quantity $\rho\, u^\mu d^3x$ is a four-vector. Then, since $\gamma(v) = dt/d\tau$, we have that $(\rho c, \rho\boldsymbol{v})d^3x\,dt$ is a four-vector as well. However, we already know that a general property of Lorentz transformations is that they leave the integration measure $d^3x\,dt$ invariant (since $|\det \Lambda| = 1$), so that also the four quantities $(\rho c, \rho v) = (\rho c, \boldsymbol{J})$ must be the components of a four-vector field, which is usually called the four-current density J^μ. That means exactly that

$$J^{\mu\prime}(\boldsymbol{x}', t') = \Lambda^\mu{}_\nu J^\nu(\boldsymbol{x}, t), \tag{1.92}$$

the four-current $J^{\mu\prime}$, measured by observer O' in the event of coordinates $\boldsymbol{x}'$ and t', is the Lorentz transformed of the four-current J^μ, measured by O in the event of coordinates $\boldsymbol{x}$ and t, where $x^{\mu\prime} = \Lambda^\mu_\nu x^\nu$. Notice that now the continuity condition, Eq. (1.85), takes a covariant form since

$$\frac{\partial \rho}{\partial t} + \boldsymbol{\nabla} \cdot \boldsymbol{J} = \frac{\partial}{\partial x^\mu} J^\mu = 0$$

Let us now consider (1.90) and (1.91): they are four equations that we require to be covariant. It is immediate the realize that the d'Alembertian differential operator is nothing but the contraction of the tensor of second derivatives, i.e.

$$\Box \equiv \frac{1}{c^2} \frac{\partial^2 \phi}{\partial t^2} - \nabla^2 = \partial_\mu \partial^\mu = \frac{\partial}{\partial x^\mu} \frac{\partial}{\partial x_\mu} .$$

Then it is clear, if we multiply (1.90) by $1/c$, that (1.90) and (1.91) form the 4 components of a covariant equation only if the auxiliary fields form the components of a four-vector field. Indeed, defining the four-potential A^μ, corresponding to $(\phi/c, \boldsymbol{A})$, the four equations can be rewritten in the form

$$\partial_\nu \partial^\nu A^\mu = \Box A^\mu = \mu_0 J^\mu . \tag{1.93}$$

The Lorenz gauge condition takes the covariant form[14] $\partial_\mu A^\mu = 0$. To summarize, we have deduced the transformation properties of the four-current on the basis of the experimental observation that electric charge is Lorentz invariant. Then, to ensure the covariance of Maxwell equations for the auxiliary fields, we have deduced the transformation properties of the latter quantities, thus introducing the four-potential A^μ. At the same time, we have noticed that the differential operator $\Box$ is Lorentz invariant, thus ensuring the covariance of the electromagnetic wave equations.

We are now in a position to determine the transformation properties of the electromagnetic field themselves. Indeed, if we define the two-index antisymmetric tensor

$$F^{\mu\nu} \equiv \partial^\mu A^\nu - \partial^\nu A^\mu \tag{1.94}$$

---

[14] Actually, it is not strictly necessary for the gauge condition to be covariant, and other gauge choices can be taken which are not so, like Coulomb gauge, in which $\boldsymbol{\nabla} \cdot \boldsymbol{A} = 0$. Covariance is therefore a virtue of Lorenz gauge, which permits to maintain the covariance of Maxwell equations for ϕ and $\boldsymbol{A}$. On the contrary, the covariant form of the equations for electric and magnetic fields, which is shown next, is guaranteed for any gauge choice, since such fields are gauge independent.

it is easy to verify, from Eqs. (1.86) and (1.87), that it corresponds to the 4×4 matrix

$$\begin{pmatrix} 0 & -E_x/c & -E_y/c & -E_z/c \\ E_x/c & 0 & -B_z & B_y \\ E_y/c & B_z & 0 & -B_x \\ E_z/c & -B_y & B_x & 0 \end{pmatrix} \tag{1.95}$$

while $F_{\mu\nu}$ is obtained by changing sign to the components with one spatial and one temporal index. $F^{\mu\nu}$ has well defined transformation properties

$$F^{\mu\nu\prime}(x', t') = \Lambda^{\mu}{}_{\rho} \Lambda^{\nu}{}_{\sigma} F^{\rho\sigma}(x, t) \tag{1.96}$$

so that, if we consider a boost along the x axis, like that represented by (1.18), we obtain the following transformation properties for E and B:

$$E'_x = E_x; \quad E'_y = \gamma(E_y - \beta c B_z); \quad E'_z = \gamma(E_z + \beta c B_y)$$
$$B'_x = B_x; \quad B'_y = \gamma(B_y + \beta E_z/c); \quad B'_z = \gamma(B_z - \beta E_y/c). \tag{1.97}$$

It is now possible to rewrite also Maxwell equations (1.81)–(1.84) in a covariant form. Indeed, one can check that (1.81) and (1.84), i.e. the ones involving four-current components, can be rewritten in the form

$$\partial_\nu F^{\nu\mu} = \mu_0 J^\mu \tag{1.98}$$

while the homogeneous equations (1.82) and (1.83) can be re-expressed by introducing the so-called dual electromagnetic tensor $\tilde{F}^{\mu\nu} = \epsilon^{\mu\nu\rho\sigma} F_{\rho\sigma}$, where $\epsilon^{\mu\nu\rho\sigma}$ is the completely antisymmetric invariant tensor,[15] whose components are

$$\begin{pmatrix} 0 & -B_x & -B_y & -B_z \\ B_x & 0 & E_z/c & -E_y/c \\ B_y & -E_z/c & 0 & E_x/c \\ B_z & E_y/c & -E_x/c & 0 \end{pmatrix} \tag{1.99}$$

---

[15] Such a tensor is defined so that $\epsilon^{\mu\nu\rho\sigma} = 0$ if any couple of indices coincide and $\epsilon^{\mu\nu\rho\sigma} = \pm 1$ if $\mu\nu\rho\sigma$ are any even/odd permutation of 0123. It is actually a pseudotensor, meaning that, for any Lorentz transformation Λ, it transforms like

$$\epsilon^{\prime\mu\nu\rho\sigma} = (\det \Lambda) \Lambda^{\mu}_{\alpha} \Lambda^{\nu}_{\beta} \Lambda^{\rho}_{\gamma} \Lambda^{\sigma}_{\delta} \epsilon^{\alpha\beta\gamma\delta}$$

Then, making use of the Leibniz formula for the determinant and of the fact that $(\det \Lambda)^2 = 1$, one proves that

$$\epsilon^{\prime\mu\nu\rho\sigma} = \det \Lambda \det \Lambda \epsilon^{\mu\nu\rho\sigma} = \epsilon^{\mu\nu\rho\sigma}$$

i.e. that it is an invariant tensor, which takes the same form in every frame.

then Eqs. (1.82) and (1.83) become

$$\partial_\nu \tilde{F}^{\nu\mu} = 0. \tag{1.100}$$

If a pointlike particle moves in the presence of an electromagnetic field, given the mass m and the charge q of the particle, Newton's law asserts that, in the instantaneous rest system of the particle, its acceleration is given by

$$m\boldsymbol{a} = q\boldsymbol{E}(\boldsymbol{r}_p, t). \tag{1.101}$$

From (1.58) and (1.95) we see that (1.101) can be written in the explicitly covariant form

$$ma^\mu = qF^{\mu\nu}u_\nu. \tag{1.102}$$

Multiplying both sides by $\gamma(v)^{-1}$ and considering the space components of (1.102) we get the well known Lorentz force equation

$$m\frac{d\boldsymbol{p}}{dt} = q(\boldsymbol{E} + \boldsymbol{v} \wedge \boldsymbol{B}). \tag{1.103}$$

It is not difficult to verify that this equation of motion for a charged particle in an external electromagnetic field corresponds to the Lagrangian

$$\mathcal{L}(\boldsymbol{r}, \boldsymbol{v}, t) = -\gamma(\boldsymbol{v})^{-1}(mc^2 + q\, u_\mu(\boldsymbol{v})A^\mu(\boldsymbol{r}, t)). \tag{1.104}$$

1.5.1 Relativistic Doppler Effect

In absence of sources, Eq. (1.93) describes the propagation of electromagnetic waves in vacuum space. Among the various possible solutions, one has monochromatic plane waves of frequency ν and wavelength $\lambda = c/\nu$, like the following

$$A^\mu(\boldsymbol{x}, t) = A_0^\mu \sin(\boldsymbol{k} \cdot \boldsymbol{x} - \omega t), \tag{1.105}$$

where A_0^μ are constants, $\omega = 2\pi\nu$ and $\boldsymbol{k}$ is the wave vector: in order for (1.105) to solve Eq. (1.93), they have to satisfy the dispersion relation $\omega = |\boldsymbol{k}|c = 2\pi c/\lambda$.

We ask now how a particular plane wave solution for an inertial frame O is described by a new inertial frame O'. Linearity implies that it is described by a plane wave too, but with a new wave vector $\boldsymbol{k}'$ and a new frequency ν'. The transformation laws for these quantities are easily found by noticing that the difference of the two phases at corresponding points in the two frames must be a space-time independent constant if the fields transforms locally. That is true only if the phase $\boldsymbol{k} \cdot \boldsymbol{x} - \omega t$ is invariant under Lorentz transformations, i.e., defining $k_0 = \omega/c$, if

$$k'_0 ct' - \mathbf{k}' \cdot \mathbf{x}' = k_0 ct - \mathbf{k} \cdot \mathbf{x}; \tag{1.106}$$

this has been implicitly assumed in writing Eq. (1.105): if by A_0^μ we mean a constant four-vector, then the oscillating function must be Lorentz invariant in order to ensure covariance.

It is easy to verify that Eq. (1.106) is true for every point $(\mathbf{x}, ct)$ in space-time if and only if[16] the four quantities $(k_0, \mathbf{k})$ transform like a new four-vector k^μ, so that (1.105) can be rewritten as $A^\mu(\mathbf{x}, t) = A_0^\mu \sin(k_\nu x^\nu)$. From $k'^\mu = \Lambda^\mu{}_\nu k^\nu$ we deduce the transformation properties of the temporal component ω. If O and O' are connected by a boost along the $\hat{\mathbf{n}}$ axis then

$$\omega' = \gamma(\omega - \boldsymbol{\beta} \cdot \mathbf{k} c) = \gamma\omega(1 - \boldsymbol{\beta} \cdot \hat{\mathbf{n}}) = \gamma\omega(1 - \beta\cos\theta) \tag{1.107}$$

where $\hat{\mathbf{n}} \equiv \mathbf{k}/|\mathbf{k}|$, θ is the angle between $\hat{\mathbf{n}}$ and $\mathbf{k}$ and the dispersion relation $\omega = |\mathbf{k}|c$ has been used.

When $\boldsymbol{\beta}$ is parallel or antiparallel to $\mathbf{k}$ we have the longitudinal relativistic Doppler effect:

$$\omega' = \gamma\omega(1 \mp \beta) = \sqrt{\frac{1 \mp \beta}{1 \pm \beta}}\, \omega. \tag{1.108}$$

The frequency is therefore reduced (increased) if the motion of O' is parallel (antiparallel) to that of the signal. The result is similar to the classical Doppler effect obtained for waves propagating in a medium, but with important differences. In particular it is impossible to distinguish the motion of the source from the motion of the observer: that is evident from (1.108), since ν and ν' can be exchanged by simply reversing $v \to -v$. This is consistent with the fact that no propagating medium (ether) exists for electromagnetic waves in vacuum.

Another relevant difference is that the frequency changes even if, in frame O, the motion of O' is orthogonal to that of the propagating signal (transverse Doppler effect): in that case Eq. (1.107) implies

$$\nu' = \gamma(v)\nu = \frac{\nu}{\sqrt{1 - v^2/c^2}}, \tag{1.109}$$

which has no analogy in non-relativistic wave propagation and can be reinterpreted as a pure time dilation effect.

---

[16]Let us put $k \equiv (k_0, \mathbf{k})$ and $x \equiv (ct, \mathbf{x})$, then from (1.106) we have

$$k'^T g x' = k'^T g \Lambda x = k^T g x \,\forall\, x \;\to\; k'^T g \Lambda = k^T g \;\to\; k' = g^{-1}(\Lambda^T)^{-1} g k = \Lambda k$$

where Λ is the Lorentz transformation bringing from O to O' and we made use of Eq. (1.42).

Suggestions for Supplementary Readings

- C. Kittel, W. D. Knight, M. A. Ruderman: *Mechanics - Berkeley Physics Course*, volume 1 (Mcgraw-Hill Book Company, New York 1965)
- L. D. Landau, E. M. Lifshitz: *The Classical Theory of Fields - Course of theoretical physics*, volume 2 (Pergamon Press, London 1959)
- J. D. Jackson: *Classical Electrodynamics*, 3d edn (John Wiley, New York 1998)
- W. K. H. Panowsky, M. Phillips: *Classical Electricity and Magnetism*, 2nd edn (Addison-Wesley Publishing Company Inc., Reading 2005).

Problems

1.1 A spaceship of length $L_0 = 150$ m is moving with respect to a space station with a speed $v = 2 \times 10^8$ m/s. What is the length L of the spaceship as measured by the space station?

Answer: $L = L_0\sqrt{1 - v^2/c^2} \simeq 112$ m .

1.2 How many years does it take for an atomic clock (with a precision of one part over 10^{15}), which is placed at rest on Earth, to lose one second with respect to an identical clock placed on the Sun?

Answer: We can take the Sun as an inertial frame, in which Earth moves on an approximately circular orbit at constant speed $v \simeq 3 \times 10^4$ m/s $\simeq 10^{-4}$ c. We must find a proper time interval on Earth, $\Delta\tau$, such that it is one second less with respect to Sun time. Adopting Eq. (1.36) with v constant and setting $\delta t = 1$ s, we have $(\gamma(v) - 1)\Delta\tau = \delta t$, hence

$$\Delta\tau = \frac{\delta t}{1/\sqrt{1 - v^2/c^2} - 1} \simeq 2\,\delta t\,\frac{c^2}{v^2} \simeq 6.34\,\text{years}.$$

Our computation is actually incomplete, since it neglects gravitational field effects: their analysis would require a treatment in the framework of general relativity.

1.3 Two spaceships, moving along the same course with the same velocity $v = 0.98\,c$, pass space station Alpha, which is placed on their course, at the same hour of two successive days. On each of the two spaceships a radar permits to know the distance from the other spaceship: what value does it measure?

Answer: In the reference frame of space station Alpha, the two spaceships stay at the two ends of a segment of length $L = vT$, where $T = 1$ day. That distance is reduced by a factor $\sqrt{1 - v^2/c^2}$ with respect to the distance L_0 among the spaceships as measured in their rest frame. We have therefore

$$L_0 = \frac{1}{\sqrt{1 - v^2/c^2}}vT \simeq 1.28 \times 10^{14}\,\text{m}.$$

1.4 Spaceship A is moving with respect to space station S with a velocity 2.7×10^8 m/s. Both A and S are placed in the origin of their respective reference frames, which are oriented so that the relative velocity of A is directed along the positive direction of both x axes; A and S meet at time $t_A = t_S = 0$. Space station S detects an event, corresponding to the emission of luminous pulse, in $x_S = 3 \times 10^{13}$ m at time $t_S = 0$. An analogous but distinct event is detected by spaceship A, with coordinates $x_A = 1.3 \times 10^{14}$ m , $t_A = 2.3 \times 10^3$ s. Is it possible that the two events have been produced by the same moving body?

Answer: The two events may have been produced by the same moving body only if they have a time-like distance, otherwise the unknown body would go faster than light. After obtaining the coordinates of the two events in the same reference frame, one obtains, e.g. in the spaceship frame, $\Delta x \simeq 6.09 \times 10^{13}$ m and $c\Delta t \simeq 6.27 \times 10^{13}$ m $> \Delta x$: the two events may indeed have been produced by the same body, moving at an average speed $\Delta x / \Delta t \simeq 0.97\, c$.

1.5 Fizeau's Experiment

In the experiment described in the figure, a light beam of frequency $\nu = 10^{15}$ Hz, produced by the source S, is split into two distinct beams which go along two different paths belonging to a rectangle of sides $L_1 = 10$ m and $L_2 = 5$ m. They recombine, producing interference in the observation point O, as illustrated in the figure. The rectangular path is contained in a tube T filled with a liquid having refraction index $n = 2$, so that the speed of light in that liquid is $v_c \simeq 1.5 \times 10^8$ m/s. If the liquid is moving counter-clockwise around the tube with a velocity 0.3 m/s, the speed of the light beams along the two different paths changes, together with their wavelength, which is constrained by the equation $v_c = \lambda \nu$ (the frequency ν instead does not change and is equal to that of the original beam). For that reason the two beams recombine in O with a phase difference $\Delta\phi$, which is different from zero (the phase accumulated by each beam is given by 2π times the number of wavelengths contained in the total path). What is the value of $\Delta\phi$? Compare the result with what would have been obtained using Galilean transformation laws.

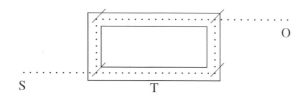

Answer: Calling $L = L_1 + L_2 = 15$ m the total path length inside the tube for each beam, and using Einstein laws for adding velocities, one finds

$$\Delta\phi = 4\pi\nu L v \frac{n^2 - 1}{c^2 - n^2 v^2} \simeq 4\pi\nu L v (n^2 - 1)/c^2 \simeq 1.89 \text{ rad}.$$

Instead, Galilean laws would lead to

$$\Delta\phi = 4\pi\nu L v \frac{n^2}{c^2 - n^2 v^2} \, ,$$

a result which does not make sense, since it is different from zero also when the tube is empty: in that case, indeed, Galilean laws would imply that the tube is still filled with a "rotating ether".

The experiment described above is very similar to the one performed by Fizeau. The only significant difference is that Fizeau made the two light beams going exactly along the same path, even if in two opposite directions, in order to eliminate any possible systematic effect related to different conditions (e.g., in temperature) in the various pieces of the tube. Fizeau observed a non-zero effect for water, and a null effect for air, for which $n \simeq 1$: this is in agreement with the relativistic prediction, but was misinterpreted at the time of Fizeau and associated with the partial drag hypothesis by Fresnel.

1.6 Relativistic Aberration of Light

In the reference frame of the Sun, Earth moves with a velocity of modulus $v = 10^{-4} c$ which forms, at a given time, an angle $\theta = 60°$ with respect to the position of a given star. Compute the variation of this angle when it is measured by a telescope placed on Earth.

Answer: By applying the relativistic transformation laws for velocities to the photons coming from the star, one easily obtains that

$$\tan\theta' = \sqrt{1-\beta^2}\,\frac{\sin\theta}{\cos\theta+\beta} \qquad \text{or equivalently} \qquad \cos\theta' = \frac{\cos\theta+\beta}{1+\beta\cos\theta}$$

to be compared with the classical expression $\tan\theta' = \sin\theta/(\cos\theta+\beta)$. In the relativistic case all angles, apart from $\theta = \pi$, get transformed into $\theta' = 0$ as $\beta \to 1$.

In the present case $\beta = v/c = 10^{-4}$ and it is sensible to expand in Taylor series obtaining, up to second order:

$$\theta' = \theta - \beta\sin\theta + \beta^2\sin\theta\cos\theta + O(\beta^3)$$

to be compared with result obtained by Galilean transformation laws:

$$\theta' = \theta - \beta\sin\theta + 2\beta^2\sin\theta\cos\theta + O(\beta^3) \, .$$

It is interesting to notice that relativistic effects show up only at the second order in β. In the given case $\delta\theta = -17.86216''$ to be compared with $\delta\theta = -17.86127''$ for the classical computation. Relativistic effects are tiny in this case and not appreciable by a usual optical telescope; an astronomical interferometer, which is able to reach resolutions of the order of few micro-arcseconds at radio wavelengths, should be used instead.

1.7 A particle is moving with a speed of modulus v and components (v_x, v_y, v_z). What is the modulus v' of the velocity for an observer moving at a speed w along the x axis? Comment the result as v and/or w approach c.

Answer: Applying the relativistic laws for the addition of velocities we find

$$v'^2 = v_x'^2 + v_y'^2 + v_z'^2 = \frac{(v_x - w)^2 + (1 - w^2/c^2)(v_z^2 + v_y^2)}{(1 - v_x w/c^2)^2}$$

and after some simple algebra

$$v'^2 = c^2 \left(1 - \frac{(1 - v^2/c^2)(1 - w^2/c^2)}{(1 - v_x w/c^2)^2}\right).$$

It is interesting to notice that as v and/or w approach c, also v' approaches c (from below): that verifies that a body moving with $v = c$ moves with the same velocity in every reference frame (invariance of the speed of light).

1.8 During a Star Wars episode, space station Alpha detects an enemy spaceship approaching it from a distance $d = 10^8$ km at a speed $v = 0.9\,c$; at the same time the station launches a missile of speed $v' = 0.95\,c$ to destroy it. As soon as the enemy spaceship detects the electromagnetic pulses emitted by the missile, it launches against space station Alpha the same kind of missile, therefore moving at a speed $0.95\,c$ in the rest frame of the spaceship. How much time do the inhabitants of space station Alpha have, after having launched their missile, to leave the station before it is destroyed by the second missile?

Answer: Let us make computations in the reference frame of the space station. Setting to zero the launching time of the first missile, the enemy spaceship detects it and launches the second missile at time $t_1 = d/(v + c)$ and when it is at a distance $x_1 = cd/(v + c)$ from the space station. The second missile approaches Alpha with a velocity $V = (v + v')/(1 + vv'/c^2) \simeq 0.9973\,c$, therefore it hits the space station at time $t_2 = t_1 + x_1/V \simeq 352$ s.

1.9 Two particles move at the same speed v along two orthogonal axes of the laboratory, x and y. Assuming that the rest frames of the two particles are connected to the laboratory frame by a pure boost, i.e. that their axes are parallel to those of the laboratory, compute the components and the modulus of the relative velocities of the two particles, i.e. the velocity of each particle in the rest frame of the other. Are the two relative velocities opposite to each other? Are their moduli equal? Do you have any explanation?

Answer: In the laboratory we have $v_1 = (v, 0)$ and $v_2 = (0, v)$. Let us call O' and O'' the rest frames of particles 1 and 2, respectively, and v_2' (v_1'') the speed of particle 2 (1) in the rest frame of particle 1 (2). Then, setting as usual $\beta = v/c$ and $\gamma = 1/\sqrt{1 - v^2/c^2}$, a simple application of the transformation law for velocities shows that $v_2' = (-v, v/\gamma)$ and $v_1'' = (v\gamma, -v)$. While the modulus is the same,

$|\boldsymbol{v}_2'| = |\boldsymbol{v}_1''| = c\sqrt{\beta^2(2-\beta^2)}$, one has that, contrary to naïve expectations, $\boldsymbol{v}_2' \neq -\boldsymbol{v}_1''$. The reason is easy to understand: while the axes of the two particle frames are both parallel to the laboratory frame, they are not parallel to each other: in order to go from reference frame O' to reference frame O'' we need the combination of two boosts along two different axes, x and y. As the reader can verify by composing the corresponding matrices, that is not a pure boost, but involves a rotation around the z axis, therefore the discrepancy is only due to a mutual rotation of the frame axes, i.e. the actual velocities are in fact opposite to each other. Such a rotation is a purely relativistic effect which leads to new phenomena, like Thomas precession.

1.10 Two particles proceed with four-velocities u^μ and w^μ in the laboratory frame. What is the relative speed v_r of the two particles?

Answer: The relative speed is defined as the velocity of one particle measured in the rest frame of the other, hence it is clear that it must be a Lorentz invariant quantity (it is, in some sense, a private fact of the two particles). Therefore we must look for an invariant quantity to be built in terms of u^μ and w^μ: the only non-trivial invariant combination of these two quantities is $u^\mu w_\mu$, since we have $u^\mu u_\mu = w^\mu w_\mu = c^2$. Since we are looking for an invariant quantity, we can work in any frame we prefer, for instance in the reference frame of particle 1, where, setting $\beta_r = v_r/c$ and $\gamma_r = 1/\sqrt{1-\beta_r^2}$, we have

$$u = (c,0); \quad w = (\gamma_r c, \gamma_r \boldsymbol{v}_r); \quad u^\mu w_\mu = c^2 \gamma_r.$$

We deduce

$$v_r = c\beta_r = c\sqrt{1 - \frac{1}{\gamma_r^2}} = c\sqrt{1 - \frac{c^2}{u^\mu w_\mu}}.$$

Last expression is explicitly Lorentz invariant and can be used to compute the relative velocity in any frame.

1.11 We are on the course of a spaceship moving with a constant speed v while emitting electromagnetic pulses which, in the rest frame of the spaceship, are equally spaced in time. We receive a pulse every second while the spaceship is approaching us, and a pulse every two seconds while the spaceship is leaving us. What is the speed of the spaceship?

Answer: Let us call $\Delta\tau$ the time spacing of pulse emissions in the rest frame of the spaceship. By time dilation, such spacing becomes $\Delta t = \gamma \Delta\tau$ in our frame. While the spaceship is approaching us, pulses travelling towards us are at a distance of $\Delta t(c-v)$ from each other: indeed, each pulse and the spaceship travel respectively $c\Delta t$ and $v\Delta t$, in the same direction, before the next pulse is emitted; on the contrary, when the spaceship is leaving, pulses towards us travel in the direction opposite to the spaceship and their mutual distance is $\Delta t(c+v)$. Therefore the arrival time intervals are respectively

$$\Delta t(c \mp v)/c = \gamma(1 \mp \beta)\Delta\tau = \sqrt{\frac{1 \mp \beta}{1 \pm \beta}}\Delta\tau,$$

in agreement with the longitudinal Doppler effect. Combining the two effects, we deduce that

$$\frac{1 + v/c}{1 - v/c} = 2$$

hence $v = 1/3\,c$.

1.12 A spacecraft is moving towards a plane mirror at rest with a constant velocity $v = 5 \times 10^7$ m/s, which is orthogonal to the mirror. The spacecraft casts a laser pulse of frequency $\nu = 10^{15}$ Hz, which is reflected by the mirror and comes back with a new frequency ν'. What is the value of ν'?

Answer: The pulse proceeds parallel to the spacecraft velocity both before and after reflection. Since reflection leaves the frequency unchanged only in the rest frame of the mirror, two different longitudinal Doppler effects have to be taken into account, that of the original pulse with respect to the mirror and that the reflected pulse with respect to the spacecraft, therefore $\nu' = \nu\,(1 + v/c)/(1 - v/c) = 1.4 \times 10^{15}$ Hz.

1.13 Consider a problem similar to the one above, with the difference that now the spacecraft is moving parallel to the mirror surface. What is the value of ν' in this case?

Answer: In this case, if we require that the pulse hits the mirror surface and catches the spacecraft on its way back, we obtain, assuming standard reflection laws, that the pulse makes an angle with respect to the spacecraft velocity, which in the mirror frame is $\pm\theta$ respectively before and after reflection, with $\cos\theta = v/c$ (the value of θ is actually irrelevant in the following).

Let us put $\omega = 2\pi\nu$, $\omega' = 2\pi\nu'$, and let us call $\tilde{\omega}$ the angular frequency of the pulse as observed in the mirror frame. We can apply equation (1.107) twice to transform frequencies, however, since the angle θ is already known in the mirror frame, the easiest way is to apply the transformation starting in both cases from the mirror frame. We then obtain:

$$\omega = \gamma(v)\tilde{\omega}(1 - \beta\cos\theta)$$

$$\omega' = \gamma(v)\tilde{\omega}(1 - \beta\cos\theta)$$

which immediately gives $\omega' = \omega$, i.e. the frequency the spacecraft gets back is unmodified in this case.

1.14 Spacecraft A passes very close to an astrophysical object S, which emits electromagnetic radiation of frequency $\nu = 10^{14}$ Hz in all directions, ν being measured in S rest frame. In our inertial rest frame, both A and S move at constant velocity

v along two orthogonal axes in the xy plane, their world lines being respectively $x_A(t) = (vt, 0)$ and $x_S(t) = (0, vt + d)$, with $v = c/5$ and $d = 10^3$ km.

What is, according to spacecraft A, the minimum distance reached from object S? What is the frequency spacecraft A receives when it sees S at such closest distance?

Answer: Let us take S world line, expressed in our reference frame, $x_S(t) = (0, vt + d)$, and transform it to reference frame A by a standard Lorentz transformation along the x axis

$$t' = \gamma t; \quad x'_S = -\gamma vt; \quad y'_S = vt + d$$

where $\gamma = 1/\sqrt{1 - v^2/c^2}$. We could re-express everything in terms of t', to find the world line of S in A, however this is not strictly necessary, since we just need to compute the point of the world line for which

$$x'^2_S + y'^2_S = \gamma^2 v^2 t^2 + (vt + d)^2 = v^2 t^2 (1 + \gamma^2) + 2vdt + d^2$$

is minimum. To that aim it is completely irrelevant whether the world line is parametrized in terms of t or t'. The condition for the minimum gives:

$$2tv^2(1 + \gamma^2) + 2vd = 0 \quad \rightarrow \quad t_{min} = -\frac{d}{v(1 + \gamma^2)}$$

from which we find

$$d^2_{min} = \frac{d^2}{1 + \gamma^2} - 2\frac{d^2}{1 + \gamma^2} + d^2 = \frac{d^2}{2 - v^2/c^2} = \frac{25d^2}{49}$$

hence $d_{min} = 5d/7 \simeq 714$ km. It is interesting to compare with the minimum distance as observed from our reference frame, which is $d/\sqrt{2} \simeq 707$ km.

In A frame, the photon leaving S when it is at minimum distance and reaching A travels orthogonal to the direction of S, hence we have to apply the trasverse Doppler effect law. However we have to be careful about the exact transformation we need to do: the photon is transverse in frame A, hence we have to transform from frame A to frame S, i.e. we can write

$$\nu = \gamma' \nu'$$

where $\gamma' \equiv 1/\sqrt{1 - v_S'^2}$ and v_S' is S velocity as seen by A. Applying the relativistic law for the transformation of velocities we obtain $v_S' = (-v, v/\gamma)$ hence

$$v_S' = \sqrt{v^2 + (1 - v^2/c^2)v^2} = \sqrt{\frac{1}{25}\left(2 - \frac{1}{25}\right)}c = \frac{7}{25}c; \quad \gamma' = 25/24$$

and

$$\nu' = \nu/\gamma' = \frac{24\nu}{25} \simeq 0.96 \times 10^{14} \text{ Hz}.$$

1.15 A particle moves in one dimension with an acceleration which is constant and equal to a in the reference frame instantaneously at rest with it. Determine its world line $x(t)$ in the reference frame where it is placed at rest in $x = 0$ at time $t = 0$.

Answer: One can make use either of the inertial frame time t or of the particle proper time τ to parametrize the world line, which are related by $d\tau = \sqrt{1 - v^2(t)/c^2}\, dt$ and will be assumed to be synchronized so that $t = 0$ when $\tau = 0$. There are various way to solve this problem, let us discuss two of them. First, to derive the equation of motion, let us notice that, in the frame instantaneously at rest with the particle, the velocity goes from 0 to $ad\tau$ in the interval $d\tau$, so that, by relativistic addition of velocities, v changes into $(v + ad\tau)/(1 + vad\tau/c^2)$ in the same interval, meaning that $dv/d\tau = a(1 - v^2/c^2)$. We can now make use of the spatial component of the four-velocity, $u = dx/d\tau$, which is related to v by the following relations:

$$u = \frac{v}{\sqrt{1 - v^2/c^2}}, \qquad v = \frac{u}{\sqrt{1 + u^2/c^2}}, \qquad \sqrt{1 + \frac{u^2}{c^2}}\sqrt{1 - \frac{v^2}{c^2}} = 1.$$

Using previous equations, it is easy to derive

$$\frac{du}{dt} = \frac{du}{dv}\frac{dv}{d\tau}\frac{d\tau}{dt} = a$$

which can immediately be integrated, setting $u(0) = 0$, as $u(t) = at$. Using the relation between u and v we have

$$v(t) = \frac{at}{\sqrt{1 + a^2t^2/c^2}}.$$

That gives the velocity, as observed in the laboratory, for a uniformly accelerated motion: for $t \ll c/a$ one recovers the non-relativistic result, while for $t \gg c/a$ the velocity reaches asymptotically that of light. The dependence of v on t can be finally integrated, using the initial condition $x(0) = 0$, giving

$$x(t) = \frac{c^2}{a}\left(\sqrt{1 + \frac{a^2t^2}{c^2}} - 1\right).$$

This is usually known as hyperbolic motion, since the world line defines an hyperbola in the xt plane:

$$\left(\frac{ax}{c^2} + 1\right)^2 - \frac{a^2t^2}{c^2} = 1.$$

A simpler way to solve the same problem makes use of the rapidity χ, which is defined by $\beta = \tanh\chi$. We have seen that, contrary to velocity, rapidity is an additive quantity for collinear boosts. By virtue of that, the infinitesimal change of rapidity in the rest

frame, $d\chi = ad\tau/c$, goes unchanged to the laboratory frame, i.e. we can write, in
the laboratory:

$$\frac{d\chi}{d\tau} = a/c \quad \rightarrow \quad \chi(\tau) = a\tau/c$$

leading immediately to $\beta(\tau) = \tanh(a\tau/c)$ and $\gamma(\tau) = \cosh(a\tau/c)$. Then $dt = \gamma d\tau$
can be immediately integrated to obtain $t = (c/a)\sinh(a\tau/c)$ and finally

$$\beta = \tanh(a\tau/c) = \frac{\sinh(a\tau/c)}{\cosh(a\tau/c)} = \frac{at/c}{\sqrt{1 + a^2 t^2/c^2}}.$$

1.16 Two identical spacecrafts of mass M and initially at rest at a relative distance L,
start moving at the same time $t = 0$ under the influence of a constant force F acting
along their relative distance, i.e. such that $dp/dt = F$ where p is the relativistic spatial
momentum. The two spacecrafts are initially connected by a thin and inextensible
cable. Determine the world lines of both spacecrafts and say whether the cable breaks
or not.

Answer: Regarding the motion of each spacecraft, this problem is identical to the
previous one, even if this is not obvious at a first sight. The equation $dp/dt = F$
can be immediately integrated yielding $p = Ft$ ($p = 0$ at $t = 0$), hence $E = \sqrt{F^2 c^2 t^2 + M^2 c^4}$, from which we obtain

$$v(t) = \frac{pc^2}{E} = \frac{Fc^2 t}{\sqrt{F^2 c^2 t^2 + M^2 c^4}} = \frac{at}{\sqrt{1 + (a/c)^2 t^2}}$$

where $a \equiv F/M$, i.e. exactly the same solution of Problem 1.15. Another way to
look at the correspondence between the two problems is to consider how the quantity
dp/dt transforms, in the case of one-dimensional motions, for longitudinal boosts.
Going to a new inertial frame O', moving with velocity V with respect to the original
one and setting $\tilde\gamma = 1/\sqrt{1 - V^2/c^2}$, we have

$$dt' = \tilde\gamma\left(dt - \frac{V}{c^2}dx\right) = \tilde\gamma dt\left(1 - \frac{Vv}{c^2}\right)$$

$$dp' = \tilde\gamma\left(dp - \frac{V}{c^2}dE\right) = \tilde\gamma dp\left(1 - \frac{Vv}{c^2}\right)$$

where in the last equation we have used the relation

$$dE = d(\sqrt{p^2 c^2 + M^2 c^4}) = \frac{pc^2}{E}dp = vdp.$$

It is now evident that $dp'/dt' = dp/dt$, i.e. such a quantity is invariant under any
longitudinal boost: that includes the boost bringing to the frame istantaneously at

rest with the spacecraft, hence a constant dp/dt implies a constant acceleration in the aircraft rest frame, which is exactly Problem 1.15.

Let us now address the cable issue. From Problem 1.15 we obtain the world lines of the two spacecrafts

$$x_1(t) = \frac{c^2}{a}\left(\sqrt{1 + \frac{a^2 t^2}{c^2}} - 1\right); \qquad x_2(t) = L + \frac{c^2}{a}\left(\sqrt{1 + \frac{a^2 t^2}{c^2}} - 1\right)$$

where integration constants have been set such that $x_1(0) = 0$ and $x_1(0) = L$. In the original frame, the two spacecrafts keep at distance L at every time: one would deduce that the cable does not break, however, since the cable is moving as well, we must compute its length in the aircraft rest frame. Let us consider, in particular, an inertial frame O' which is istantaneously at rest with aircraft 1, and let us fix its origin so that $x_1' = t_1' = 0$ corresponds to the event in which aircraft 1 reaches speed v in the original frame O; for convenience, from now on we translate the origin of O so that also in this frame $x_1 = t_1 = 0$, while $x_2 = L$ and $t_2 = 0$ corresponds to the event for which also spacecraft 2 reaches the same speed v. After a Lorentz transformation, we obtain that, according to O', spacecraft 2 is at rest in $x_2' = \gamma L$ at $t_2' = -\beta\gamma L$. In order to understand where spacecraft 2 is at time $t' = 0$, thus obtaining the relative distance as seen from aircraft 1, we need to integrate again the equation of the hyperbolic motion, which tells us that, in a time interval $\beta\gamma L$ and starting from rest, aircraft 2 will travel a distance

$$\frac{c}{a}\sqrt{c^2 + a^2(\gamma\beta L)^2} - \frac{c^2}{a}$$

so that the total distance will be

$$\gamma L + \frac{c}{a}\left(\sqrt{c^2 + (a\gamma\beta L)^2} - 1\right)$$

which is surely greater than L, hence the cable will break.

1.17 A muon, which is a particle (usually indicated by the Greek letter μ) of mass $m = 1.89 \times 10^{-28}$ kg and carrying the same electric charge as the electron, has a mean life time $\tau = 2.2 \times 10^{-6}$ s when it is at rest. The particle is accelerated instantaneously through a potential gap $\Delta V = 10^8$ V. What is the expected life time t of the particle in the laboratory after the acceleration? What is the expected distance D traveled by the particle before decaying?

Answer:

$$t = \tau \frac{(mc^2 + e\Delta V)}{mc^2} \simeq 4.28 \times 10^{-6} \text{ s}$$

$$D = \frac{\sqrt{(e\Delta V)^2 + 2emc^2\Delta V}}{mc^2} c\tau = 1.1 \times 10^3 \text{ m.}$$

Notice that the average traveled distance D is larger than what expected in absence of time dilation, which is limited to $c\tau$ due to the finiteness of the speed of light. The increase in the traveled distance for relativistic unstable particles is one of the best experimental proofs of time dilation, think e.g. of the muons created when cosmic rays collide with the upper regions of the atmosphere: a large fraction of them reaches Earth's surface and that is possible only since their life times appear dilated in Earth's frame.

1.18 The energy of a particle is equal to 2.5×10^{-12} J, its momentum is 7.9×10^{-21} N s. What are its mass m and velocity v?

Answer: $m = \sqrt{E^2 - c^2 p^2}/c^2 \simeq 8.9 \times 10^{-30}$ kg, $v = pc^2/E \simeq 2.84 \times 10^8$ m/s.

1.19 A particle of mass $M = 10^{-27}$ kg decays, while at rest, into a particle of mass $m = 4 \times 10^{-28}$ kg plus a photon. What is the energy E of the photon?

Answer: The two outgoing particles must have opposite momenta with an equal modulus p to conserve total momentum. Energy conservation is then written as $Mc^2 = \sqrt{m^2 c^4 + p^2 c^2} + pc$, so that

$$E = pc = \frac{M^2 - m^2}{2M} c^2 = 0.42\, Mc^2 \simeq 3.78 \times 10^{-11} \text{ J} \simeq 2.36 \times 10^8 \text{ eV},$$

while the energy of the massive particle is $c^2 (M^2 + m^2)/(2M)$.

1.20 Consider a system composed of two photons of momenta k_1 and k_2. Determine its invariant mass M_{inv} and the velocity v_{cm} of its center of mass frame.

Answer: Since $E_1 = k_1 c$ ed $E_2 = k_2 c$, where $k_1 = |k_1|$ and $k_2 = |k_2|$, we have

$$M_{inv}^2 c^2 = (k_1 + k_2)^2 - |k_1 + k_2|^2 = 2k_1 k_2 (1 - \cos\theta)$$

where θ is the angle formed by the directions of the two photons. Moreover

$$v_{CM} = \frac{k_1 + k_2}{k_1 + k_2} c; \quad v_{CM} = c \sqrt{1 - (1 - \cos\theta)\frac{2k_1 k_2}{(k_1 + k_2)^2}}.$$

Therefore only in the case of collinear photons ($\theta = 0$) we have a vanishing invariant mass and $v_{cm} = c$.

1.21 A spaceship of initial mass $M = 10^4$ kg is boosted by a photonic engine: a light beam is emitted opposite to the direction of motion, with a power $W = 10^{13}$ W, as measured in the spaceship rest frame. What is the derivative of the spaceship rest mass with respect to its proper time? What is the spaceship acceleration a in the frame instantaneously at rest with it?

Answer: The engine power must be subtracted from the spaceship energy, which is Mc^2 in its rest frame, hence $dM/dt = -W/c^2 \simeq 1.1 \times 10^{-4}$ kg/s. Since the particles emitted by the engine are photons, they carry a momentum equal to $1/c$ times their energy, hence

$$a(\tau) = \frac{W}{c(M - W\tau/c^2)}.$$

1.22 Consider again Problem 1.21. If the spaceship moves along the positive x direction and leaves the space station at $\tau = 0$, compute its velocity with respect to the station reference frame (which is assumed to be inertial) as a function of the spaceship proper time.

Answer: According to the solution of previous Problem 1.21, the spaceship acceleration a in the frame instantaneously at rest with it is

$$a(\tau) = \frac{W}{c(M - \tau W/c^2)} = \frac{\alpha c}{1 - \alpha\tau}; \qquad \alpha = \frac{W}{Mc^2} \simeq 1.1 \times 10^{-8} \text{ s}^{-1}.$$

Recalling from the solution of Problem 1.15 that

$$v = \frac{u}{\sqrt{1 + \frac{u^2}{c^2}}}, \qquad \frac{du}{d\tau} = a(\tau)\sqrt{1 + \frac{u^2}{c^2}},$$

where u is the x-component of the four-velocity we can easily integrate last equation

$$\frac{du}{\sqrt{1 + \frac{u^2}{c^2}}} = \frac{\alpha c\, d\tau}{1 - \alpha\tau}$$

obtaining

$$\frac{u}{c} = \sinh\left(-\ln(1 - \alpha\tau)\right).$$

Expressing v/c as a function of u/c we finally get

$$\frac{v}{c} = \tanh\left(-\ln(1 - \alpha\tau)\right) = \frac{1 - (1 - \alpha\tau)^2}{1 + (1 - \alpha\tau)^2}.$$

A much simpler way to obtain the same result is the following. Since all photons are emitted by the engine in the same direction (they are collinear), from Problem 1.20 we know that, independently of their energies, they can be considered as a single photon, i.e. a system of zero invariant mass. Therefore, from the point of view of the spaceship, what happens after a proper time τ can be described by the result of Problem 1.19: a spaceship of initial mass M and which is initially at rest decays into a photon plus a residual spaceship of residual mass $M(1 - \alpha\tau)$; the fact that photons are emitted continuously over a finite time interval is irrelevant, since, from the kinematic point of view, we are only interested in the initial and final states. Then, from the solution to Problem 1.19 and setting $m = M(1 - \alpha\tau)$, we easily obtain:

$$\frac{v(\tau)}{c} = \frac{p(\tau)c}{E(\tau)} = \frac{M^2 - M^2(1 - \alpha\tau)^2}{M^2 + M^2(1 - \alpha\tau)^2} = \frac{1 - (1 - \alpha\tau)^2}{1 + (1 - \alpha\tau)^2}$$

which coincides with previous result. The solution is defined, of course, only for $\alpha\tau <$ 1 since after that the spaceship mass vanishes. Notice that, for $\alpha\tau = 1$, the spaceship does not disappear, since that would not be consistent with kinematic constraints (the total invariant mass must stay M forever): it becomes a photon of momentum $Mc/2$. However, from $\tau = 1/\alpha$ on, the engine must stop, since a photon cannot emit another photon in the opposite direction (that would violate four-momentum conservation, leading in particular to a different invariant mass).

One can also obtain time t in the station frame as a function of proper time τ:

$$t = \int_0^\tau d\tau' \gamma(\tau') = \int_0^\tau d\tau' \frac{1 + (1 - \alpha\tau)^2}{2(1 - \alpha\tau)} = \frac{1}{2\alpha} \left[\frac{1 - (1 - \alpha\tau)^2}{2} - \ln(1 - \alpha\tau) \right]$$

which diverges as $\alpha\tau \to 1$.

1.23 A spaceprobe of mass $M = 10$ kg is boosted by a laser beam of frequency $\nu = 10^{15}$ Hz and power $W = 10^{12}$ W, which is directed from Earth against an ideal reflecting mirror (i.e. reflecting all incoming photons) placed in the back of the probe. Assuming that the probe is initially at rest with respect to Earth, and that the laser beam is always parallel to the spaceprobe velocity and orthogonal to the mirror, determine the evolution of the spaceprobe position in the Earth frame and compute the total time Δt for which the laser must be kept switched on, in order for the spaceprobe to reach a velocity $v = 0.5\,c$.

Answer: In the reference frame of the spaceship, every photon gets reflected from the mirror with a negligible change in frequency ($\Delta\lambda'/\lambda' \sim h\nu'/(Mc^2) < 10^{-37}$, see Problem 1.33), hence it transfers a momentum $2h\nu'/c$ to the spaceprobe, where ν' is the photon frequency in the spaceprobe frame. The acceleration of the probe in its proper frame is therefore

$$a' = \frac{2h\nu'}{Mc} \frac{dN'_\gamma}{dt'} = \frac{2W'}{Mc}$$

where W' is the power of the beam measured in the proper frame, which is given by the photon energy, $h\nu'$, times the rate at which photons arrive, i.e. the number of photons hitting the mirror per unit time, dN'_γ/dt'. The above result is twice as large as that obtained if the light beam is emitted directly by the spaceprobe, as in Problem 1.21, since in this case each reflected photon transfers twice its momentum.

Both ν' and the rate of arriving photons are frequencies, hence they get transformed by Doppler effect and we can write

$$W' = h\nu' \frac{dN'_\gamma}{dt'} = \sqrt{\frac{1-\beta}{1+\beta}} h\nu \sqrt{\frac{1-\beta}{1+\beta}} \frac{dN_\gamma}{dt} = \frac{1-\beta}{1+\beta} W$$

which gives us the transformation law for the beam power. From the acceleration in the proper frame, $a' = 2W(1-\beta)/[(1+\beta)Mc]$, one obtains the derivative of β with respect to the proper time τ (see Problem 1.15)

$$\frac{d\beta}{d\tau} = \frac{dv}{cd\tau} = \frac{a'}{c}\left(1 - \frac{v^2}{c^2}\right) = \alpha(1-\beta)^2$$

where we have set $\alpha = 2W/(Mc^2)$. Last equation, after integration with the initial condition $\beta(0) = 0$, leads to $\alpha\tau = \beta/(1-\beta)$, i.e.

$$\beta(\tau) = \frac{\alpha\tau}{1+\alpha\tau}.$$

Regarding the position of the probe, we have

$$dx = c\beta dt = c\beta \frac{d\tau}{\sqrt{1-\beta^2}} = \frac{\alpha\tau}{\sqrt{1+2\alpha\tau}} d\tau$$

which gives, after integration and using $x(0) = 0$

$$x = \frac{c}{3\alpha}\left((\alpha\tau - 1)\sqrt{2\alpha\tau + 1} + 1\right).$$

Setting $\beta = 0.5$ we obtain $\tau = \alpha^{-1} = Mc^2/(2W) = 9 \times 10^5$ s. The corresponding time in the Earth frame can be obtained by integrating the relation

$$dt = \frac{d\tau}{\sqrt{1-\beta^2}} = \frac{1+\alpha\tau}{\sqrt{1+2\alpha\tau}} d\tau$$

yielding

$$t = \frac{1}{3\alpha}\left((\alpha\tau + 2)\sqrt{2\alpha\tau + 1} - 2\right).$$

In order to compute the total time Δt that the laser must be kept switched on, we have to consider that a photon reaching the spaceprobe at time t has left Earth at a time $t - x(t)/c$, where x is the probe position, hence the emission time is

$$t_{em} = t - x/c = \frac{1}{\alpha}\left(\sqrt{2\alpha\tau + 1} - 1\right).$$

Setting $\alpha\tau = 1$, we obtain $\Delta t = (\sqrt{3} - 1)/\alpha \simeq 6.59 \times 10^5$ s.

It is interesting to notice that, analogously to Problem 1.22, also in this case the solution can be obtained in a much simpler way by mapping the problem onto a photon-particle diffusion one. Indeed, all the photons emitted by the laser are collinear to each other and can be considered as one single incident photon; the same applies to reflected photons, which can be treated as one single reflected photon. Then the problem reduces to a one-dimensional problem in which a photon of momentum $k = Wt_{em}/c$ hits a particle of mass M which is initially at rest: we have to tune k so that after the collision the particle has a velocity $v = c/2$, i.e. a momentum $p = Mv/\sqrt{1 - v^2/c^2} = Mc/\sqrt{3}$. If we call $-k'$ the momentum of the reflected photon, conservation of four-momentum implies

$$k = p - k'; \qquad kc + Mc^2 = k'c + \sqrt{p^2c^2 + M^2c^4}$$

which combined together lead, after some algebra, to

$$k = \frac{Mc}{2}\left(\frac{p}{Mc} - 1 + \sqrt{1 + \left(\frac{p}{Mc}\right)^2}\right) = \frac{Mc}{2}\left(\sqrt{3} - 1\right)$$

hence to $t_{em} = Mc^2(\sqrt{3}-1)/(2W)$ which coincides with the result found previously.

1.24 An electron–proton collision can give rise to a fusion process in which all available energy is transferred to a neutron. As a matter of fact, there is a neutrino emitted whose energy and momentum in the present situation can be neglected. The proton rest energy is 0.938×10^9 eV, while those of the neutron and of the electron are respectively 0.940×10^9 eV and 5×10^5 eV. What is the velocity of the electron which, knocking into a proton at rest, may give rise to the process described above?

Answer: Notice that we are not looking for a minimum electron energy: since, neglecting the final neutrino, the final state is a single particle state, its invariant mass is fixed and equal to the neutron mass. That must be equal to the invariant mass of the initial system of two particles, leaving no degrees of freedom on the possible values of the electron energy: only for one particular value v_e of the electron velocity the reaction can take place.

A rough estimate of v_e can be obtained by considering that the electron energy must be equal to the rest energy difference $(m_n - m_p)\,c^2 = (0.940-0.938) \times 10^9$ eV plus the kinetic energy of the final neutron. Therefore the electron is surely relativistic and $(m_n - m_p)c$ is a reasonable estimate of its momentum: it coincides with the neutron momentum which is instead non-relativistic $((m_n - m_p)c \ll m_n c)$. The kinetic energy of the neutron is thus roughly $(m_n - m_p)^2 c^2/(2m_n)$, hence negligible with respect to $(0.940 - 0.938) \times 10^9$ eV. The total electron energy is therefore, within a good approximation, $E_e \simeq 2 \times 10^6$ eV, and its velocity is $v_e = c\sqrt{1 - m_e^2/E_e^2} \simeq 2.9 \times 10^8$ m/s. The exact result is obtained by writing $E_e = (m_n^2 - m_p^2 - m_e^2)c^2/(2m_p)$, which differs by less than 0.1 % from the approximate result.

1.25 A system made up of an electron and a positron, which is an exact copy of the electron but with opposite charge (i.e. its antiparticle), annihilates, while both particles are at rest, into two photons. The mass of the electron is $m_e \simeq 0.9 \times 10^{-30}$ kg: what is the wavelength of each outgoing photon? Explain why the same system cannot decay into a single photon.

Answer: The two photons carry momenta of modulus mc which are opposite to each other in order to conserve total momentum: their common wavelength is therefore, as we shall see in next Chapter, $\lambda = h/(m_e c) \simeq 2.4 \times 10^{-12}$ m. A single photon should carry zero momentum since the initial system is at rest, but then energy could not be conserved; more in general the initial invariant mass of the system, which is $2m_e$, cannot fit the invariant mass of a single photon, which is always zero.

1.26 A piece of copper of mass $M = 1$ g, is heated from $0\,°C$ up to $100\,°C$. What is the mass variation ΔM if the copper specific heat is $C_{Cu} = 0.4$ J/g°C?

Answer: The piece of copper is actually a system of interacting particles whose mass is defined as the invariant mass of the system. That is proportional to the total energy if the system is globally at rest, see Eq. (1.75). Therefore $\Delta M = C_{Cu}\,\Delta T/c^2 \simeq 4.45 \times 10^{-16}$ kg .

1.27 Consider a system made up of two point-like particles of equal mass $m = 10^{-20}$ kg, bound together by a rigid massless rod of length $L = 2 \times 10^{-4}$ m. The center of mass of the system lies in the origin of the inertial frame O, in the same frame the rod rotates in the x–y plane with an angular velocity $\omega = 3 \times 10^{10}$ s^{-1}. A second inertial reference frame O' moves with respect to O with velocity $v = 4c/5$ parallel to the x axis. Compute the sum of the kinetic energies of the particles at the same time in the frame O', disregarding corrections of order ω^3.

Answer: There are two independent ways of computing the sum of the kinetic energies of the particles. The first way, which is the simplest one, is based on the assumption that the kinetic energy of the system coincides with the sum of the kinetic energies of the particles, hence one can compute the total energy of the system in the reference O, which, in the chosen approximation, is: $E_t = m\,(2c^2 + \omega^2 L^2/4)$. In O' the total energy is computed by a Lorentz transformation $E'_t = E_t/\sqrt{1 - v^2/c^2} = (5/3)m\,(2c^2 + \omega^2 L^2/4)$, hence the sum of the kinetic energies is $E'_t - 2mc^2 = 4/3mc^2 + 5m\omega^2 L^2/12$. Numerically one has $12 \times 10^{-4}(1 + 1.25 \times 10^{-4})$ J. Alternatively, we can compute the velocities of both particles in a situation corresponding to equal time in O'. The simplest such situation is when the rod is parallel to the y axis and the particles move with velocity $v_\pm = \pm\omega L/2$ parallel to the x axis. Using Einstein formula one finds in O' the velocities $v'_\pm = c(4/5 \pm \omega L/(2c))/(1 \pm 2\omega L/(5c))$ and hence the kinetic energies $E'_\pm = mc^2(1/\sqrt{1 - (v'_\pm/c)^2} - 1) = mc^2\left((5 \pm 2\omega L/c)/(3\sqrt{1 - (\omega L/(2c))^2}) - 1\right)$. It is apparent that the sum $E'_+ + E'_-$ gives the known result up to corrections of order $(\omega L/c)^4$.

1.28 A spinning top, which can be described as a rigid disk of mass $M = 10^{-1}$ kg, radius $R = 5 \times 10^{-2}$ m and uniform density, starts rotating with angular velocity $\Omega = 10^3$ rad/s. What is the energy variation of the spinning top due to rotation, as seen from a reference frame moving with a relative speed $v = 0.9\,c$ with respect to it?

Answer: The speed of the particles composing the spinning top is surely non-relativistic in their frame, since it is limited by $\Omega R = 50\,\text{m/s} \simeq 1.67 \times 10^{-7}\,c$. In that frame the total energy is therefore, apart from corrections of order $(\Omega R/c)^2$, $E_{tot} = Mc^2 + I\Omega^2/2$, where I is the moment of inertia, $I = MR^2/2$. Lorentz transformations yield in the moving frame

$$E'_{tot} = \frac{1}{\sqrt{1 - v^2/c^2}} E_{tot} = \frac{1}{\sqrt{1 - v^2/c^2}} \left(Mc^2 + \frac{1}{2} I\Omega^2 \right),$$

to be compared with the energy observed in absence of rotation, $Mc^2/\sqrt{1 - v^2/c^2}$. Therefore, the energy variation due to rotation in the moving frame is $\Delta E' = I\Omega^2/(2\sqrt{1 - v^2/c^2}) \simeq 143\,\text{J}$.

1.29 A photon of energy E knocks into an electron at rest producing a final state composed of an electron–positron pair plus the initial electron: all three final particles have the same momentum. What is the value of E and the common momentum p of the final particles?

Answer: $E = 4\,mc^2 \simeq 3.3 \times 10^{-13}\,\text{J}, \quad p = E/3c = 4/3\,mc \simeq 3.6 \times 10^{-22}\,\text{N/m}.$

1.30 A particle of mass $M = 1\,\text{GeV}/c^2$ and energy $E = 10\,\text{GeV}$ decays into two particles of equal mass $m = 490\,\text{MeV}$. What is the maximum angle that each of the two outgoing particles may form, in the laboratory, with the trajectory of the initial particle?

Answer: Let $\hat{x}$ be the direction of motion of the initial particle, and x–y the decay plane: this is defined as the plane containing both the initial particle momentum and the two final momenta, which are indeed constrained to lie in the same plane by total momentum conservation. Let us consider one of the two outgoing particles: in the center of mass frame it has energy $\epsilon = Mc^2/2 = 0.5\,\text{GeV}$ and a momentum $p_x = p\cos\theta$, $p_y = p\sin\theta$, with θ being the decay angle in the center of mass frame and

$$p = c\sqrt{\frac{M^2}{4} - m^2} \simeq \frac{0.1\,\text{GeV}}{c}.$$

The momentum components in the laboratory are obtained by Lorentz transformations with parameters $\gamma = (\sqrt{1 - v^2/c^2})^{-1} = E/(Mc^2) = 10$ and $\beta = v/c = \sqrt{1 - 1/\gamma^2} \simeq 0.995$,

$$p'_y = p_y = p\sin\theta$$

$$p'_x = \gamma(p\cos\theta + \beta\epsilon).$$

It can be easily verified that, while in the center of mass frame the possible momentum components lie on a circle of radius p centered in the origin, in the laboratory p'_x and p'_y lie on an ellipse of axes γp and p, centered in $(\gamma\beta\epsilon, 0)$. If θ' is the angle formed in the laboratory with respect to the initial particle trajectory and defining $\alpha \equiv \beta\epsilon/p \simeq 5$, we can write

$$\tan\theta' = \frac{p'_y}{p'_x} = \frac{1}{\gamma}\frac{\sin\theta}{\cos\theta + \alpha}.$$

If $\alpha > 1$ the denominator is always positive, $\tan\theta'$ is limited and $|\theta'| < \pi/2$, i.e. the particle is always forward emitted, in the laboratory, with a maximum possible angle which can be computed by solving $d\tan\theta'/d\theta = 0$; that can also be appreciated pictorially by noticing that, if $\alpha > 1$, the ellipse containing the possible momentum components does not contain the origin. Finally one finds

$$\theta'_{max} = \tan^{-1}\left(\frac{1}{\gamma}\frac{1}{\sqrt{\alpha^2 - 1}}\right) \simeq 0.02 \text{ rad.}$$

1.31 A particle of mass $M = 10^{-27}$ kg, which is moving in the laboratory with a speed $v = 0.99\,c$, decays into two particles of equal mass $m = 3 \times 10^{-28}$ kg. What is the possible range of energies (in GeV) which can be detected in the laboratory for each of the outgoing particles? Supposing that in the center of mass frame the final particles are emitted isotropically, i.e. with equal probability in all directions, what is the probability, in the laboratory, of detecting a particle of energy in the range $[E, E + dE]$?

Answer: In the CM frame the outgoing particles have equal energy and modulus of the momentum completely fixed by the kinematic constraints: $\epsilon = Mc^2/2$ and $p = c\sqrt{M^2/4 - m^2}$. The only free variable is the decaying angle θ, measured with respect to the initial particle trajectory, which however results in a variable energy E in the laboratory frame. Indeed by Lorentz transformations $E = \gamma(\epsilon + vp\cos\theta)$, with $\gamma = (1 - v^2/c^2)^{-1/2}$, hence ignorance about θ in the CM frame results in ignorance about E in the laboratory. The maximum/mininum values of E are obtained as $\cos\theta$ becomes maximum/minimum, hence

$$E_{max/min} = \gamma\left(\frac{Mc^2}{2} \pm vp\right) \quad \Rightarrow \quad E_{max} \simeq 3.56 \text{ GeV}, \ E_{min} \simeq 0.41 \text{ GeV.}$$

Since the infinitesimal solid angle in spherical coordinates is $\sin\theta\,d\theta\,d\phi$, where ϕ is the azimuthal angle, an isotropic distribution in the CM frame means that the probability for one of the two final particles to be emitted with an angle in the range $[\theta, \theta + d\theta]$ is $P_\theta(\theta)d\theta = \sin\theta\,d\theta/2$. The energy in the laboratory is a function of θ, hence, calling $P_E dE$ the probability for the energy to be in the range $[E, E + dE]$, we have $P_E dE = P_\theta(\theta)d\theta = \sin\theta\,d\theta/2$. Since, by differentiating $E(\theta)$, we get

$dE = \gamma v p \sin \theta \, d\theta$, we finally obtain

$$P_E dE = \frac{dE}{2\gamma v p}$$

i.e. a flat distribution between the minimum and maximum possible values.

1.32 A particle of rest energy $Mc^2 = 10^9$ eV, which is moving in the laboratory with momentum $p = 5 \times 10^{-18}$ N s, decays into two particles of equal mass $m = 2 \times 10^{-28}$ kg. In the center of mass frame the decay direction is orthogonal to the trajectory of the initial particle. What is the angle between the trajectories of the outgoing particles in the laboratory?

Answer: Let $\hat{x}$ be the direction of the initial particle and $\hat{y}$ the decay direction in the center of mass frame, $\hat{y} \perp \hat{x}$. For the process described in the text, $\hat{x}$ is a symmetry axis, hence the outgoing particles will form the same angle θ also in the laboratory. For one of the two particles we can write $p_x = p/2$ by momentum conservation in the laboratory, and $p_y = c\sqrt{M^2/4 - m^2}$ by energy conservation. Finally, the angle between the two particles is $2\theta = 2 \operatorname{atan}(p_y/p_x) \simeq 0.207$ rad.

1.33 Compton Effect

A photon of wavelength λ knocks into an electron at rest. After the elastic collision, the photon moves in a direction forming an angle θ with respect to its original trajectory. What is the change $\Delta\lambda \equiv \lambda' - \lambda$ of its wavelength as a function of θ?

Answer: Let q and q' be respectively the initial and final momentum of the photon, and p the final momentum of the electron. As we shall discuss in next chapter, the photon momentum is related to its wavelength by the relation $q \equiv |q| = h/\lambda$, where h is Planck's constant. Total momentum conservation implies that q, q' and p must lie in the same plane, which we choose to be the x–y plane, with the x-axis parallel to the photon initial trajectory. Momentum and energy conservation lead to:

$$p_x = q - q' \cos \theta,$$

$$p_y = q' \sin \theta,$$

$$qc + m_e c^2 - q'c = \sqrt{m_e c^4 + p_x^2 c^2 + p_y^2 c^2}.$$

Substituting the first two equations into the third and squaring both sides of the last we easily arrive, after some trivial simplifications, to $m_e(q - q') = qq'(1 - \cos\theta)$, which can be given in terms of wavelengths as follows

$$\Delta\lambda \equiv \lambda' - \lambda = \frac{h}{m_e c}(1 - \cos\theta); \qquad \frac{\Delta\lambda}{\lambda} = \frac{h\nu}{m_e c^2}(1 - \cos\theta).$$

The difference is always positive, since part of the photon energy, depending on the diffusion angle θ, is always transferred to the electron. This phenomenon, known as

Compton effect, is not predicted by the classical theory of electromagnetic waves and is an experimental proof of the corpuscular nature of radiation. Notice that, while the angular distribution of outgoing photons can only be predicted on the basis of the quantum relativistic theory, i.e. Quantum Electrodynamics, the dependence of $\Delta\lambda$ on θ that we have found is only based on relativistic kinematics and can be used to get an experimental determination of h. The coefficient $h/(m_e c)$ is known as *Compton wavelength,* which for the electron is of the order of 10^{-12} m, so that the effect is not detectable $(\Delta\lambda/\lambda \simeq 0)$ for visible light.

1.34 A particle of mass M decays, while at rest, into three particles of equal mass m. What is the maximum and minimum possible energy for each of the outgoing particles?

Answer: Let E_1, E_2, E_3 and p_1, p_2, p_3 be respectively the energies and the momenta of the three outgoing particles. We have to find, for instance, the maximum and minimum value of E_1 which are compatible with the constraints $p_1 + p_2 + p_3 = 0$ and $E_1 + E_2 + E_3 = M c^2$. The minimum value is realized when the particle is produced at rest, $E_{1min} = m c^2$, implying that the other two particles move with equal and opposite momenta. Finding the maximum value requires some more algebra. From $E_1^2 = m^2 c^4 + p_1^2 c^2$ and momentum conservation we obtain

$$E_1^2 = m^2 c^4 + |p_2 + p_3|^2 c^2 = m^2 c^4 + (E_2 + E_3)^2 - \mu^2 c^4$$

where μ is the invariant mass of particles 2 and 3, $\mu^2 c^4 = (E_2 + E_3)^2 - |p_2 + p_3|^2 c^2$. Applying energy conservation, $E_2 + E_3 = (M c^2 - E_1)$, last equation leads to

$$E_1 = \frac{1}{2M c^2} \left(m^2 c^4 + M^2 c^4 - \mu^2 c^4 \right).$$

We have written E_1 as a function of μ^2: F_{1max} corresponds to the minimum possible value for the invariant mass of the two remaining particles. On the other hand it can be easily checked (see Eq. (1.79)) that, for a system made up of two or more massive particles, the minimum possible value of the invariant mass is equal to the sum of the masses and is attained when the particles are at rest in their center of mass frame, meaning that the particles move with equal velocities in any other reference frame. Therefore $\mu_{min} = 2m$ and $E_{1max} = (M^2 - 3m^2) c^2/(2M)$: this value is obtained in particular for $p_2 = p_3$.

1.35 A particle of mass M decays, while at rest, into N particles of masses m_i, $i = 1, N$ ($N \geq 3$). What is the maximum and minimum possible energy for each of the outgoing particles?

Answer: The problem is very similar to the previous one, we work it out for particle 1 but the solution can be trivially generalized. The minimum energy is always $E_{1min} = m_1 c^2$, since the particle can be produced at rest with the remaining particles taking care of momentum conservation. This is true even if $m_1 = 0$, like for a photon: in this case the minimum energy is just a lower bound, since a photon with exactly zero energy means no photon at all.

As for the maximum energy, let us call p_i the momenta of the various emitted particles, q the total momentum of all particles except particle 1 and ϵ their energy ($q \equiv \sum_{i>1} p_i$, $\epsilon = \sum_{i>1} E_i$). Then $p_1 = -q$ and we can write

$$E_1^2 = m_1^2 c^4 + |q|^2 c^2 = m_1^2 c^4 + \epsilon^2 - \mu^2 c^4$$

where μ is the invariant mass of particles $2, 3, \ldots N$. Applying energy conservation, $\epsilon = (M c^2 - E_1)$, we have again

$$E_1 = \frac{1}{2M c^2} \left(m_1^2 c^4 + M^2 c^4 - \mu^2 c^4 \right)$$

which is maximum when μ attains its minimum, which by Eq. (1.79) is $\mu_{min} = \sum_{i>1} m_i$. Finally we can write

$$E_{1max} = \frac{M^2 + m_1^2 - \left(\sum_{i=2}^{N} m_i \right)^2}{2M} c^2 .$$

1.36 A particle at rest, whose mass is $M = 750$ MeV/c^2, decays into a photon and a second, lighter, particle of mass $m = 135$ MeV/c^2. Subsequently the lighter particle decays into two further photons. Considering all the possible decay angles of the second particle, compute the maximum and the minimum values of the possible energies of the three final photons.

One can ask the same question in the case of a direct decay of the first particle into three photons whose energies are constrained only by energy-momentum conservation. What are the maximum and minimum energies of the final photons in the direct decay?

Answer: This problem is analogous to Problem 1.34. We first consider the direct decay. If E_i with $i = 1, 2$ are the energies of two, arbitrarily chosen, final photons, $p_i = E_i/c$ are their momenta. If θ is the angle between these momenta, on account of momentum conservation, we can compute the momentum of the third photon as the length of the third side of a triangle whose other sides have lengths p_1 and p_2 and form an angle $\pi - \theta$. Therefore energy conservation gives: $\sqrt{E_1^2 + E_2^2 + 2E_1 E_2 \cos\theta} + E_1 + E_2 = Mc^2$ from which we get: $M^2 c^4 - 2Mc^2(E_1 + E_2) = -2E_1 E_2(1 - \cos\theta)$. Since $0 \le 1 - \cos\theta \le 2$ one has the inequalities:

$$Mc^2/2 \le (E_1 + E_2) \quad \text{and} \quad (Mc^2 - 2E_1)(Mc^2 - 2E_2) \ge 0 .$$

If we interpret E_1 and E_2 as cartesian coordinates of a point in a plane, we find that the point must lie in a triangle with vertices in the points of coordinates $(0, Mc^2/2)$, $(Mc^2/2, 0)$ and $(Mc^2/2, Mc^2/2)$. This shows that each of the three photon energies can range between 0 and $Mc^2/2$. In Particle Physics the distribution of points associated with a sample of decay events (in this case the distribution of points inside the triangle described above) is called Dalitz plot.

Now consider the indirect decay and assume that photons 1 and 2 are the decay products of the second particle with mass m. In this case there is a constraint for the third photon with energy $E_3 = Mc^2 - E_1 - E_2$. Indeed its momentum $p_3 = E_3/c$ must be equal, and opposite, to that of the light particle whose energy is $E_1 + E_2$. Thus we have $(E_1 + E_2)^2 - (Mc^2 - E_1 - E_2)^2 = m^2c^4$ which implies $E_1 + E_2 = (M^2+m^2)c^2/(2M)$, which can be read as the equation of a line intersecting the above mentioned triangle. It is apparent that the boundaries of the intersection segment give the maximum and minimum possible values for E_1 and E_2, which are respectively $m^2c^2/(2M) = 12.15$ MeV and $Mc^2/2 = 375$ MeV. The energy of the third photon is instead fixed and equal to $E_3 = (M^2 - m^2)c^2/(2M)$: we have $E_3 < Mc^2/2$ and, for the given values of M and m, also $E_3 > m^2c^2/(2M)$.

Another significant difference for the indirect decay is that two of the emitted photons will always form a system with an invariant mass equal to m. Therefore, if one observes several examples of such decays and measures in each case all kinematic parameters of the emitted photons, thus reconstructing the invariant mass for each couple of photons and making an histogram for its probability distribution, one would obtain a smooth distribution for the direct decay, and instead a smooth distribution plus a sharp peak located at m for the indirect decay, thus "discovering" the presence of the intermediate particle m, even if it is not directly revealed by the detector. This is a common strategy for the discovery of new particles, the last renowned example being the Higgs boson.

1.37 A proton beam is directed against a laser beam coming from the opposite direction and having wavelength 0.5×10^{-6} m. Determine what is the minimum value needed for the kinetic energy of the protons in order to produce the reaction (proton + photon $\rightarrow$ proton + π), where the π particle has mass $m \simeq 0.14\,M$, the proton mass being $M \simeq 0.938$ GeV$/c^2$.

Answer: Let p and k be the momenta of the proton and of the photon respectively, $k = h/\lambda \simeq 2.48$ eV$/c$. The reaction can take place only if the invariant mass of the initial system is larger or equal to $(M + m)$: that is most easily seen in the center of mass frame, where the minimal energy condition corresponds to the two final particles being at rest. In particular, if E is the energy of the proton, we can write

$$(E + kc)^2 - (p - k)^2 c^2 \geq (M + m)^2 c^4, \quad \text{hence} \quad E + pc \geq \frac{mc^2}{kc}(M + m/2)c^2.$$

Taking into account that $mc^2 \simeq 0.13$ GeV and $kc \simeq 2.48$ eV we deduce that $E + pc \sim 5 \times 10^7\,Mc^2$, so that the proton is ultra-relativistic and $E \simeq pc$. The minimal kinetic energy of the proton is therefore $p_{min}c \simeq (mc/k)(M + m/2)c^2/2 \simeq 2.6 \times 10^7$ GeV.

1.38 A particle of mass M decays into two particles of masses m_1 and m_2. A detector reveals the energies and momenta of the outgoing particles to be $E_1 = 2.5$ GeV,

$E_2 = 8\,\text{GeV}$, $p_{1x} = 1\,\text{GeV}/c$, $p_{1y} = 2.25\,\text{GeV}/c$, $p_{2x} = 7.42\,\text{GeV}/c$ and $p_{2y} = 2.82\,\text{GeV}/c$. Determine the masses of all involved particles, as well as the velocity v of the initial particle.

Answer: $M \simeq 3.69\,\text{GeV}/c^2$, $m_1 \simeq 0.43\,\text{GeV}/c^2$, $m_2 \simeq 1\,\text{GeV}/c^2$, $v_x = 0.802\ c$, $v_y = 0.483\ c$.

1.39 A particle of mass $\mu = 0.14\,\text{GeV}/c^2$ and momentum directed along the positive z axis, knocks into a particle at rest of mass M. The final state after the collision is made up of two particles of mass $m_1 = 0.5\,\text{GeV}/c^2$ and $m_2 = 1.1\,\text{GeV}/c^2$ respectively. The momenta of the two outgoing particles form an equal angle $\theta = 0.01\,\text{rad}$ with the z axis and have equal magnitude $p = 10^4\,\text{GeV}/c$. What is the value of M?

Answer: Momentum conservation gives the momentum of the initial particle, $k = 2p\cos\theta$. The initial energy is therefore $E_{in} = \sqrt{\mu^2 c^4 + k^2 c^2} + M\,c^2$ and must be equal to the final energy $E_{fin} = \sqrt{m_1^2 c^4 + p^2 c^2} + \sqrt{m_2^2 c^4 + p^2 c^2}$, hence

$$Mc^2 = \sqrt{m_1^2 c^4 + p^2 c^2} + \sqrt{m_2^2 c^4 + p^2 c^2} - \sqrt{\mu^2 c^4 + 4p^2 \cos\theta^2 c^2}.$$

The very high value of p makes it sensible to apply the ultra-relativistic approximation,

$$Mc^2 \simeq pc\left(1 + \frac{m_1^2 c^2}{2p^2}\right) + pc\left(1 + \frac{m_2^2 c^2}{2p^2}\right) - 2pc\cos\theta\left(1 + \frac{\mu^2 c^2}{8p^2\cos\theta^2}\right)$$

$$\simeq pc\left(\theta^2 + (2m_1^2 + 2m_2^2 - \mu^2)c^2/(4p^2)\right) \simeq pc\,\theta^2 = 1\,\text{GeV}.$$

1.40 A flux of particles, each carrying an electric charge $q = 1.6 \times 10^{-19}$ C, is moving along the x axis with a constant velocity $v = 0.9\ c$. If the total carried current is $I = 10^{-9}$ A, what is the linear density of particles, as measured in the reference frame at rest with them?

Answer: If d_0 is the distance among particles in their rest frame, the distance measured in the laboratory appears contracted and equal to $d = \sqrt{1 - v^2/c^2}\ d_0$. The electric current is given by $I = d^{-1} vq$, hence the particle density in the rest frame is

$$d_0^{-1} = \sqrt{1 - v^2/c^2}\,\frac{I}{vq} \simeq 10.1\ \text{particles/m}.$$

An alternative approach is to apply the transformation properties of the four-current, whose components are $(\rho c, I)$ in the laboratory and $(\rho_0 c, 0)$ in the rest frame of the particles, then from Eq. (1.92):

$$\rho_0 c = \gamma\left(\rho c - \frac{v}{c}I\right)$$

which coincides with the result above considering that $I = \rho v$ and $\rho_0 = q/d_0$.

1.41 Transformation Laws for Electromagnetic Fields

Our inertial reference frame moves with respect to a conducting rectilinear wire with velocity $v = 0.9\,c$ parallel to the wire. In its reference frame the wire appears neutral and one has an electric current $I = 1$ A through the wire in the direction of our velocity. We adopt a simplified scheme in which the current is carried by electrons with a linear density $\rho_{wire} = 6 \times 10^{16}$ m^{-1}, moving with an average uniform velocity $V = 10^2$ m/s in the opposite direction with respect to our velocity. The wire is made neutral by protons at rest, having the same linear density as the electrons. Coming back to our reference frame, do we detect any electric field? If the answer to our question is positive, what is the absolute value of the electric field at a distance $r = 1$ cm from the wire?

Answer: The answer could be obtained straightforwardly by applying the transformation laws of electromagnetic fields that we have derived (see Eq. (1.97)). However, let us proceed in a different way, starting from the sources observed in the different inertial frames: the reader is invited to verify that results coincide with those obtainable from Eq. (1.97).

If the electrons in the wire are uniformly distributed the distance between two neighboring electrons is $d = \rho_{wire}^{-1} = 1.66 \times 10^{-17}$ m in the wire frame. Due to the length contraction the same distance is $d/\sqrt{1 - (V/c)^2}$ in the electron frame, that is, in a frame moving with respect to the wire with the average electron velocity. Using Einstein formula we compute our velocity with respect to the electron frame $v' = (v + V)/(1 + vV/c^2)$ and the electron density in our frame: $\rho_{moving,e} = \sqrt{1 - (V/c)^2}/(d\sqrt{1 - (v'/c)^2}) = (1 + vV/c^2)/(d\sqrt{1 - (v/c)^2})$ while the proton density is $\rho_{moving,p} = 1/(d\sqrt{1 - (v/c)^2}) \equiv \gamma/d$. Thus the resulting charge density is $\rho_{moving,tot} = -evV/(c^2d\sqrt{1 - (v/c)^2}) = -I\beta\gamma/c$ where we have set $\beta = v/c$. Since Maxwell equations are the same in every inertial frame, we conclude that in our frame we have an electric field with absolute value $E_{moving} = \beta\gamma I/(2\pi\epsilon_0 rc) = \beta\gamma c B_{wire} = 1.23 \times 10^4$ V/m and directed towards the wire. B_{wire} is the absolute value of the magnetic induction at the same distance from the wire in the wire frame; the electric field measured in our frame is orthogonal with respect to the original magnetic field (in particular it is directed like $v \wedge B_{wire}$). It is also easy to verify that the electric current in our frame is γI, so that we measure a magnetic field of absolute value $B_{moving} = \gamma B_{wire}$ and parallel to the original magnetic field.

The complete set of field transformation rules, Eq. (1.97), could be obtained through the analysis of similar *gedanken* experiments. To that purpose, the reader is invited to compute the electric and magnetic fields felt by an observer moving: (a) parallel to a wire having a uniform charge density and zero electric current; (b) parallel to an infinite plane carrying zero charge density and a uniform current density orthogonal to the observer velocity; (c) orthogonal to an infinite plane carrying uniform charge density and zero electric current.

1.42 A relativistic particle with mass m and charge q moves in a time independent magnetic field B. In cylindrical coordinates z, ρ, ϕ the magnetic field components are $B_\phi = B_\rho = 0$ and $B_z = f(\rho)$, those of the vector potential are $A_z = A_\rho = 0$ and

$A_\phi = \rho a(\rho)$ with $a(\rho)$ a positive increasing function of ρ. The Lagrangian equations of the particle (see (1.104)) are easily reduced to three prime integrals. Compute them.

Answer: The Lagrangian is $\mathcal{L} = -mc^2\sqrt{1 - v^2/c^2} + q\rho^2 a\dot\phi$, we denote the time derivative by a dot and $v^2 = (\dot z)^2 + (\dot\rho)^2 + \rho^2(\dot\phi)^2$. We have three prime integrals because the Lagrangian is independent of z, ϕ and t. The corresponding prime integrals are: the energy $mc^2/\sqrt{1 - v^2/c^2} \equiv E$, the z-momentum component $P_z = m\dot z/\sqrt{1 - v^2/c^2}$ and the z-angular momentum component $L_z = m\rho^2\dot\phi/\sqrt{1 - v^2/c^2} + q\rho^2 a(\rho)$. Therefore we have

$$\dot z = P_z c^2/E \ , \ \dot\phi = (L_z/\rho^2 - qa(\rho))c^2/E$$

$$\dot\rho = \pm c\sqrt{1 - c^2(m^2c^2 + P_z^2 + (L_z/\rho^2 - qa(\rho))^2)/E^2}.$$

These equations must be integrated starting from suitable initial coordinates. A typical ultra-relativistic case corresponds to $(P_z c/E)^2 = 1 - 2\epsilon^2$ and $m^2 c^4/E^2 = \epsilon^2$, this implies $\epsilon^2 \geq c^2(L_z/\rho^2 - qa(\rho))^2)/E^2$. This sets bounds on the range of ρ which cannot vanish if L_z does not, while is bound by the increasing behavior of $a(\rho)$. The values of ρ for which one has equality correspond to spiral orbits of the particle.

Chapter 2
Introduction to Quantum Physics

The gestation of Quantum Physics has been very long and its phenomenological foundations were various. Historically, the original idea came from the analysis of the black body spectrum. This is not surprising since the black body, in fact an oven in thermal equilibrium with the electromagnetic radiation, is a simple and fundamental system once the law of electrodynamics are established. As a matter of fact many properties of the spectrum can be deduced starting from the general laws of electrodynamics and thermodynamics; the crisis came from the violation of the equipartition of energy. That suggested to Planck the idea of *quantum*, from which everything originated. Of course a long sequence of different discoveries, first of all the photoelectric effect, the line spectra for atomic emission/absorption, the Compton effect and so on, gave a compelling evidence for the new theory.

Due to the particular limits of the present notes, an exhaustive analysis of the whole phenomenology is impossible. Even a clear discussion of the black body problem needs an exceeding amount of space. Therefore we have chosen a particular line, putting major emphasis on the photoelectric effect and on the inadequacy of a classical approach based on Thomson's model of the atom, followed by Bohr's analysis of the quantized structure of Rutherford's atom and by the construction of Schrödinger's theory. This does not mean that we have completely overlooked the remaining phenomenology; we have just presented it in the light of the established quantum theory. Thus, for example, Chap. 3 deals with the analysis of the black body spectrum in the light of quantum theory.

2.1 The Photoelectric Effect

The photoelectric effect was discovered by H. Hertz in 1887. As sketched in Fig. 2.1, two electrodes are placed in a vacuum cell; one of them (C) is hit by monochromatic light of variable frequency, while the second (A) is set to a negative potential with respect to the first, as determined by a generator G and measured by a voltmeter V.

© Springer International Publishing Switzerland 2016
C.M. Becchi and M. D'Elia, *Introduction to the Basic Concepts of Modern Physics*, Undergraduate Lecture Notes in Physics, DOI 10.1007/978-3-319-20630-1_2

Fig. 2.1 A sketch of Hertz's
photoelectric effect
apparatus

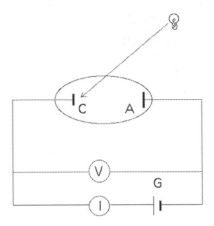

By measuring the electric current going through the amperometer I, one observes
that, if the light frequency is higher than a given threshold ν_V, determined by the
potential difference V between the two electrodes, the amperometer reveals a flux
of current i going from A to C which is proportional to the flux of luminous energy
hitting C. The threshold ν_V is a linear function of the potential difference V

$$\nu_V = a + bV . \tag{2.1}$$

The reaction time of the apparatus to light is substantially determined by the (RC)
time constant of the circuit and can be reduced down to values of the order of 10^{-8} s.
The theoretical interpretation of this phenomenon remained an open issue for about
14 years, because of the following reasons.

The direction of the current and the possibility to stop it by increasing the potential
difference clearly show that the electric flux is made up of electrons pulled out from
the atoms of electrode C by the luminous radiation.

A reasonable model for this process, which was inspired by Thomson's atomic
model, assumed that electrons, which are particles of mass $m = 9 \times 10^{-31}$ kg and
electric charge $-e \simeq -1.6 \times 10^{-19}$ C, were elastically bound to atoms of size
$R_A \sim 3 \times 10^{-10}$ m and subject to a viscous force of constant η. The value of η is
determined as a function of the atomic relaxation time, $\tau = 2m/\eta$, that is the time
needed by the atom to release its energy through radiation or collisions, which is
of the order of 10^{-8} s. Let us confine ourselves to considering the problem in one
dimension and write the equation of motion for an electron

$$m\ddot{x} = -kx - \eta\dot{x} - eE , \tag{2.2}$$

where E is an applied electric field and k is determined on the basis of atomic
frequencies. In particular we suppose the presence of many atoms with different
frequencies continuously distributed around

$$\sqrt{\frac{k}{m}} = \omega_0 = 2\pi\nu_0 \sim 10^{15}\ \mathrm{s}^{-1}\,. \tag{2.3}$$

If we assume an oscillating electric field $E = E_0 \cos(\omega t)$ with $\omega \sim 10^{15}\ \mathrm{s}^{-1}$, corresponding to visible light, then a general solution to (2.2) is given by

$$x = x_0 \cos(\omega t + \phi) + A_1 e^{-\alpha_1 t} + A_2 e^{-\alpha_2 t}\,, \tag{2.4}$$

where the second and third term satisfy the homogeneous equation associated with (2.2), so that $\alpha_{1/2}$ are the solutions of the following equation

$$m\alpha^2 - \eta\alpha + k = 0\,,$$

$$\alpha = \frac{\eta \pm \sqrt{\eta^2 - 4km}}{2m} = \frac{1}{\tau} \pm \sqrt{\frac{1}{\tau^2} - \omega_0^2} \simeq \frac{1}{\tau} \pm i\,\omega_0\,, \tag{2.5}$$

where last approximation is due to the assumption $\tau \gg \omega_0^{-1}$.

Regarding the particular solution $x_0 \cos(\omega t + \phi)$, we obtain by substitution:

$$-m\omega^2 x_0 \cos(\omega t + \phi) = -k x_0 \cos(\omega t + \phi) + \eta \omega x_0 \sin(\omega t + \phi) - e E_0 \cos(\omega t) \tag{2.6}$$

hence

$$(k - m\omega^2) x_0 \,(\cos(\omega t)\cos\phi - \sin(\omega t)\sin\phi)$$
$$= \eta\omega x_0 \,(\sin(\omega t)\cos\phi + \cos(\omega t)\sin\phi) - e E_0 \cos(\omega t)$$

from which, by taking alternatively $\omega t = 0, \pi/2$, we obtain the following system

$$\left(m\left(\omega_0^2 - \omega^2\right)\cos\phi - \eta\omega\sin\phi\right) x_0 = -e E_0\,,$$
$$m\left(\omega_0^2 - \omega^2\right) x_0 \sin\phi + \eta\omega\, x_0 \cos\phi = 0 \tag{2.7}$$

which can be solved for ϕ

$$\tan\phi = \frac{2\omega}{\tau\left(\omega^2 - \omega_0^2\right)}\,,$$

$$\cos\phi = \frac{\omega^2 - \omega_0^2}{\sqrt{\left(\omega_0^2 - \omega^2\right)^2 + \frac{4\omega^2}{\tau^2}}}\,, \qquad \sin\phi = \frac{(2\omega/\tau)}{\sqrt{\left(\omega_0^2 - \omega^2\right)^2 + \frac{4\omega^2}{\tau^2}}} \tag{2.8}$$

and finally for x_0, for which we obtain the well known resonant form

$$x_0 = \frac{eE_0/m}{\sqrt{\left(\omega_0^2 - \omega^2\right)^2 + \frac{4\omega^2}{\tau^2}}} \, . \tag{2.9}$$

To complete our computation, we must determine A_1 and A_2. On the other hand, taking into account (2.5) and the fact that x is real, we can rewrite the general solution in the following equivalent form:

$$x = x_0 \cos(\omega t + \phi) + Ae^{-t/\tau} \cos(\omega_0 t + \phi_0) \, . \tag{2.10}$$

If we assume that the electron is initially at rest, we can determine A and ϕ_0 by taking $x = \dot{x} = 0$ for $t = 0$, i.e.

$$x_0 \cos \phi + A \cos \phi_0 = 0 \, , \tag{2.11}$$

$$x_0 \, \omega \sin \phi = -A \left(\frac{\cos \phi_0}{\tau} + \omega_0 \sin \phi_0 \right) \, , \tag{2.12}$$

hence in particular

$$\tan \phi_0 = \frac{\omega}{\omega_0} \tan \phi - \frac{1}{\omega_0 \tau} \, . \tag{2.13}$$

These equations give us enough information to discuss the photoelectric effect without explicitly substituting A in (2.10).

Indeed in our simplified model the effect, i.e. the liberation of the electron from the atomic bond, happens as the amplitude of the electron displacement x is greater than the atomic radius. In Eq. (2.10) x is the sum of two parts, the first corresponding to stationary oscillations, the second to a transient decaying with time constant τ. In principle, the maximum amplitude could take place during the transient or later: to decide which is the case we must compare the value of A with that of x_0. It is apparent from (2.11) that the magnitude of A is of the same order as x_0 unless $\cos \phi_0$ is much less than $\cos \phi$. On the other hand, Eq. (2.13) tells us that, if $\tan \phi_0$ is large, then $\tan \phi$ is large as well, since $(\omega_0 \tau)^{-1} \sim 10^{-7}$ and $\omega/\omega_0 \sim 1$. Therefore, the order of magnitude of the maximum displacement is given by x_0, and can be sensitive to the electric field frequency. That happens in the resonant regime, where ω differs from ω_0 by less than $2\sqrt{\omega/\tau}$.

Let us consider separately the generic case from the resonant one. In the first case the displacement is of the order of $eE_0/(\omega^2 m)$, since the square root of the denominator in (2.9) has the same order of magnitude as ω^2. In order to induce the photoelectric effect it is therefore necessary that

$$\frac{eE_0}{\omega^2 m} \sim R_A \, ,$$

from which we can compute the power density needed for the luminous beam which hits electrode C:

$$P = c\epsilon_0 E_0^2 \sim c\epsilon_0 \left(\frac{R_A \omega^2 m}{e}\right)^2,$$

where c is the speed of light and ϵ_0 is the vacuum dielectric constant. P comes out to be of the order of 10^{15} W/m^2, a power density which is difficult to realize in practice and which would anyway be enough to vaporize any kind of electrode. We must conclude that our model cannot explain the photoelectric effect if ω is far from resonance. Let us consider therefore the resonant case and set $\omega = \omega_0$. On the basis of (2.9), (2.11) and (2.13), that implies:

$$\phi = \phi_0 = \frac{\pi}{2}, \quad A = -x_0,$$

hence

$$x = \frac{-eE_0\tau}{2m\omega_0} \left(1 - e^{-t/\tau}\right) \sin(\omega_0 t). \tag{2.14}$$

In order for the photoelectric effect to take place, the oscillation amplitude must be greater than the atomic radius:

$$\frac{eE_0\tau}{2m\omega_0} \left(1 - e^{-t/\tau}\right) \geq R_A.$$

That sets the threshold field to $2m\omega_0 R_A/(e\tau)$ and the power density of the beam to

$$P = c\epsilon_0 \left(\frac{4\omega_0 m R_A}{\tau e}\right)^2 \sim 100 \text{ W/m}^2,$$

while the time required to reach the escape amplitude is of the order of τ.

In conclusion, our model predicts a threshold for the power of the beam, but not for its frequency, which however must be tuned to the resonance frequency: the photoelectric effect would cease both below and above the typical resonance frequencies of the atoms in the electrode. Moreover the expectation is that the electron does not gain any further appreciable energy from the electric field once it escapes the atomic bond: hence the emission from the electrode could be strong, but made up of electrons of energy equal to that gained during the last atomic oscillation. Equation (2.14) shows that, during the transient ($t \ll \tau$), the oscillation amplitude grows roughly by $eE_0/(m\omega_0^2)$ in one period, so that the energy of the escaped electron would be of the order of magnitude of $kR_AeE_0/(m\omega_0^2) = eE_0R_A$, corresponding also to the energy acquired by the electron from the electric field E_0 when crossing the atom. It is easily computed that for a power density of the order of $10-100$ W/m^2, the electric field E_0 is roughly 100 V/m, so that the final kinetic energy of the electron would be 10^{-8} eV $\sim 10^{-27}$ J: this value is much smaller than the typical thermal energy at room temperature ($3kT/2 \sim 10^{-1}$ eV).

The prediction of the model is therefore in clear contradiction with the experimental results described above. In particular, the very small energy of the emitted electrons implies that the electric current I should vanish even for small negative potential differences.

Einstein proposed a description of the effect based on the hypothesis that the energy be transferred from the luminous radiation to the electron in a single *elementary* (i.e. no further separable) process, instead than through a gradual excitation. Moreover, he proposed that the transferred energy be equal to $h\nu = h\omega/(2\pi) \equiv \hbar\omega$, a quantity called *quantum* by Einstein himself. The constant h had been introduced by Planck several years before to describe the radiation emitted by an oven and its value is 6.63×10^{-34} J s.

If the quantum of energy is enough for electron liberation, i.e. according to our model if it is larger than $E_t \equiv kR_A^2/2 = \omega_0^2 R_A^2 m/2 \sim 10^{-19}$ J ~ 1 eV, and the frequency exceeds 1.6×10^{14} Hz (corresponding to ω in our model), then the electron is emitted keeping the energy exceeding the threshold in the form of kinetic energy. The number of emitted electrons, hence the intensity of the process, is proportional to the flux of luminous energy, i.e. to the number of quanta hitting the electrode.

Since $E = h\nu$ is the energy gained by the electron, which spends a part E_t to get free from the atom, the final electron kinetic energy is $T = h\nu - E_t$, so that the electric current can be interrupted by placing the second electrode at a negative potential

$$V = \frac{h\nu - E_t}{e},$$

thus reproducing (2.1).

The most important point in Einstein's proposal, which was already noticed by Planck, is that a physical system of typical frequency ν can exchange only quanta of energy equal to $h\nu$. The order of magnitude in the atomic case is $\omega \sim 10^{15}$ s^{-1}, hence $\hbar\omega \equiv (h/2\pi)\,\omega \sim 1$ eV.

2.2 Bohr's Quantum Theory

After the introduction of the concept of a quantum of energy, *quantum theory* was developed by N. Bohr in 1913 and then perfected by A. Sommerfeld in 1916: they gave a precise proposal for multi-periodic systems, i.e. systems which can be described in terms of periodic components.

The main purpose of their studies was that of explaining, in the framework of Rutherford's atomic model, the light spectra emitted by gases (in particular mono-atomic ones) excited by electric discharges. The most simple and renowned case is that of the mono-atomic hydrogen gas (which can be prepared with some difficulties since hydrogen tends to form diatomic molecules). It has a discrete spectrum, i.e. the emitted frequencies can assume only some discrete values, in particular:

$$\nu_{n,m} = R \left(\frac{1}{n^2} - \frac{1}{m^2} \right) \tag{2.15}$$

for all possible positive integer pairs with $m > n$: this formula was first proposed by J. Balmer in 1885 for the case $n = 2$, $m \geq 3$, and then generalized by J. Rydberg in 1888 for all possible pairs (n, m). The emission is particularly strong for $m = n + 1$.

Rutherford had shown that the positive charge in an atom is localized in a practically point-like nucleus, which also contains most of the atomic mass. In particular the hydrogen atom can be described as a two-body system: a heavy and positively charged particle, which nowadays is called proton, bound by Coulomb forces to a light and negatively charged particle, the electron.

We will confine our discussion to the case of circular orbits of radius r, covered with uniform angular velocity ω, and will consider the proton as if it were infinitely heavy (its mass is about 2×10^3 times that of the electron). In this case we have

$$m\omega^2 r = \frac{e^2}{4\pi\epsilon_0 r^2} ,$$

where m is the electron mass. Hence the orbital frequencies, which in classical physics correspond to those of the emitted radiation, are continuously distributed as a function of the radius

$$\nu = \frac{\omega}{2\pi} = \frac{e}{\sqrt{16\pi^3\epsilon_0 m r^3}} ; \tag{2.16}$$

this is in clear contradiction with (2.15). Based on Einstein's theory of the photoelectric effect, Bohr proposed to interpret (2.15) by assuming that only certain orbits be allowed in the atom, which are called *levels*, and that the frequency $\nu_{n,m}$ correspond to the transition from the m-th level to n-th one. In that case

$$h\nu_{n,m} - E_m - E_n , \tag{2.17}$$

where the atomic energies (which are negative since the atom is a bound system) would be given by

$$E_n = -\frac{hR}{n^2} . \tag{2.18}$$

Since, according to classical physics for the circular orbit case, the atomic energy is given by

$$E_{\text{circ}} = -\frac{e^2}{8\pi\epsilon_0 r} ,$$

Bohr's hypothesis is equivalent to the assumption that the admitted orbital radii be

$$r_n = \frac{e^2 n^2}{8\pi\epsilon_0 hR} . \tag{2.19}$$

It is clear that Bohr's hypothesis seems simply aimed at reproducing the observed experimental data; it does not permit any particular further development, unless further conditions are introduced. The most natural, which is called *correspondence principle*, is that the classical law, given in (2.16), be reproduced by (2.15) for large values of r, hence of n, and at least for the strongest emissions, i.e. those with $m = n + 1$, for which we can write

$$\nu_{n,n+1} = R\frac{2n+1}{n^2(n+1)^2} \rightarrow \frac{2R}{n^3} \, , \tag{2.20}$$

these frequencies should be identified in the above mentioned limit with what resulting from the combination of (2.16) and (2.19):

$$\nu = \frac{e}{\sqrt{16\pi^3\epsilon_0 m r_n^3}} = \frac{\epsilon_0\sqrt{32(hR)^3}}{e^2\sqrt{m}n^3} \, . \tag{2.21}$$

By comparing last two equations we get the value of the coefficient R in (2.15), which is called *Rydberg constant*:

$$R = \frac{me^4}{8\epsilon_0^2 h^3}$$

and is in excellent agreement with experimental determinations. We have then the following *quantized* atomic energies

$$E_n = -\frac{me^4}{8\epsilon_0^2 h^2 n^2} \, , \quad n = 1, 2, \ldots$$

while the quantized orbital radii are

$$r_n = \frac{\epsilon_0 h^2 n^2}{\pi m e^2} \, . \tag{2.22}$$

In order to give a numerical estimate of our results, it is convenient to introduce the ratio $e^2/(2\epsilon_0 hc) \equiv \alpha \simeq 1/137$, which is dimensionless and is known as the *fine structure constant*. The energy of the state with $n = 1$, which is called the *ground state*, is

$$E_1 = -hR = -\frac{mc^2}{2}\alpha^2 \, ;$$

noticing that $mc^2 \sim 0.51$ MeV, we have $E_1 \simeq -13.6$ eV. The corresponding atomic radius (Bohr radius) is $r_1 \simeq 0.53 \times 10^{-10}$ m.

Notwithstanding the excellent agreement with experimental data, the starting hypothesis, to be identified with (2.18), looks still quite conditioned by the particular form of Balmer law given in (2.15). For that reason Bohr tried to identify a physical observable to be quantized according to a simpler and more fundamental law. He proceeded according to the idea that such observable should have the same dimensions of the Planck constant, i.e. those of an action, or equivalently of an angular momentum. In the particular case of quantized circular orbits this last quantity reads:

$$L = pr = m\omega r^2 = \frac{e}{\sqrt{4\pi\epsilon_0}}\sqrt{mr_n} = \frac{h}{2\pi}n \equiv n\hbar , \quad n = 1, 2, \ldots . \quad (2.23)$$

2.3 de Broglie's Interpretation

In this picture of partial results, even if quite convincing from the point of view of the phenomenological comparison, the real progress towards understanding quantum physics came as L. de Broglie suggested the existence of a *universal wave-like behavior* of material particles and of energy quanta associated to force fields. As we have seen in the case of electromagnetic waves, when discussing the Doppler effect, a phase can always be associated with a wave-like process, which is variable both in space and in time (e.g. given by $2\pi (x/\lambda - \nu t)$ in the case of waves moving parallel to the x axis). The assumption that quanta can be interpreted as real particles and that Einstein's law $E = h\nu$ be universally valid, would correspond to identifying the wave phase with $2\pi (x/\lambda - Et/h)$. If we further assume the phase to be relativistically invariant, then it must be expressed in the form $(p\,x - E\,t)/\hbar$, where E and p are identified with relativistic energy and momentum, i.e. in the case of material particles:

$$E = \frac{mc^2}{\sqrt{\left(1 - \frac{v^2}{c^2}\right)}} , \quad p = \frac{mv}{\sqrt{\left(1 - \frac{v^2}{c^2}\right)}} .$$

In order to simplify the discussion as much as possible, we will consider here and in most of the following a one-dimensional motion (parallel to the x axis). In conclusion, by comparing last two expressions given for the phase, we obtain de Broglie's equation:

$$p = \frac{h}{\lambda} ,$$

which is complementary to Einstein's law, $E = h\nu$.

These formulae give an idea of the scale at which quantum effects are visible. For an electron having kinetic energy $E_k = 10^2$ eV $\simeq 1.6 \times 10^{-17}$ J, quantum effects show up at distances of the order of $\lambda = h/p = h/\sqrt{2mE_k} \sim 10^{-10}$ m, corresponding to atomic or slightly subatomic distances; that confirms the importance of quantum effects for electrons in condensed matter and in particular in solids, where typical energies are of the order of a few electron-volts. For a gas of light atoms in equilibrium at temperature T, the kinetic energy predicted by the equipartition theorem is $3\,kT/2$, where k is Boltzmann's constant. At a temperature $T = 300\,°$K (room temperature) the kinetic energy is roughly 2.5×10^{-2} eV, corresponding to wavelengths of about 10^{-10} m for atom masses of the order of 10^{-26} kg. However at those distances the picture of a non-interacting (perfect) gas does not apply because of strong repulsive forces coming into play: in order to gain a factor ten over distances, it is necessary to reduce the temperature by a factor 100, going down to a few Kelvin degrees, at which quantum effects are manifest. For a macroscopic body of mass 1 kg and kinetic energy 1 J, quantum effects would show up at distances roughly equal to 3×10^{-34} m, hence completely negligible with respect to the thermal oscillation amplitudes of atoms, which are proportional to the square root of the absolute temperature, and are in particular of the order of a few nanometers at $T = 10^3\,°$K, where the solid melts.

On the other hand, Einstein's formula gives us information about the scale of times involved in quantum processes, which is of the order of $h/\Delta E$, where ΔE corresponds to the amount of exchanged energy. For $\Delta E \sim 1$ eV, times are roughly 4×10^{-15} s, while for thermal interactions at room temperature time intervals increase by a factor 40.

In conclusion, in the light of de Broglie's formula, quantum effects are not visible for macroscopic bodies and at macroscopic energies. For atoms in matter they show up after condensation, or anyway at very low temperatures, while electrons in solids or in atoms are fully in the quantum regime.

In Rutherford's atomic model illustrated in the previous Section, the circular motion of the electron around the proton must be associated, according to de Broglie, with a wave closed around a circular orbit. That resembles wave-like phenomena analogous to the oscillations of a ring-shaped elastic string or to air pressure waves in a toroidal reed pipe. That implies well tuned wavelengths, as in the case of musical instruments (which are not ring-shaped for obvious practical reasons). The need for tuned wavelength can be easily understood in the case of the toroidal reed pipe: a complete round of the ring must bring the phase back to its initial value, so that the total length of the pipe must be an integer multiple of the wavelength.

Taking into account previous equations regarding circular atomic orbits, we have the following electron wavelength:

$$\lambda = \frac{h}{p} = \frac{h}{e}\sqrt{\frac{4\pi\epsilon_0 r}{m}},$$

so that the tuning condition reads

$$2\pi r = n\lambda = \frac{nh}{e}\sqrt{\frac{4\pi\epsilon_0 r}{m}}$$

giving

$$r = \frac{n^2 h^2 \epsilon_0}{\pi e^2 m},$$

which confirms (2.22) and gives support to the picture proposed by Bohr and Sommerfeld. De Broglie's hypothesis, which was formulated in 1924, was confirmed in 1926 by Davisson and Gerner by measuring the intensity of an electron beam reflected by a nickel crystal. The apparatus used in the experiment is sketched in Fig. 2.2. The angular distribution of the electrons, reflected in conditions of normal incidence, shows a strongly anisotropic behavior with a marked dependence on the beam accelerating potential. In particular, an accelerating potential equal to 48 V leads to a quite pronounced peak at a reflection angle $\phi = 55.3°$. An analogous X-ray diffraction experiment permits to interpret the nickel crystal as an atomic lattice of spacing 0.215×10^{-9} m. The comparison between the angular distributions obtained for X-rays and for electrons shows relevant analogies, suggesting a diffractive interpretation also in the case of electrons. Bragg's law, giving the n-th maximum in the diffraction figure, is $d \sin \phi_n = n\lambda$.

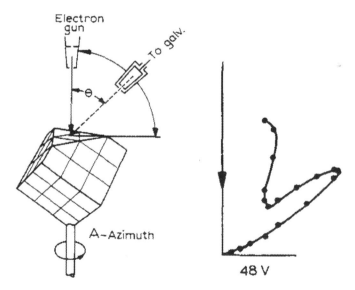

Fig. 2.2 A schematic description of Davisson-Gerner apparatus and a polar coordinate representation of the results obtained at 48 V electron energy, as they appear in Davisson's Noble Price Lecture, from *Nobel Lectures*, Physics 1922–1941 (Elsevier Publishing Company, Amsterdam 1965)

For the peak corresponding to the principal maximum at 55.3° we have

$$d \, \sin \phi = \lambda \simeq 0.175 \times 10^{-9} \, \text{m} \, .$$

On the other hand the electrons in the beam have a kinetic energy

$$E_k \simeq 7.68 \times 10^{-18} \, \text{J} \, ,$$

hence a momentum $p \simeq 3.7 \times 10^{-24}$ N s, in excellent agreement with de Broglie's formula $p = h/\lambda$. In the following years analogous experiments were repeated using different kinds of material particles, in particular neutrons.

Once established the wave-like behavior of propagating material particles, it must be clarified what is the physical quantity the phenomenon refers to, i.e. what is the physical meaning of the oscillating quantity (or quantities) usually called *wave function*, for which a *linear* propagating equation will be supposed, in analogy with mechanical or electromagnetic waves. It is known that, in the case of electromagnetic waves, the quantities measuring the amplitude are electric and magnetic fields. Our question regards exactly the analogous of those fields in the case of de Broglie's waves. The experiment by Davisson and Gerner gives an answer to this question. Indeed, as illustrated in Fig. 2.2, the detector reveals the presence of one or more electrons at a given angle; if we imagine to repeat the experiment several times, with a single electron in the beam at each time, and if we measure the frequency at which electrons are detected at the various angles, we get the *probability* of having the electron in a given site covered by the detector.

In the case of an optical measure, what is observed is the interference effect in the energy deposited on a plate; that is proportional to the square of the electric field on the plate. Notice that the linearity in the wave equation and the quadratic relation between the measured quantity and the wave amplitude are essential conditions for the existence of interference and diffractive phenomena. We must conclude that also in the case of material particles some *positive quadratic form* of the de Broglie wave function gives the probability of having the electron in a given point.

We have quite generically mentioned a quadratic form, since at the moment it is still not clear if the wave function has one or more components, i.e. if it corresponds to one or more real functions. By a positive quadratic form we mean a homogeneous second order polynomial in the wave function components, which is positive for real and non-vanishing values of its arguments. In the case of a single component, we can say without loss of generality that the probability density is the wave function squared, while in the case of two or more components it is always possible, by suitable linear transformations, to reduce the quadratic form to a sum of squares.

We are now going to show that the hypothesis of a single component must be discarded. Let us indicate by $\rho(\boldsymbol{r}, t)d^3r$ the probability of the particle being in a region of size d^3r around $\boldsymbol{r}$ at time t, and by $\psi(\boldsymbol{r}, t)$ the wave function, which for the moment is considered as a real valued function, defined so that

$$\rho(\boldsymbol{r}, t) = \psi^2(\boldsymbol{r}, t) \, . \tag{2.24}$$

If Ω indicates the whole region accessible to the particle, the probability density must satisfy the natural constraint:

$$\int_\Omega d^3r\, \rho\,(\boldsymbol{r}, t) = 1, \qquad (2.25)$$

which implies the condition:

$$\int_\Omega d^3r\, \dot{\rho}(\boldsymbol{r}, t) \equiv \int_\Omega d^3r\, \frac{\partial \rho(\boldsymbol{r}, t)}{\partial t} = 0. \qquad (2.26)$$

This expresses the fact that, if the particle cannot escape Ω, the probability of finding it in that region must always be one. This condition can be given in mathematical terms analogous to those used to express electric charge conservation: the charge contained in a given volume, i.e. the integral of the charge density, may change only if the charge flows through the boundary surface. The charge flux through the boundaries is expressed in terms of the current density flow and can be rewritten as the integral of the divergence of the current density itself by using Gauss–Green theorem

$$\int_\Omega \dot{\rho} = -\Phi_{\partial\Omega}(J) = -\int_\Omega \nabla \cdot \boldsymbol{J}.$$

Finally, by reducing the equation from an integral form to a differential one, we can set the temporal derivative of the charge density equal to minus the divergence of the current density. Based on this analogy, let us introduce the *probability current density* $\boldsymbol{J}$ and write

$$\dot{\rho}(\boldsymbol{r}, t) = -\frac{\partial J_x(\boldsymbol{r}, t)}{\partial x} - \frac{\partial J_y(\boldsymbol{r}, t)}{\partial y} - \frac{\partial J_z(\boldsymbol{r}, t)}{\partial z} \equiv -\nabla \cdot \boldsymbol{J}(\boldsymbol{r}, t). \qquad (2.27)$$

The conservation equation must be automatically satisfied as a consequence of the propagation equation of de Broglie's waves, which we write in the form:

$$\dot{\psi} = L\left(\psi, \nabla\psi, \nabla^2\psi, \dots\right), \qquad (2.28)$$

where L indicates a generic linear function of ψ and its derivatives like:

$$L\left(\psi, \nabla\psi, \nabla^2\psi, \dots\right) = \alpha\psi + \beta\nabla^2\psi. \qquad (2.29)$$

Notice that if L were not linear the interference mechanism upon which quantization is founded would soon or later fail. Furthermore we assume invariance under the reflection of the coordinates, at least in the free case, so that terms proportional to first derivatives are excluded.

From Eq. (2.24) we have $\dot\rho = 2\psi\dot\psi$, which can be rewritten, using (2.28), as:

$$\dot\rho = 2\psi L\left(\psi, \nabla\psi, \nabla^2\psi, \dots\right) . \tag{2.30}$$

The right-hand side of last equation must be identified with $-\nabla \cdot \boldsymbol{J}(\boldsymbol{r}, t)$. Moreover $\boldsymbol{J}$ must necessarily be a bilinear function of ψ and its derivatives, exactly like $\dot\rho$. Therefore, since $\boldsymbol{J}$ is a vector-like quantity, it must be expressible as

$$\boldsymbol{J} = c\ \psi\nabla\psi + d\ \nabla\psi\nabla^2\psi + \cdots$$

from which it appears that $\nabla \cdot \boldsymbol{J}(\boldsymbol{r}, t)$ must necessarily contain bilinear terms in which both functions are derived, like $\nabla\psi \cdot \nabla\psi$. However, such terms are clearly missing in (2.30).

We come to the conclusion that the description of de Broglie's waves requires at least two wave functions ψ_1 and ψ_2, defined so that $\rho = \psi_1^2 + \psi_2^2$. In an analogous way we can introduce the complex valued function:

$$\psi = \psi_1 + i\psi_2 , \tag{2.31}$$

defined so that

$$\rho = |\psi|^2 ; \tag{2.32}$$

this choice implies:

$$\dot\rho = \psi^*\dot\psi + \psi\dot\psi^* .$$

If we assume, for instance, the wave equation corresponding to (2.29):

$$\dot\psi = \alpha\psi + \beta\nabla^2\psi , \tag{2.33}$$

we obtain:

$$\dot\rho = \psi^*\left(\alpha\psi + \beta\nabla^2\psi\right) + \psi\left(\alpha^*\psi^* + \beta^*\nabla^2\psi^*\right) .$$

If we also assume that the probability current density be

$$\boldsymbol{J} = i k\left(\psi^*\nabla\psi - \psi\nabla\psi^*\right) , \tag{2.34}$$

with k real so as to make $\boldsymbol{J}$ real as well, we easily derive

$$\nabla \cdot \boldsymbol{J} = i k\left(\psi^*\nabla^2\psi - \psi\nabla^2\psi^*\right) .$$

It can be easily verified that the continuity equation (2.27) is satisfied if

$$\alpha + \alpha^* = 0 , \qquad \beta = -i k . \tag{2.35}$$

It is of great physical interest to consider the case in which the wave function has more than two real components. In particular, the wave function of electrons has four components or, equivalently, two complex components. In general, the multiplicity of the complex components is linked to the existence of an intrinsic angular momentum, which is called *spin*. The various complex components are associated with the different possible spin orientations. In the case of particles with non-vanishing mass, the number of components is $2S + 1$, where S is the spin of the particle. In the case of the electron, $S = 1/2$.

For several particles, as for the electron, spin is associated with a magnetic moment which is inherent to the particle: it behaves as a microscopic magnet with various possible orientations, corresponding to those of the spin, which can be selected by placing the particle in a non-uniform magnetic field and measuring the force acting on the particle.

2.4 Schrödinger's Equation

The simplest case to which our considerations can be applied is that of a non-relativistic free particle of mass m. To simplify notations and computations, we will confine ourselves to a one-dimensional motion, parallel, for instance, to the x axis; if the particle is not free, forces will be parallel to the same axis as well. The obtained results will be extensible to three dimensions by exploiting the vector formalism. In practice, we will systematically replace ∇ by its component $\nabla_x = \partial/\partial x \equiv \partial_x$ and the Laplacian operator $\nabla^2 = \partial^2/\partial x^2 + \partial^2/\partial y^2 + \partial^2/\partial z^2$ by $\partial^2/\partial x^2 \equiv \partial_x^2$; the probability current density $\boldsymbol{J}$ will be replaced by J_x (J) as well. The inverse replacement will suffice to get back to three dimensions.

The energy of a non-relativistic free particle is

$$E = c\sqrt{m^2c^2 + p^2} \simeq mc^2 + \frac{p^2}{2m} + O\left(\frac{p^4}{m^3c^2}\right),$$

where we have explicitly declared our intention to neglect terms of the order of $p^4/(m^3c^2)$. Assuming de Broglie's interpretation, we write the wave function:

$$\psi_P(x,t) \sim e^{2\pi i(x/\lambda - \nu t)} = e^{i(px - Et)/\hbar} \tag{2.36}$$

(we are considering a motion in the positive x direction). Our choice implies the following wave equation

$$\dot{\psi}_P = -\frac{iE}{\hbar}\psi_P = -\frac{i}{\hbar}\left(mc^2 + \frac{1}{2m}p^2\right)\psi_P. \tag{2.37}$$

We have also

$$\partial_x \psi_P = \frac{i}{\hbar} p \, \psi_P \,, \tag{2.38}$$

from which we deduce

$$i\hbar \dot{\psi}_P = mc^2 \psi_P - \frac{\hbar^2}{2m} \partial_x^2 \psi_P \,. \tag{2.39}$$

Our construction can be simplified by multiplying the initial wave function by the phase factor $e^{imc^2 t/\hbar}$, i.e. defining

$$\psi \equiv e^{imc^2 t/\hbar} \psi_P \sim \exp\left(\frac{i}{\hbar} \left(px - \frac{p^2}{2m} t \right) \right) \,. \tag{2.40}$$

Since the dependence on x is unchanged, ψ still satisfies (2.38) and has the same probabilistic interpretation as ψ_P. Indeed both ρ and J are unchanged. The wave equation instead changes:

$$i\hbar \dot{\psi} = -\frac{\hbar^2}{2m} \partial_x^2 \psi \equiv T \psi \,. \tag{2.41}$$

This is the *Schrödinger equation* for a free (non-relativistic) particle, in which the right-hand side has a natural interpretation in terms of the particle energy, which in the free case is only of kinetic type.

In the case of particles under the influence of a force field corresponding to a potential energy $V(x)$, the equation can be generalized by adding $V(x)$ to the kinetic energy:

$$i\hbar \dot{\psi} = -\frac{\hbar^2}{2m} \partial_x^2 \psi + V(x)\psi \,. \tag{2.42}$$

This is the one-dimensional Schrödinger equation that we shall apply to various cases of physical interest.

Equations (2.34) and (2.35) show that the probability density current does not depend on V and is given by:

$$J = -\frac{i\hbar}{2m} \left(\psi^* \partial_x \psi - \psi \partial_x \psi^* \right) \,. \tag{2.43}$$

Going back to the free case and considering the *plane* wave function given in (2.36), it is interesting to notice that the corresponding probability density, $\rho = |\psi|^2$, is a constant function. This result is paradoxical since, by reducing (2.25) to one dimension, we obtain

$$\int_{-\infty}^{\infty} dx \, \rho(x, t) = \int_{-\infty}^{\infty} dx \, |\psi(x, t)|^2 = 1 \,, \tag{2.44}$$

which cannot be satisfied in the examined case since the integral of a constant function is divergent. We must conclude that our interpretation excludes the possibility that a particle has a well defined momentum.

We are left with the hope that this difficulty may be overcome by admitting some (small) uncertainty on the knowledge of momentum. This possibility can be easily analyzed thanks to the linearity of the Schrödinger equation. Indeed Eq. (2.41) admits other different solutions besides the simple plane wave, in particular the *wave packet* solution, which is constructed as a linear superposition of many plane waves according to the following integral:

$$\psi(x,t) = \frac{1}{\sqrt{2\pi\hbar}} \int_{-\infty}^{\infty} dp \; \tilde{\psi}(p) \exp\left(\frac{i}{\hbar}\left(px - \frac{p^2}{2m}t\right)\right). \tag{2.45}$$

Considering the expression for $\psi(x,0)$ it is easy to deduce the expression[1]:

$$\tilde{\psi}(p) = \frac{1}{\sqrt{2\pi\hbar}} \int_{-\infty}^{\infty} dx \; \psi(x,0) \exp\left(-\frac{i}{\hbar}px\right). \tag{2.46}$$

The squared modulus of the superposition coefficients, $|\tilde{\psi}(p)|^2$, can be naturally interpreted as the probability density in terms of momentum, exactly in the same way as $\rho(x)$ is interpreted as a probability density in terms of position.

Let us choose in particular a Gaussian distribution:

$$\tilde{\psi}(p) = \frac{1}{\sqrt{\sqrt{2\pi}\,\Delta}} e^{-(p-p_0)^2/(4\Delta^2)}, \tag{2.47}$$

corresponding to

$$\psi_\Delta(x,t) = \frac{1}{\sqrt{\sqrt{(2\pi)^3}\,\Delta\hbar}} \int_{-\infty}^{\infty} dp \; e^{-(p-p_0)^2/(4\Delta^2)} \; e^{i(px-p^2t/2m)/\hbar} \tag{2.48}$$

where the coefficients in (2.45) and (2.47) are determined in such a way that

$$\int_{-\infty}^{\infty} dx |\psi_\Delta(x,t)|^2 = 1. \tag{2.49}$$

---

[1]Using the following formulae it is easy to verify (2.45) and (2.46) in the case of Gaussian wave packets. By linearity this proves the validity of the same equations in the case of fast decreasing, infinitely differentiable (C^∞) functions. A further extension of the validity is shown in the framework of distribution theory. In our analysis the restriction to fast decreasing C^∞ functions is understood.

The integral in (2.48) can be computed by recalling that, if α is a complex number with positive real part (Re $(\alpha) > 0$), then

$$\int_{-\infty}^{\infty} dp\, e^{-\alpha p^2} = \sqrt{\frac{\pi}{\alpha}}$$

and that the Riemann integral measure dp is left invariant by translations in the complex plane,

$$\int_{-\infty}^{\infty} dp\, e^{-\alpha p^2} \equiv \int_{-\infty}^{\infty} d(p+\gamma)\, e^{-\alpha(p+\gamma)^2}$$

$$= \int_{-\infty}^{\infty} dp\, e^{-\alpha(p+\gamma)^2} = e^{-\alpha\gamma^2} \int_{-\infty}^{\infty} dp\, e^{-\alpha p^2} e^{-2\alpha\gamma p},$$

for every complex number γ. Therefore we have

$$\int_{-\infty}^{\infty} dp\, e^{-\alpha p^2} e^{\beta p} = \sqrt{\frac{\pi}{\alpha}} e^{\beta^2/4\alpha}. \tag{2.50}$$

Developing (2.48) with the help of (2.50) we can write

$$\psi_\Delta(x,t) = \frac{1}{\sqrt{\sqrt{(2\pi)^3}\Delta\hbar}} e^{-\frac{p_0^2}{4\Delta^2}} \int_{-\infty}^{\infty} dp\, e^{-\left[\frac{1}{4\Delta^2}+\frac{it}{2m\hbar}\right]p^2}\, e^{\left[\frac{p_0}{2\Delta^2}+\frac{ix}{\hbar}\right]p}$$

$$= \sqrt{\frac{1}{\sqrt{2\pi}(\frac{\hbar}{2\Delta}+\frac{i\Delta t}{m})}} \exp\left(\frac{\left[\frac{p_0}{2\Delta^2}+\frac{ix}{\hbar}\right]^2}{\frac{1}{\Delta^2}+\frac{2it}{m\hbar}} - \frac{p_0^2}{4\Delta^2}\right). \tag{2.51}$$

We are interested in particular in the x dependence of the probability density $\rho(x)$: that is solely related to the real part of the exponent of the rightmost term in (2.51), which can be expanded as follows:

$$\frac{\frac{p_0^2}{4\Delta^4}+\frac{ip_0x}{\Delta^2\hbar}-\frac{x^2}{\hbar^2}}{\frac{1}{\Delta^2}+\frac{2it}{m\hbar}} - \frac{p_0^2}{4\Delta^2} = -\frac{p_0^2}{4\Delta^2}\frac{\frac{4t^2\Delta^4}{m^2\hbar^2}+\frac{2it\Delta^2}{m\hbar}}{1+\frac{4t^2\Delta^4}{m^2\hbar^2}} - \left(\frac{\Delta^2x^2}{\hbar^2}-\frac{ip_0x}{\hbar}\right)\frac{1-\frac{2it\Delta^2}{m\hbar}}{1+\frac{4t^2\Delta^4}{m^2\hbar^2}}$$

the real part being

$$-\frac{\Delta^2\left(x-\frac{p_0t}{m}\right)^2}{\hbar^2\left(1+\frac{4t^2\Delta^4}{m^2\hbar^2}\right)} \equiv -\frac{\Delta^2\left(x-v_0t\right)^2}{\hbar^2\left(1+\frac{4t^2\Delta^4}{m^2\hbar^2}\right)}.$$

Since p_0 is clearly the average momentum of the particle, we have introduced the corresponding average velocity $v_0 = p_0/m$. Recalling the definition of ρ as well as its normalization constraint, we finally find

$$\rho(x,t) = \frac{\Delta}{\hbar} \sqrt{\frac{2}{\pi \left(1 + \frac{4t^2 \Delta^4}{m^2 \hbar^2}\right)}} \, \exp\left(-\frac{2\Delta^2}{\hbar^2} \frac{(x - v_0 t)^2}{1 + \frac{4t^2 \Delta^4}{m^2 \hbar^2}}\right), \qquad (2.52)$$

while the probability distribution in terms of momentum reads

$$\tilde{\rho}(p) = \frac{1}{\sqrt{2\pi}\Delta} e^{-(p-p_0)^2/(2\Delta^2)} . \qquad (2.53)$$

Given a Gaussian distribution $\rho(x) = 1/(\sqrt{2\pi}\sigma)e^{-(x-x_0)^2/(2\sigma^2)}$, it is a well known fact, which anyway can be easily derived from previous formulae, that the mean value $\bar{x}$ is x_0 while the mean quadratic deviation $\overline{(x - \bar{x})^2}$ is equal to σ^2. Hence, in the examined case, we have an average position $\bar{x} = v_0 t$ with a mean quadratic deviation equal to $\hbar^2/(4\Delta^2) + t^2 \Delta^2/m^2$, while the average momentum is p_0 with a mean quadratic deviation Δ^2. The mean values represent the kinematic variables of a free particle, while the mean quadratic deviations are roughly inversely proportional to each other: if we improve the definition of one observable, the other becomes automatically less defined.

The distributions given in (2.52) and (2.53), even if derived in the context of a particular example, permit us to reach important general conclusions which, for the sake of clarity, are listed in the following as distinct points.

2.4.1 The Uncertainty Principle

While the mean quadratic deviation relative to the momentum distribution

$$\overline{(p - \bar{p})^2} = \Delta^2$$

has been fixed a priori, by choosing $\tilde{\psi}(p)$, and is independent of time, thus confirming that momentum is a constant of motion for a free particle, that relative to the position

$$\overline{(x - \bar{x})^2} = \left(1 + \frac{4t^2 \Delta^4}{m^2 \hbar^2}\right) \frac{\hbar^2}{4\Delta^2}$$

does not contain further free parameters and does depend on time. Indeed, Δ_x grows significantly for $2t\Delta^2/(m\hbar) > 1$, hence for times greater than $t_s = m\hbar/(2\Delta^2)$. Notice that t_s is nothing but the time needed for a particle of momentum Δ to cover a distance $\hbar/(2\Delta)$, therefore this spreading has a natural interpretation also from a classical point of view: a set of independent particles having momenta distributed according to a width Δ_p, spreads with velocity $\Delta_p/m = v_s$; if the particles are statistically distributed in a region of size initially equal to Δ_x, the same size will grow significantly after times of the order of Δ_x/v_s.

What is new in our results is, first of all, that they refer to a single particle, meaning that uncertainties in position and momentum are not avoidable; secondly, these uncertainties are strictly interrelated. Without considering the spreading in time, it is evident that the uncertainty in one variable can be diminished only as the other uncertainty grows. Indeed, Δ can be eliminated from our equations by writing the inequality:

$$\Delta_x \Delta_p \equiv \sqrt{\overline{(x - \bar{x})^2}\,\overline{(p - \bar{p})^2}} \geq \frac{\hbar}{2}, \qquad (2.54)$$

which is known as the *Heinsenberg uncertainty principle*. The case of a real Gaussian packet corresponds to the minimal possible value $\Delta_x \Delta_p = \hbar/2$.

We have discussed Heisenberg's uncertainty principle using Gaussian wave packets and understanding that the results have general validity. It is not difficult to prove this generality. Indeed, let us consider a generic wave packet $\psi(x)$ satisfying the normalization condition (2.49), and let us denote by $\bar{x}$ the average value of the particle position

$$\bar{x} = \int_{-\infty}^{\infty} dx\, x |\psi(x)|^2 . \qquad (2.55)$$

The mean quadratic deviation in position can be easily computed by:

$$\Delta_x^2 = \int_{-\infty}^{\infty} dx\, (x - \bar{x})^2 |\psi(x)|^2 = \int_{-\infty}^{\infty} dx\, x^2 |\psi(x + \bar{x})|^2 . \qquad (2.56)$$

Completely analogous formulae in terms of $\tilde{\psi}$ hold true for $\bar{p}$ and Δ_p. However, using (2.45) and (2.46), it is possible to compute $\bar{p}$ and Δ_p directly from ψ. Indeed one has:

$$\begin{aligned}
\bar{p} &= \int_{-\infty}^{\infty} dp\, \tilde{\psi}^*(p)\, p\, \tilde{\psi}(p) = \frac{1}{\sqrt{2\pi\hbar}} \int_{-\infty}^{\infty} dp\, \tilde{\psi}^*(p) \int_{-\infty}^{\infty} dx\, \psi(x) i\hbar \frac{d}{dx} e^{-\frac{ipx}{\hbar}} \\
&= \frac{1}{\sqrt{2\pi\hbar}} \int_{-\infty}^{\infty} dx \int_{-\infty}^{\infty} dp\, \tilde{\psi}^*(p) e^{-\frac{ipx}{\hbar}} (-i\hbar) \frac{d}{dx} \psi(x) \\
&= \int_{-\infty}^{\infty} dx\, \psi^*(x)(-i\hbar) \frac{d}{dx} \psi(x) .
\end{aligned} \qquad (2.57)$$

This shows that the same results are obtained replacing at the same time $\tilde{\psi}$ by ψ and the multiplication of $\tilde{\psi}$ by p by the action of $(-i\hbar)\frac{d}{dx}$ on ψ. Both the multiplication by the variable, be it x or p, and the action of the derivative on the wave packets, in much the same way as the Laplacian, are operations which transform linearly wave packets of a certain class (e.g. with a certain number of continuous derivatives) into wave packets of another class. They are called linear *operators* since they act linearly on the space of wave functions.[2] Linear operators can be combined into linear combinations and ordered *products*. A product of operators represents the combined action on the wave function of the factors of the operator product, beginning from the first operator on the right and ending with the last one on the left. The resulting action depends on the order of the factors, for this reason one says that linear operators form a *non-commutative algebra*. In general the above mentioned properties allow defining functions of operators. We shall use operators in Sects. 2.7 and 2.9.

Based on this comment, we can write a formula for the mean quadratic deviation in momentum. For this it is convenient to introduce the modified wave packet:

$$\hat{\psi}(x) \equiv e^{-\frac{i\bar{p}x}{\hbar}} \psi(x + \bar{x}) . \tag{2.58}$$

Notice that

$$(\tilde{\hat{\psi}})(p) = \frac{1}{\sqrt{2\pi\hbar}} \int_{-\infty}^{\infty} dx \, \psi(x + \bar{x}) e^{-\frac{i(p+\bar{p})x}{\hbar}} = \tilde{\psi}(p + \bar{p}) e^{-\frac{i(p+\bar{p})\bar{x}}{\hbar}} . \tag{2.59}$$

Therefore, using (2.46) and (2.56) we have:

$$\Delta_x^2 = \int_{-\infty}^{\infty} dx \, |x \, \hat{\psi}(x)|^2, \quad \Delta_p^2 = \hbar^2 \int_{-\infty}^{\infty} dx \, |\frac{d}{dx} \, \hat{\psi}(x)|^2 , \tag{2.60}$$

indeed, the first equation is identical to Eq. (2.56), while the second one is the same equation written in terms of the variable p.

Now we can consider the product $\Delta_x^2 \Delta_p^2$. Before that, let us introduce a general inequality called Cauchy-Schwarz inequality. Consider two wave packets ψ_1 and ψ_2 not satisfying the normalization constraint (2.49), one has the inequality:

$$\int_{-\infty}^{\infty} dx_1 |\psi_1(x_1)|^2 \int_{-\infty}^{\infty} dx_2 |\psi_2(x_2)|^2 \geq \left| \int_{-\infty}^{\infty} dx \, \psi_1^*(x)\psi_2(x) \right|^2 , \tag{2.61}$$

which is analogous to the triangular inequality: the square scalar product of two vectors cannot exceed the product of the square lengths of the same vectors. The product of the integrals on the left-hand side of the inequality corresponds to that of the square lengths of the vectors. The integral on the right-hand side corresponds to the scalar product. This analogy justifies the general validity of the inequality.

Let us now replace into Eq. (2.61) $\psi_1(x)$ by $x \, \psi(x)$ and $\psi_2(x)$ by $\hbar\frac{d}{dx}\psi(x)$. Taking into account Eq. (2.60), we have

---

[2]In general, given a normalized wave function, e.g. $\psi(x)$, and an operator O acting on it, $\int dx\psi^*(x)O\psi(x) \equiv \langle O \rangle$ is the average value of the physical quantity associated with O in the state described by the wave function $\psi(x)$.

$$\Delta_x^2 \Delta_p^2 \geq \hbar^2 \left| \int_{-\infty}^{\infty} dx\, x\psi^*(x)\frac{d}{dx}\psi(x) \right|^2$$

$$\geq \frac{\hbar^2}{4} \left| \int_{-\infty}^{\infty} dx (x\psi^*(x)\frac{d}{dx}\psi(x) + x\psi(x)\frac{d}{dx}\psi^*(x)) \right|^2$$

$$= \frac{\hbar^2}{4} \left| \int_{-\infty}^{\infty} dx\, \psi^*(x)(x\frac{d}{dx} - \frac{d}{dx}x)\psi(x)) \right|^2 = \frac{\hbar^2}{4}. \qquad (2.62)$$

It appears clearly that the uncertainty relation (2.54) holds true for any wave packet, and that it is due to the lack of commutativity of the operators corresponding to multiplication by x and to x-derivative. In more physical terms, the origin of the uncertainty relation is due to the lack of commutativity of the operators associated with position (x) and momentum $(-i\hbar\frac{d}{dx})$.

This result obviously generalizes to any pair of quantities (observables) whose corresponding operators do not commute. We shall use this generalization in Sect. 2.9, considering pairs of components of the angular momentum.

From a phenomenological point of view this principle originates from the universality of diffractive phenomena. Indeed, diffractive effects are those which prevent the possibility of a simultaneous measurement of position and momentum with arbitrarily good precision for both quantities. Let us consider for instance the case in which the measurement is performed through optical instruments; in order to improve the resolution it is necessary to make use of radiation of shorter wavelength, thus increasing the momenta of photons, which hitting the object under observation change its momentum in an unpredictable way. If instead position is determined through mechanical instruments, like slits, then the uncertainty in momentum is caused by diffractive phenomena.

It is important to evaluate the order of magnitude of quantum uncertainty in cases of practical interest. Let us consider for instance a beam of electrons emitted by a cathode at a temperature $T = 1000\,°K$ and accelerated through a potential difference equal to 10^4 V. The order of magnitude of the kinetic energy uncertainty Δ_E is kT, where $k = 1.381 \times 10^{-23}$ J/°K is the Boltzmann constant (alternatively one can use $k = 8.617 \times 10^{-5}$ eV/°K). Therefore $\Delta_E = 1.38 \times 10^{-20}$ J while $E = 1.6 \times 10^{-15}$ J, corresponding to a quite precise determination of the beam energy $(\Delta_E/E \sim 10^{-5})$. We can easily compute the momentum uncertainty by using error propagation $(\Delta_p/p = \frac{1}{2}\Delta_E/E)$ and computing $p = \sqrt{2m_e E} = 5.6 \times 10^{-23}$ N s; we thus obtain $\Delta_p = 2.8 \times 10^{-28}$ N s, hence, making use of (2.54), $\Delta_x \geq 2 \times 10^{-7}$ m. It is clear that the uncertainty principle does not place significant constraints in the case of particle beams.

A macroscopic body of mass $M = 1$ kg placed at room temperature $(T \simeq 300\,°K)$ has an average thermal momentum, caused by collisions with air molecules, which is equal to $\Delta_p \sim \sqrt{2M\,3kT/2} \simeq 9 \times 10^{-11}$ N s, so that the minimal quantum uncertainty on its position is $\Delta_x \sim 10^{-24}$ m, hence not appreciable.

The uncertainty principle is instead quite relevant at the atomic level, where it is the stabilizing mechanism which prevents the electron from collapsing onto the nucleus. We can think of the electron orbital radius as a rough estimate of its position uncertainty ($\Delta_x \sim r$) and evaluate the kinetic energy deriving from the momentum uncertainty; we have $E_k \sim \Delta_p^2/(2m) \sim \hbar^2/(2mr^2)$. Taking into account the binding Coulomb energy, the total energy is

$$E(r) \sim \frac{\hbar^2}{2mr^2} - \frac{e^2}{4\pi\epsilon_0 r} .$$

We infer that the system is stable, since the total energy $E(r)$ has an absolute minimum. The stable radius r_m corresponding to this minimum can be computed through the equation

$$\frac{e^2}{4\pi\epsilon_0 r_m^2} - \frac{\hbar^2}{mr_m^3} = 0 ,$$

hence

$$r_m \sim \frac{4\pi\epsilon_0 \hbar^2}{me^2} ,$$

which nicely reproduces the value of the atomic radius for the ground level in Bohr's model, see (2.22).

2.4.2 The Speed of Waves

It is known that electromagnetic waves move without distortion at a speed $c = 1/\sqrt{\epsilon_0 \mu_0}$ and that, for a harmonic wave, c is given by the wavelength multiplied by the frequency.

In the case of de Broglie's waves introduced in (2.40), we have $\nu = p^2/(2mh)$ and $\lambda = h/p$; therefore the velocity of harmonic waves is given by $v_F \equiv \lambda\nu = p/(2m)$. If we consider instead the wave packet given in (2.51) and its corresponding probability density given in (2.52), we clearly see that it moves with a velocity $v_G \equiv p_0/m$, which is equal to the classical velocity of a particle with momentum p_0. We have used different symbols to distinguish the velocity of plane waves v_F, which is called *phase velocity*, from v_G, which is the speed of the packet and is called *group velocity*. Previous equations lead to the result that, contrary to what happens for electromagnetic waves propagating in vacuum, the two velocities are different for de Broglie's waves, and in particular the group velocity does not coincide with the average value of the phase velocities of the different plane waves making up the packet. Moreover, the phase velocity depends on the wavelength ($v_F = h/(2m\lambda)$). The relation between frequency and wavelength is given by $\nu = c/\lambda$ for electromagnetic waves, while for de Broglie's waves it is $\nu = h/(2m\lambda^2)$.

There is a very large number of examples of wave-like propagation in physics: electromagnetic waves, elastic waves, gravity waves in liquids and several other ones. In each case the frequency presents a characteristic dependence on the wavelength, $\nu(\lambda)$. Considering as above the propagation of gaussian wave packets, it is always possible to define the phase velocity, $v_F = \lambda\,\nu(\lambda)$, and the group velocity, which in general is defined by the relation:

$$v_G = -\lambda^2 \frac{d\nu(\lambda)}{d\lambda} \, . \tag{2.63}$$

Last equation can be verified by considering that, for a generic dependence of the wave phase on the wave number $\exp(ikx - i\omega(k)t)$ and for a generic wave packet described by superposition coefficients strongly peaked around a given value $k = k_0$, the resulting wave function

$$\psi(x) \propto \int_{-\infty}^{\infty} dk \; f(k - k_0) \, e^{i(kx - \omega(k)t)}$$

will be peaked around an x_0 such that the phase factor is stationary, hence almost constant, for $k \sim k_0$, leading to $x_0 \sim \omega'(k_0)t$.

In the case of de Broglie's waves, Eq. (2.63) reproduces the result found previously. Media where the frequency is inversely proportional to the wavelength, as for electromagnetic waves in vacuum, are called *non-dispersive media*, and in that case the two velocities coincide.

It may be interesting to notice that, if we adopt the relativistic form for the plane wave, we have $\nu(\lambda) = \sqrt{m^2 c^4 / h^2 + c^2 / \lambda^2}$, hence

$$v_F = \lambda \sqrt{\frac{m^2 c^4}{h^2} + \frac{c^2}{\lambda^2}} = \frac{E}{p} > c \, ,$$

$$v_G = \frac{c^2}{\lambda} \left(\frac{m^2 c^4}{h^2} + \frac{c^2}{\lambda^2} \right)^{-1/2} = \frac{p c^2}{E} < c \, .$$

In particular v_G, which describes the motion of wave packets, satisfies the constraint of being less than c and coincides with the relativistic expression for the speed of a particle in terms of momentum and energy given in Chap. 1.

2.4.3 The Collective Interpretation of de Broglie's Waves

The description of single particles as wave packets is at the basis of a rigorous formulation of Schrödinger's theory. There is however an alternative interpretation of the wave function, which is of much simpler use and can be particularly useful to describe average properties, like a particle flow in the free case.

Let us consider the plane wave in (2.40): $\psi = \exp\left(i\left(p\,x - p^2 t/(2m)\right)/\hbar\right)$ and compute the corresponding current density J:

$$J = -\frac{i\hbar}{2m}\left(\psi^* \partial_x \psi - \psi \partial_x \psi^*\right) = -\frac{i\hbar}{2m}\left(\psi^* \frac{ip}{\hbar}\psi - \psi \frac{-ip}{\hbar}\psi^*\right) = \frac{p}{m}, \qquad (2.64)$$

while $\rho = \psi^* \psi = 1$. On the other hand we notice that given a distribution of classical particles with density ρ and moving with velocity v, the corresponding current density is $J = \rho v$.

That suggests to go beyond the problem of normalizing the probability distribution in (2.44), relating instead the wave function in (2.40) not to a single particle, as we have done till now, but to a stationary flux of independent particles, which are uniformly distributed with unitary density and move with the same velocity v.

It should be clear that in this way we are a priori giving up the idea of particle localization, however we obtain in a much simpler way information about the group velocity and the flux. We will thus be able, in the following Section, to easily and clearly interpret the effects of a potential barrier on a particle flux.

2.5 The Potential Barrier

The most interesting physical situation is that in which particles are not free, but subject to forces corresponding to a potential energy $V(x)$. In these conditions the Schrödinger equation in the form given in (2.42) has to be used. Since the equation is linear, the study can be limited, without loss of generality, to solutions which are periodic in time, like:

$$\psi(x, t) = e^{-i\,Et/\hbar}\psi_E(x). \qquad (2.65)$$

Indeed the general time dependent solution can always be decomposed in periodic components through a Fourier expansion, so that its knowledge is equivalent to that of $\psi_E(x)$ plus the expansion coefficients.

Furthermore, according to the collective interpretation of de Broglie waves presented in the last section, the wave function in (2.65) describes either a stationary flow or a *stationary state* of particles. In particular we shall begin studying a stationary flow hitting a potential barrier.

The function $\psi_E(x)$ is a solution of the equation obtained by replacing (2.65) into (2.42), i.e.

$$i\hbar\,\partial_t e^{-i\,Et/\hbar}\psi_E(x) = E e^{-i\,Et/\hbar}\psi_E(x) = e^{-i\,Et/\hbar}\left[-\frac{\hbar^2}{2m}\partial_x^2 \psi_E + V(x)\psi_E\right]$$

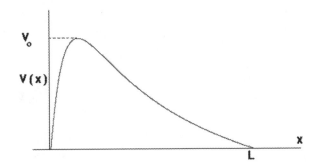

Fig. 2.3 A typical example of a potential barrier, referring in particular to that due to Coulomb repulsion that will be used when discussing Gamow's theory of nuclear α-emission

hence

$$E\psi_E(x) = -\frac{\hbar^2}{2m}\partial_x^2\psi_E(x) + V(x)\psi_E(x),\qquad(2.66)$$

which is known as the *time-independent* or *stationary* Schrödinger equation.

We shall consider at first the case of a *potential barrier*, in which $V(x)$ vanishes for $x < 0$ and $x > L$, and is positive in the segment $[0, L]$, as shown in Fig. 2.3. A flux of classical particles hitting the barrier from the left will experience slowing forces as $x > 0$. If the starting kinetic energy, corresponding in this case to the total energy E in (2.66), is greater than the barrier height V_0, the particles will reach the point where V has a maximum, being accelerated from there forward till they pass point $x = L$, where the motion gets free again. Therefore the flux is completely transmitted, the effect of the barrier being simply a slowing down in the segment $[0, L]$. If instead the kinetic energy is less than V_0, the particles will stop before they reach the point where V has a maximum, reversing their motion afterwards: the flux is completely reflected in this case. Quantum Mechanics gives a completely different result.

In order to analyze the differences from a qualitative point of view, it is convenient to choose a barrier which makes the solution of (2.66) easier: that is the case of a potential which is piecewise constant, like the square given below. The choice is motivated by the fact that, if V is constant, then (2.66) can be rewritten as follows:

$$\partial_x^2\psi_E(x) + \frac{2m}{\hbar^2}(E - V)\psi_E(x) = 0,\qquad(2.67)$$

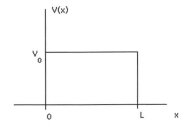

and has the general solution:

$$\psi_E(x) = a_+ \exp\left(i \,\frac{\sqrt{2m(E-V)}}{\hbar}x\right) + a_- \exp\left(-i\,\frac{\sqrt{2m(E-V)}}{\hbar}x\right) , \quad (2.68)$$

if $E > V$, while

$$\psi_E(x) = a_+ \exp\left(\frac{\sqrt{2m(V-E)}}{\hbar}x\right) + a_- \exp\left(-\frac{\sqrt{2m(V-E)}}{\hbar}x\right) , \quad (2.69)$$

in the opposite case. The problem is then to establish how the solution found in a definite region can be connected to those found in the nearby regions. In order to solve this kind of problem we must be able to manage differential equations in the presence of discontinuities in their coefficients, and that requires a brief *mathematical interlude*.

2.5.1 Mathematical Interlude: Differential Equations with Discontinuous Coefficients

Differential equations with discontinuous coefficients can be treated by smoothing the discontinuities, then solving the equations in terms of functions which are derivable several times, and finally reproducing the correct solutions in the presence of discontinuities through a limit process. In order to do so, let us introduce the function $\varphi_\epsilon(x)$, which is defined as

$$\varphi_\epsilon(x) = 0 \quad \text{if} \quad |x| > \epsilon ,$$

$$\varphi_\epsilon(x) = \frac{\epsilon^2 + x^2}{2\left(\epsilon^2 - x^2\right)^2} \frac{1}{\cosh^2\left(x/(\epsilon^2 - x^2)\right)} \quad \text{if} \quad |x| < \epsilon .$$

This function, as well as all of its derivatives, is continuous and it can be easily shown that

$$\int_{-\infty}^{\infty} \varphi_\epsilon(x)dx = 1 .$$

Based on this property we conclude that if $f(x)$ is locally integrable, i.e. if it admits at most isolated singularities where the function may diverge with a degree less than one, like for instance $1/|x|^{1-\delta}$ when $\delta > 0$, then the integral

$$\int_{-\infty}^{\infty} \varphi_\epsilon(x - y)f(y)dy \equiv f_\epsilon(x)$$

defines a function which can be derived in x an infinite number of times; the derivatives of f_ϵ tend to those of f in the limit $\epsilon \to 0$ and in all points where the latter are defined. We have in particular, by part integration,

$$\frac{d^n}{dx^n} f_\epsilon(x) = \int_{-\infty}^{\infty} \varphi_\epsilon(x - y) \frac{d^n}{dy^n} f(y) dy ; \qquad (2.70)$$

f_ϵ is called *regularized function*. If for instance we consider the case in which f is the step function in the origin, i.e. $f(x) = 0$ for $x < 0$ and $f(x) = 1$ for $x > 0$, we have for $f_\epsilon(x)$, $\partial_x f_\epsilon(x) = f'_\epsilon(x)$ and $\partial_x^2 f_\epsilon(x) = f''_\epsilon(x)$ the behaviors showed in the figures given below. Notice in particular that since

$$f_\epsilon(x) = \int_0^{\infty} \varphi_\epsilon(x - y) dy = \int_{-\infty}^{x} \varphi_\epsilon(z) dz$$

we have $\partial_x f_\epsilon(x) = \varphi_\epsilon(x)$. By looking at the three figures it is clear that $f_\epsilon(x)$ continuously interpolates between the two values, zero and one, which the function assumes respectively to the left of $-\epsilon$ and to the right of ϵ, staying less than 1 for every value of x. It is important to notice that instead the second figure, showing $\partial_x f_\epsilon(x)$, i.e. $\varphi_\epsilon(x)$, has a maximum of height proportional to $1/\epsilon^2$, hence diverging as $\epsilon \to 0$.

The third figure, showing the second derivative $\partial_x^2 f_\epsilon(x)$, has an oscillation of amplitude proportional to $1/\epsilon^4$ around the discontinuity point. Since, for small ϵ, the regularized function depends, close to the discontinuity, on the nearby values of the original function, it is clear that the qualitative behaviors showed in the figures are valid, close to discontinuities of the first kind (i.e. where the function itself has a discontinuous gap), for every starting function f.

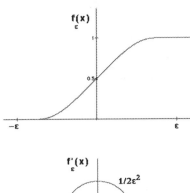

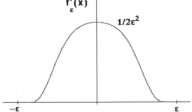

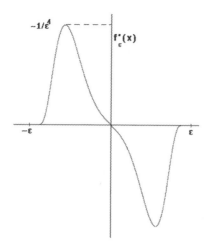

Let us now consider (2.67) close to a discontinuity point of the first kind (step function) for V, and suppose we regularize both terms on the left hand side. Assuming that the wave function does not present discontinuities worse than first kind, the second term in the equation may present only steps so that, once regularized, it is limited independently of ϵ. However the first term may present oscillations of amplitude $\sim 1/\epsilon^4$ if ψ_E has a first kind discontinuity, or a peak of height $\sim +1/\epsilon^2$ if ψ_E is continuous but its first derivative has such discontinuity: in each case the modulus of the first regularized term would diverge faster than the second in the limit $\epsilon \to 0$. That shows that in the presence of a first kind discontinuity in V, both the wave function ψ_E and its derivative must be continuous.

In order to simply deal with barriers of length L much smaller than the typical wavelengths of the problem, it is useful to introduce infinitely thin barriers: that can be done by choosing a potential energy which, once regularized, is equal to $V_\epsilon(x) = \mathcal{V} \varphi_\epsilon(x)$, i.e.

$$V(x) = \mathcal{V} \lim_{\epsilon \to 0} \varphi_\epsilon(x) \equiv \mathcal{V}\delta(x) . \tag{2.71}$$

Equation (2.71) defines the so-called *Dirac's delta function* as a limit of φ_ϵ.

When studying Schrödinger equation regularized as done above, it is possible to show, by integrating the differential equation between $-\epsilon$ and ϵ, that in the presence of a potential barrier proportional to the Dirac delta function the wave function stays continuous, but its derivative has a first kind discontinuity of amplitude

$$\lim_{\epsilon \to 0} (\psi'_E(\epsilon) - \psi'_E(-\epsilon)) = \frac{2m}{\hbar^2} \mathcal{V} \psi_E(0) . \tag{2.72}$$

Notice that a potential barrier proportional to the Dirac delta function can be represented equally well by a square barrier of height $\mathcal{V}/L$ and width L, in the limit as $L \to 0$ with $\int_{-\infty}^{\infty} dx\, V(x) = \mathcal{V}$ kept constant.

2.5.2 The Square Barrier

Let us consider the stationary Schrödinger equation (2.66) with a potential corresponding to the square barrier described above, that is $V(x) = V$ for $0 < x < L$ and vanishing elsewhere. As in the classical case we can distinguish two different regimes:

(a) the case $E > V$, in which classically the flux would be entirely transmitted;
(b) the opposite case, $E < V$, in which classically the flux would be entirely reflected.

Let us start with case (a) and distinguish three different regions:
(1) the region $x < 0$, in which the general solution is

$$\psi_E(x) = a_+ e^{i\sqrt{2mE}\, x/\hbar} + a_- e^{-i\sqrt{2mE}\, x/\hbar}; \tag{2.73}$$

this wave function corresponds to two opposite fluxes, the first moving rightwards and equal to $|a_+|^2\sqrt{2E/m}$, the other opposite to the first and equal to $-|a_-|^2\sqrt{2E/m}$. Since we want to study a quantum process analogous to that described classically, we arbitrarily choose $a_+ = 1$, thus fixing the incident flux to $\sqrt{2E/m}$, hence

$$\psi_E(x) = e^{i\sqrt{2mE}\, x/\hbar} + a\, e^{-i\sqrt{2mE}\, x/\hbar}; \tag{2.74}$$

a takes into account the possible reflected flux, $|a|^2\sqrt{2E/m}$. The physically interesting quantity is the fraction of the incident flux which is reflected, which is called the *reflection coefficient* of the barrier and, with our normalization for the incident flux, is $R = |a|^2$;
(2) the region $0 < x < L$, where the general solution is

$$\psi_E(x) = b\, e^{i\sqrt{2m(E-V)}\, x/\hbar} + c\, e^{-i\sqrt{2m(E-V)}\, x/\hbar}; \tag{2.75}$$

(3) the region $x > L$, where the general solution is given again by (2.73). However, since we want to study reflection and transmission through the barrier, we exclude the possibility of a backward flux, i.e. coming from $x = \infty$, thus assuming that the only particles present in this region are those going rightwards after crossing the barrier. Therefore in this region we write

$$\psi_E(x) = d\, e^{i\sqrt{2mE}\, x/\hbar}. \tag{2.76}$$

The potential has two discontinuities in $x = 0$ and $x = L$, therefore we have the following conditions for the continuity of the wave function and its derivative:

$$1 + a = b + c \,,$$

$$1 - a = \sqrt{\frac{E - V}{E}} \, (b - c) \,,$$

$$b \, e^{i \sqrt{2m(E-V)} L/\hbar} + c \, e^{-i \sqrt{2m(E-V)} L/\hbar} = d \, e^{i \sqrt{2m E} L/\hbar} \,, \qquad (2.77)$$

$$\sqrt{\frac{E - V}{E}} \left[b \, e^{i \sqrt{2m(E-V)} L/\hbar} - c \, e^{-i \sqrt{2m(E-V)} L/\hbar} \right] = d \, e^{i \sqrt{2m E} L/\hbar} \,.$$

We have thus a linear system of 4 equations with 4 unknown variables which, for a generic choice of parameters, should univocally identify the solution. However our main interest is the determination of $|a|^2$. Dividing side by side the first two as well as the last two equations, we obtain after simple algebra:

$$\frac{1 - a}{1 + a} = \sqrt{\frac{E - V}{E}} \frac{\frac{b}{c} - 1}{\frac{b}{c} + 1} \,,$$

$$\frac{\frac{b}{c} - e^{-2i \sqrt{2m(E-V)} L/\hbar}}{\frac{b}{c} + e^{-2i \sqrt{2m(E-V)} L/\hbar}} = \sqrt{\frac{E}{E - V}} \,. \qquad (2.78)$$

Solving the second equation for b/c and the first for a we obtain:

$$\frac{b}{c} = e^{-2i \sqrt{2m(E-V)} L/\hbar} \frac{\sqrt{\frac{E-V}{E}} + 1}{\sqrt{\frac{E-V}{E}} - 1} \,,$$

$$a = \frac{1 + \sqrt{\frac{E-V}{E}} + \frac{b}{c} \left(1 - \sqrt{\frac{E-V}{E}} \right)}{1 - \sqrt{\frac{E-V}{E}} + \frac{b}{c} \left(1 + \sqrt{\frac{E-V}{E}} \right)} \qquad (2.79)$$

and finally, by substitution:

$$a = \frac{\left(1 - \frac{E-V}{E} \right) \left(e^{i \sqrt{2m(E-V)} L/\hbar} - e^{-i \sqrt{2m(E-V)} L/\hbar} \right)}{\left(1 - \sqrt{\frac{E-V}{E}} \right)^2 e^{i \sqrt{2m(E-V)} L/\hbar} - \left(1 + \sqrt{\frac{E-V}{E}} \right)^2 e^{-i \sqrt{2m(E-V)} L/\hbar}} \,,$$

so that

$$a = \frac{V}{E} \frac{\sin \left(\frac{\sqrt{(2m(E-V))}}{\hbar} L \right)}{\frac{2E-V}{E} \sin \left(\frac{\sqrt{(2m(E-V))}}{\hbar} L \right) + 2i \sqrt{\frac{E-V}{E}} \cos \left(\frac{\sqrt{(2m(E-V))}}{\hbar} L \right)} \,, \qquad (2.80)$$

which clearly shows that $0 \le |a| < 1$ and that, for $V > 0$, a vanishes only when $\sqrt{(2m(E - V)}L/\hbar = n\pi$.

This is a clear interference effect, showing that reflection by the barrier is a wave-like phenomenon. For those knowing the physics of coaxial cables there should be a clear analogy between our result and the reflection happening at the junction of two cables having mismatching impedances: television set technicians well known that as a possible origin of failure.

The quantum behavior in case (b), i.e. when $E < V$, is more interesting and important for its application to microscopic physics. In this case the wave functions in regions 1 and 3 do not change, while for $0 < x < L$ the general solution is:

$$\psi_E(x) = b\, e^{\sqrt{2m(V-E)}\, x/\hbar} + c\, e^{-\sqrt{2m(V-E)}\, x/\hbar}\,, \tag{2.81}$$

so that the continuity conditions become:

$$
\begin{aligned}
&1 + a = b + c\,, \\
&1 - a = -i\sqrt{\frac{V - E}{E}}\,(b - c)\,, \\
&b\, e^{\sqrt{2m(V-E)}L/\hbar} + c\, e^{-\sqrt{2m(V-E)}L/\hbar} = d\, e^{i\sqrt{2mE}L/\hbar}\,, \\
&-i\sqrt{\frac{V - E}{E}}\left[b\, e^{\sqrt{2m(V-E)}L/\hbar} - c\, e^{-\sqrt{2m(V-E)}L/\hbar}\right] = d\, e^{i\sqrt{2mE}L/\hbar}\,.
\end{aligned}
\tag{2.82}
$$

Dividing again side by side we have:

$$
\begin{aligned}
&\frac{\frac{b}{c} - e^{-2\sqrt{2m(V-E)}L/\hbar}}{\frac{b}{c} + e^{-2\sqrt{2m(V-E)}L/\hbar}} = i\sqrt{\frac{E}{V - E}}\,, \\
&\frac{1 - a}{1 + a} = i\sqrt{\frac{V - E}{E}}\,\frac{1 - \frac{b}{c}}{1 + \frac{b}{c}}\,,
\end{aligned}
\tag{2.83}
$$

which can be solved as follows:

$$a = -\frac{1 - \frac{b}{c} + i\sqrt{\frac{E}{V-E}}\left(1 + \frac{b}{c}\right)}{1 - \frac{b}{c} - i\sqrt{\frac{E}{V-E}}\left(1 + \frac{b}{c}\right)}\,,$$

$$\frac{b}{c} = e^{-2\sqrt{2m(V-E)}L/\hbar}\,\frac{1 + i\sqrt{\frac{E}{V-E}}}{1 - i\sqrt{\frac{E}{V-E}}}\,. \tag{2.84}$$

We can get the expression for a, hence the reflection coefficient $R \equiv |a|^2$, by replacing b/c in the first equation. The novelty is that R is not equal to one since, as it is clear from (2.84), b/c is a complex number. Therefore a fraction $1 - R \equiv T$ of the incident flux is transmitted through the barrier, in spite of the fact that, classically, the particles do not have enough energy to reach the top of it. That is known as *tunnel effect* and plays a very important role in several branches of modern physics, from radioactivity to electronics.

Instead of giving a complete solution for a, hence for the *transmission coefficient* T, and in order to avoid too complex and unreadable formulae, we will confine the discussion to two extreme cases, which however have a great phenomenological interest. We consider in particular:

(a) the case in which $e^{-2\sqrt{2m(V-E)}L/\hbar} \ll 1$, with a generic value for $\frac{E}{V-E}$, i.e. $L \gg \hbar/\sqrt{2m(V-E)}$, which is known as the thick barrier case;
(b) the case in which the barrier is thin, corresponding in particular to the limit $L \to 0$ with $VL \equiv \mathcal{V}$ kept constant.

The thick barrier

In this case $|b/c|$ is small, so that it could be neglected in a first approximation, however it is clear from (2.84) that if $b/c = 0$ then $|a| = 1$, so that there is actually no tunnel effect. For this reason we must compute the Taylor expansion in the expression of a as a function of b/c up to the first order:

$$
\begin{aligned}
a &= -\frac{1+i\sqrt{\frac{E}{V-E}}}{1-i\sqrt{\frac{E}{V-E}}} \frac{1 - \frac{b}{c}\frac{1-i\sqrt{E/(V-E)}}{1+i\sqrt{E/(V-E)}}}{1 - \frac{b}{c}\frac{1+i\sqrt{E/(V-E)}}{1-i\sqrt{E/(V-E)}}} \\[2mm]
&\sim -\frac{1+i\sqrt{\frac{E}{V-E}}}{1-i\sqrt{\frac{E}{V-E}}}\left[1 - \frac{b}{c}\left(\frac{1-i\sqrt{\frac{E}{V-E}}}{1+i\sqrt{\frac{E}{V-E}}} - \frac{1+i\sqrt{\frac{E}{V-E}}}{1-i\sqrt{\frac{E}{V-E}}}\right)\right] \\[2mm]
&= -\frac{1+i\sqrt{\frac{E}{V-E}}}{1-i\sqrt{\frac{E}{V-E}}}\left[1 + 4i\frac{b}{c}\frac{\sqrt{E(V-E)}}{V}\right] \\[2mm]
&= -\frac{1+i\sqrt{\frac{E}{V-E}}}{1-i\sqrt{\frac{E}{V-E}}}\left[1 + 4i\frac{\sqrt{E(V-E)}}{V}e^{-2\sqrt{2m(V-E)}L/\hbar}\frac{1+i\sqrt{\frac{E}{V-E}}}{1-i\sqrt{\frac{E}{V-E}}}\right].
\end{aligned}
\tag{2.85}
$$

In the last line we have replaced b/c by the corresponding expression in (2.84). Neglecting terms of the order of $e^{-4\sqrt{2m(V-E)}L/\hbar}$ or smaller we obtain

$$
|a|^2 = R = 1 - 16\frac{E(V-E)}{V^2}e^{-2\sqrt{2m(V-E)}L/\hbar}.
\tag{2.86}
$$

Therefore the transmission coefficient, which measures the probability for a particle hitting the barrier to cross it, is given by:

$$
T \equiv 1 - R = 16\frac{E(V-E)}{V^2}e^{-2\sqrt{2m(V-E)}L/\hbar}.
\tag{2.87}
$$

Notice that the result seems to vanish for $V = E$, but this is not true since in this case the terms neglected in our approximation come into play.

This formula was first applied in nuclear physics, and more precisely to study α emission, a phenomenon in which a heavy nucleus breaks up into a lighter nucleus plus a particle carrying twice the charge of the proton and roughly four times its mass, which is known as α particle. The decay can be simply described in terms of particles of mass $\sim 0.66 \times 10^{-26}$ kg and energy $E \simeq 4\text{--}8$ MeV$\simeq 10^{-12}$ J, hitting barriers of width roughly equal to 3×10^{-14} m; the difference $V-E$ is of the order of 10 MeV $\simeq 1.6 \times 10^{-12}$ J.

In these conditions we have $2\sqrt{2m(V - E)}L/\hbar \simeq 83$ and therefore $T \sim e^{-2\sqrt{2m(V-E)}L/\hbar} \sim 10^{-36}$. Given the order of magnitude of the energy E and of the mass of the particle, we infer that it moves with a velocity of the order of 10^7 m/s: since the radius R_0 of heavy nuclei is roughly 10^{-14} m, the frequency of collisions against the barrier is $\nu_u \sim 10^{21}$ Hz. That indicates that, on average, the time needed for the α particle to escape the nucleus is of the order of $1/(\nu_u T)$, i.e. about 10^{15} s, equal to 10^8 years. However, if the width of the barrier is only 4 times smaller, the decay time goes down to about 100 years. That shows a great sensitivity of the result to the parameters and justifies the fact that we have neglected the pre-factor in front of the exponential in (2.87). On the other hand that also shows that, for a serious comparison with the actual mean lives of nuclei, an accurate analysis of parameters is needed, but it is also necessary to take into account the fact that we are not dealing with a true square barrier, since the repulsion between the nucleus and the α particle is determined by Coulomb forces, i.e. $V(x) = 2Ze^2/(4\pi\epsilon_0 x)$ for x greater than a given threshold, see Fig. 2.3.

As a consequence, the order of magnitude of the transmission coefficient given in (2.87), i.e.

$$T \simeq e^{-2\sqrt{2m(V-E)}L/\hbar} \tag{2.88}$$

must be replaced by[3]

$$T \simeq \exp\left(-2\int_{R_0}^{R_1} dx \,\frac{\sqrt{2m(V(x) - E)}}{\hbar}\right) \equiv e^{-G}, \tag{2.89}$$

where R_0 is the already mentioned nuclear radius and $R_1 = 2Ze^2/(4\pi E\epsilon_0)$ is the solution of the equation $V(R_1) = E$. We have then

---

[3]One can think of a thick, but not square, barrier as a series of thick square barriers of different heights.

$$G = 2\frac{\sqrt{2m}}{\hbar} \int_{R_0}^{R_1} dx \sqrt{\frac{2Ze^2}{4\pi\epsilon_0 x} - E} = 2\frac{\sqrt{2mE}}{\hbar} \int_{R_0}^{R_1} dx \sqrt{\frac{R_1}{x} - 1}$$

$$= 2\frac{\sqrt{2mE}\,R_1}{\hbar} \int_{\frac{R_0}{R_1}}^{1} dy \sqrt{\frac{1}{y} - 1} = 2\sqrt{\frac{2m}{E}}\frac{Ze^2}{\pi\epsilon_0\hbar} \int_{\sqrt{\frac{R_0}{R_1}}}^{1} dz \sqrt{1 - z^2}$$

$$= \sqrt{\frac{2m}{E}}\frac{Ze^2}{\pi\epsilon_0\hbar} \left[\mathrm{acos}\sqrt{\frac{R_0}{R_1}} - \sqrt{\frac{R_0}{R_1} - \left(\frac{R_0}{R_1}\right)^2} \right]. \tag{2.90}$$

In the approximation $R_0/R_1 \ll 1$ we have

$$G \simeq \frac{2\pi Ze^2}{\epsilon_0 h v}, \tag{2.91}$$

where v is the velocity of the alpha particle. Hence, if we assume like above that the collision frequency be $\nu_u \sim 10^{21}$ Hz, the mean life is

$$\tau = 10^{-21} \exp\left(\frac{2\pi Ze^2}{\epsilon_0 h v}\right). \tag{2.92}$$

If we instead make use of the last expression in (2.90), with $R_0 = 1.1 \times 10^{-14}$ m, we infer for $\ln \tau$ the behavior shown in Fig. 2.4, where the crosses indicate experimental values for the mean lives of various isotopes: ^{232}Th, ^{238}U, ^{230}Th, ^{241}Am, ^{230}U, ^{210}Rn, ^{220}Rn, ^{222}Ac, ^{215}Po, ^{218}Th. Taking into account that the figure covers 23 orders of magnitude, the agreement is surely remarkable. Indeed Gamow's first presentation of these results in 1928 made a great impression.

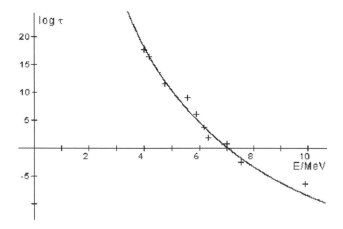

Fig. 2.4 The mean lives of a sample of α-emitting isotopes plotted against the corresponding α-energies. The *solid line* shows the values predicted by Gamow's theory

The thin barrier

In the case of a thin barrier we can neglect E with respect to V, so that $\sqrt{E/(V-E)} \simeq \sqrt{E/V}$ and $e^{-\sqrt{2m(V-E)}L/\hbar} \simeq 1 - \sqrt{2mV}L/\hbar$. We also remind that $\sqrt{2mV}L/\hbar$ is infinitesimal, since $L \to 0$ with $VL \equiv \mathcal{V}$ fixed, so that $e^{i\sqrt{2mE}L/\hbar}$ can be put equal to 1. Therefore Eq. (2.83) becomes

$$1 + a = b + c \,,$$

$$1 - a = -i\sqrt{\frac{V}{E}}(b - c) \,,$$

$$b + c + \frac{\sqrt{2mV}}{\hbar}L(b - c) = d \,,$$

$$b - c + \frac{\sqrt{2mV}}{\hbar}L(b + c) = i\sqrt{\frac{E}{V}}d \,, \tag{2.93}$$

and substituting $b \pm c$ we obtain:

$$d = 1 + a + i\sqrt{\frac{E}{V}}\frac{\sqrt{2mV}}{\hbar}L(1 - a) \simeq 1 + a \,,$$

$$i\sqrt{\frac{E}{V}}(1 - a) + \frac{\sqrt{2mV}}{\hbar}L(1 + a) = i\sqrt{\frac{E}{V}}d \,, \tag{2.94}$$

in its simplest form. Taking further into account our approximation, the system can be rewritten as

$$1 + a = d \,,$$

$$1 - a = \left(i\sqrt{\frac{2m}{E}}\frac{VL}{\hbar} + 1\right)d \equiv \left(1 + i\sqrt{\frac{2m}{E}}\frac{\mathcal{V}}{\hbar}\right)d \,. \tag{2.95}$$

Finally we find, by eliminating a, that

$$d = \frac{1}{1 + i\sqrt{\frac{m}{2E}}\frac{\mathcal{V}}{\hbar}} \,, \tag{2.96}$$

hence

$$T = \frac{1}{1 + \frac{m}{2E}\frac{\mathcal{V}^2}{\hbar^2}} \tag{2.97}$$

and

$$R = \frac{1}{1 + \frac{2E}{m}\frac{\hbar^2}{\mathcal{V}^2}} \,. \tag{2.98}$$

Notice that the system (2.95) confirms what predicted about the continuity conditions for the wave function in the presence of a potential energy equal to $V\delta(x)$, i.e. that the wave function is continuous $(1 + a = d)$ while its derivative has a discontinuity $(i\sqrt{2mE}(1 - a - d)/\hbar)$ equal to $2mV/\hbar^2$ times the value of the wave function (d in our case).

2.6 Quantum Wells and Energy Levels

Having explored the tunnel effect in some details, let us now discuss the solutions of the Schrödinger equation in the case of binding potentials. For bound states, i.e. for solutions with wave functions localized in the neighborhood of a potential well, we expect computations to lead to energy quantization, i.e. to the presence of discrete *energy levels*. Let us start our discussion from the case of a square well

$$V(x) = -V \quad \text{for} \quad |x| < \frac{L}{2}, \qquad V(x) = 0 \quad \text{for} \quad |x| > \frac{L}{2}. \qquad (2.99)$$

Notice that the origin of the coordinate has been chosen in order to emphasize the *symmetry* of the system, corresponding in this case to the invariance of Schrödinger equation under axis reflection $x \to -x$. In general, the symmetry of the potential allows us to find new solutions of the equation starting from known solutions, or to simplify the search for solutions by a priori fixing some of their features. In this case it can be noticed that if $\psi_E(x)$ is a solution, $\psi_E(-x)$ is a solution too, so that, by linearity of the differential equation, any linear combination (with complex coefficients) of the two wave functions is a good solution corresponding to the same value of the energy E, in particular the combinations $\psi_E(x) \pm \psi_E(-x)$, which are even/odd under reflection of the x axis. Naturally one of the two solutions may well vanish, but it is clear that all possible solutions can be described in terms of (i.e. they can be written as linear combinations of) functions which are either even or odd under x-reflection.

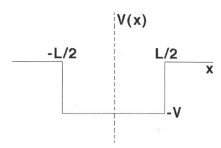

To better clarify the point, let us notice that, since the Schrödinger equation is linear, the set of all possible solutions having the same energy constitutes what is usually called a *linear space*, which is completely fixed once we know one particular basis for it. What we have learned is that in the present case even/odd functions are a good basis, so that the search for solutions can be solely limited to them. This is probably the simplest example of the application of a *symmetry principle* asserting that, if the Schrödinger equation is invariant under a coordinate transformation, it is always possible to choose its solutions so that the transformation does not change them but for a constant phase factor, which in the present case is ± 1.

We will consider in the following only bound solutions which, assuming that the potential energy vanishes for $|x| \rightarrow \infty$, correspond to a negative total energy E and are therefore the analogous of bound states in classical mechanics. Solutions with positive energy, instead, present reflection and transmission phenomena, as in the case of barriers. We notice that, in the case of bound states, the collective interpretation of the wave function does not apply, since these are states involving a single particle: that is in strict relation with the fact that bound state solutions vanish rapidly enough as $|x| \rightarrow \infty$, so that the probability distribution in (2.44) can be properly normalized.

Let us start by considering even solutions: it is clear that we can limit our study to the positive x axis, with the additional constraint of a vanishing first derivative in the origin, as due for an even function (whose derivative is odd). We can divide the positive x axis into two regions where the potential is constant:

(a) that corresponding to $x < L/2$, where the general solution is:

$$\psi_E(x) = a_+ e^{i\sqrt{2m(E+V)}\, x/\hbar} + a_- e^{-i\sqrt{2m(E+V)}\, x/\hbar},$$

which is even for $a_+ = a_-$, so that

$$\psi_E(x) = a \cos\left(\frac{\sqrt{2m(E+V)}}{\hbar} x\right) ; \qquad (2.100)$$

(b) that corresponding to $x > L/2$, where the general solution is:

$$\psi_E(x) = b_+ e^{\sqrt{2m|E|}\, x/\hbar} + b_- e^{-\sqrt{2m|E|}\, x/\hbar}.$$

The condition that $|\psi|^2$ be an integrable function constrains $b_+ = 0$, otherwise the probability density would unphysically diverge for $|x| \rightarrow \infty$; therefore we can write

$$\psi_E(x) = b\, e^{-\sqrt{2m|E|}\, x/\hbar}. \qquad (2.101)$$

Notice that we have implicitly excluded the possibility $E < -V$, the reason being that in this case (2.100) would be replaced by

$$\psi_E(x) = a \cosh\left(\frac{\sqrt{2m|E+V|}}{\hbar} x\right)$$

which for $x > 0$ has a positive logarithmic derivative $(\partial_x \psi_E(x)/\psi_E(x))$ which cannot continuously match the negative logarithmic derivative of the solution in the second region given in (2.101). Therefore quantum theory is in agreement with classical mechanics about the impossibility of having states with total energy less than the minimum of the potential energy.

The solutions to the Schrödinger equation on the whole axis can be found by solving the system:

$$a \cos \frac{\sqrt{2m(E+V)}L}{2\hbar} = b\, e^{-\sqrt{2m|E|}L/(2\hbar)}, \tag{2.102}$$

$$\frac{\sqrt{2m(E+V)}}{\hbar} a \sin \frac{\sqrt{2m(E+V)}L}{2\hbar} = \frac{\sqrt{2m|E|}}{\hbar} b\, e^{-\sqrt{2m|E|}L/(2\hbar)}$$

dividing previous equations side by side we obtain the continuity condition for the logarithmic derivative:

$$\tan \frac{\sqrt{2m(E+V)}L}{2\hbar} = \sqrt{\frac{|E|}{E+V}}. \tag{2.103}$$

In order to discuss last equation let us introduce the variable

$$x \equiv \frac{\sqrt{2m(E+V)}L}{2\hbar} \tag{2.104}$$

and the parameter

$$y \equiv \sqrt{2mV}L/2\hbar, \tag{2.105}$$

and let us plot together the behavior of the two functions $\tan x$ and $\sqrt{(y^2 - x^2)/x^2} = \sqrt{|E|/(E+V)}$. In the figure we show the case $y^2 = 20$. From a qualitative point of view the figure shows that energy levels, corresponding to the intersection points of the two functions, are quantized, thus confirming also for the case of potential wells the discrete energy spectrum predicted by Bohr's theory. In particular the plot shows two intersections, the first for $x - x_1 < \pi/2$, the second for $\pi < x = x_2 < 3\pi/2$.

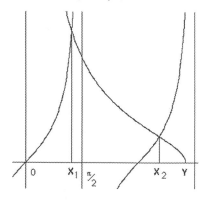

Notice that quantization of energy derives from the physical requirement of having a bound state solution which does not diverge but instead vanishes outside the well:

for this reason the external solution is parametrized in terms of only one parameter. The reduced number of available parameters allows for non-trivial solutions of the homogeneous linear system (2.103) only if the energy quantization condition (2.103) is satisfied.

The number of possible solutions increases as y grows and since $y > 0$ it is anyway greater than zero. Therefore the square potential well in one dimension has always at least one bound state corresponding to an even wave function. It can be proved that the same is true for every symmetric well in one dimension (i.e. such that $V(-x) = V(x) \leq 0$). On the contrary, an extension of this analysis (see in particular the discussion about the spherical well in Sect. 2.9) shows that in the three-dimensional case the existence of at least one bound state is not guaranteed any more.

Let us now consider the case of odd solutions: we must choose a wave function which vanishes in the origin, so that the cosine must be replaced by a sine in (2.100). Going along the same lines leading to (2.103) we arrive to the equation

$$\cot \frac{\sqrt{2m(E + V)}L}{2\hbar} = -\sqrt{\frac{|E|}{E + V}} . \qquad (2.106)$$

Using the same variables x and y as above, we have the corresponding figure given below, which shows that intersections are present only if $y > \pi/2$, i.e. if $V > \pi^2\hbar^2/(2mL^2)$ (which by the way is also the condition for the existence of at least one bound state in three dimensions). Notice that the energy levels found in the odd case are different from those found in the even case. In particular any possible negative energy level can be put in correspondence with only one wave function (identified by neglecting a possible irrelevant constant phase factor): this implies that, in the present case, dealing with solutions having a definite transformation property under the symmetry of the problem (i.e. even or odd) is not a matter of choice, as it is in the general case, but a necessity, since those are the only possible solutions. Indeed a different kind of solution could only be constructed in the presence of two solutions, one even and the other odd, corresponding to the same energy level.

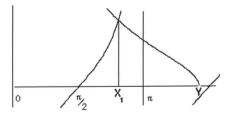

The number of independent solutions corresponding to a given energy level is usually called the *degeneracy* of the level. We have therefore demonstrated that, for the potential square well in one dimension, the discrete energy levels have always degeneracy equal to one or, stated otherwise, that they are *non-degenerate*. This is in fact a general property of bound states in one dimension, which can be demonstrated for any kind of potential well.

It is interesting to apply our analysis to the case of an infinitely deep well. Obviously, if we want to avoid dealing with divergent negative energies as we deepen the well, it is convenient to shift the zero of the energy so that the potential energy vanishes inside the well and is V outside. That is equivalent to replacing in previous formulae $E + V$ by E and $|E|$ by $V - E$; moreover, bound states will now correspond to energies $E < V$. Taking the limit $V \to \infty$ in the quantization conditions given in (2.103) and (2.106), we obtain respectively $\tan \sqrt{2mE}L/(2\hbar) = +\infty$ and $-\cot \sqrt{2mE}L/(2\hbar) = +\infty$, so that $\sqrt{2mE}\,L/(2\hbar) = (2n - 1)\pi/2$ and $\sqrt{2mE}\,L/(2\hbar) = n\pi$ with $n = 1, 2, \ldots$ Finally, combining odd and even states, we have

$$\frac{\sqrt{2mE}}{\hbar}L = n\pi : \qquad n = 1, 2, \ldots$$

and the following energy levels

$$E_n = \frac{n^2\pi^2\hbar^2}{2mL^2}. \tag{2.107}$$

The corresponding wave functions vanish outside the well while in the region $|x| < L/2$ the even functions are $\sqrt{2/L}\cos(2n - 1)\pi x/L$ and the odd ones are $\sqrt{2/L}\sin 2n\pi x/L$, with the coefficients fixed in order to satisfy (2.44). It is also possible to describe all wave functions by a unique formula:

$$\psi_{E_n}(x) = \sqrt{\frac{2}{L}}\sin\frac{n\pi(x + \frac{L}{2})}{L} \quad \text{for} \ |x| < \frac{L}{2},$$

$$\psi_{E_n}(x) = 0 \quad \text{for} \ |x| > \frac{L}{2}. \tag{2.108}$$

While all wave functions are continuous in $|x| = L/2$, their derivatives are not, as in the case of the potential barrier proportional to the Dirac delta function. The generic solution ψ_{E_n} has the behavior showed in the figure, where the analogy with the electric component of an electromagnetic wave reflected between two mirrors clearly appears. Therefore the infinitely deep well can be identified as the region between two reflecting walls.

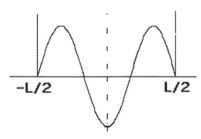

If the wave amplitude vanishes over the mirrors, the distance between them must necessarily be an integer multiple of half the wavelength; this is the typical tuning condition for a musical instrument and implies wavelength and energy quantization. The exact result agrees with that of Problem 2.4.

Going back to the analogy with electromagnetic waves, the present situation corresponds to a one-dimensional *resonant cavity*. In the cavity the field can only oscillate according to the permitted wavelengths, which are $\lambda_n = 2L/n$ for $n = 1, 2, \ldots$ corresponding to the frequencies $\nu_n = c/\lambda_n = nc/(2L)$, which are all multiple of the *fundamental* frequency of the cavity.

Our results regarding the infinitely deep well can be easily generalized to three dimensions. To that purpose, let us introduce a cubic box of side L with reflecting walls. The condition that the wave function vanishes over the walls is equivalent, inside the box and choosing solutions for which the dependence on x, y and z is factorized, to:

$$\psi_{n_x,n_y,n_z} = \sqrt{\frac{8}{L^3}} \sin \frac{n_x \pi (x + \frac{L}{2})}{L} \sin \frac{n_y \pi (y + \frac{L}{2})}{L} \sin \frac{n_z \pi (z + \frac{L}{2})}{L}, \quad (2.109)$$

where we have assumed the origin of the coordinates to be placed in the center of the box. The corresponding energy coincides with the kinetic energy inside the box and can be obtained by writing the Schrödinger equation in three dimensions:

$$-\frac{\hbar^2}{2m} \left(\partial_x^2 + \partial_y^2 + \partial_z^2 \right) \psi_{n_x,n_y,n_z} = E_{n_x,n_y,n_z} \psi_{n_x,n_y,n_z}, \quad (2.110)$$

leading to

$$E_{n_x,n_y,n_z} = \frac{\pi^2 \hbar^2}{2mL^2} \left[n_x^2 + n_y^2 + n_z^2 \right]. \quad (2.111)$$

This result will be useful for studying the properties of a gas of non-interacting particles (perfect gas) contained in a box with reflecting walls. Following the same analogy as above one can study in a similar way the oscillations of an electromagnetic field in a three-dimensional cavity, with proper frequencies given by $\nu_{n_x,n_y,n_z} = (c/2L)\sqrt{n_x^2 + n_y^2 + n_z^2}$.

2.7 The Harmonic Oscillator

The one-dimensional harmonic oscillator can be identified with the mechanical system formed by a particle of mass m bound to a fixed point (taken as the origin of the coordinate) by an ideal spring of elastic constant k and vanishing length at rest. This is equivalent to a potential energy $V(x) = kx^2/2$. In classical mechanics the corresponding equation of motion is

$$m\ddot{x} + kx = 0,$$

whose general solution is

$$x(t) = X \cos(\omega t + \phi),$$

where $\omega = \sqrt{k/m} = 2\pi\nu$ and ν is the proper frequency of the oscillator.

At the quantum level we must solve the following stationary Schrödinger equation:

$$-\frac{\hbar^2}{2m}\partial_x^2 \psi_E(x) + \frac{m\omega^2}{2} x^2 \psi_E(x) = E\psi_E(x). \tag{2.112}$$

In order to solve this equation we can use the identity

$$\left(\sqrt{\frac{m\omega^2}{2}}x - \frac{\hbar}{\sqrt{2m}}\partial_x\right)\left(\sqrt{\frac{m\omega^2}{2}}x + \frac{\hbar}{\sqrt{2m}}\partial_x\right) f(x)$$

$$= \frac{m\omega^2}{2}x^2 f(x) + \frac{\hbar\omega}{2}x\partial_x f(x) - \frac{\hbar\omega}{2}\partial_x(xf(x)) - \frac{\hbar^2}{2m}\partial_x^2 f(x)$$

$$= -\frac{\hbar^2}{2m}\partial_x^2 f(x) + \frac{m\omega^2}{2}x^2 f(x) - \frac{\hbar\omega}{2}f(x)$$

$$= \left(-\frac{\hbar^2}{2m}\partial_x^2 + \frac{m\omega^2}{2}x^2 - \frac{\hbar\omega}{2}\right) f(x), \tag{2.113}$$

which is true for any function f which is derivable at least two times.

It is important to notice the *operator* notation used in last equation, where we have introduced some specific symbols, $(\sqrt{m\omega^2/2}\, x \pm (\hbar/\sqrt{2m})\,\partial_x)$ or $(-(\hbar^2/2m)\,\partial_x^2 + (m\omega^2/2)\,x^2 - \hbar\omega/2)$, to indicate operations in which derivation and multiplication by some variable are combined together. As already mentioned, these are usually called operators, meaning that they give a correspondence law between functions belonging to some given class (for instance those which can be derived n times) and other functions belonging, in general, to a different class.

In this way, leaving aside the specific function f, Eq. (2.113) can be rewritten as an operator relation

$$\left(\sqrt{\frac{m\omega^2}{2}}x - \frac{\hbar}{\sqrt{2m}}\partial_x\right)\left(\sqrt{\frac{m\omega^2}{2}}x + \frac{\hbar}{\sqrt{2m}}\partial_x\right) = \left(-\frac{\hbar^2}{2m}\partial_x^2 + \frac{m\omega^2}{2}x^2 - \frac{\hbar\omega}{2}\right) \tag{2.114}$$

and equations of similar nature can be introduced, like for instance:

$$\left(\sqrt{\frac{m\omega^2}{2}}x + \frac{\hbar}{\sqrt{2m}}\partial_x\right)\left(\sqrt{\frac{m\omega^2}{2}}x - \frac{\hbar}{\sqrt{2m}}\partial_x\right)$$

$$- \left(\left(\sqrt{\frac{m\omega^2}{2}}x - \frac{\hbar}{\sqrt{2m}}\partial_x\right)\left(\sqrt{\frac{m\omega^2}{2}}x + \frac{\hbar}{\sqrt{2m}}\partial_x\right)\right) = \hbar\omega. \tag{2.115}$$

In order to shorten formulae, it is useful to introduce the two symbols:

$$X_{\pm} \equiv \left(\sqrt{\frac{m\omega^2}{2}} x \pm \frac{\hbar}{\sqrt{2m}} \partial_x \right) = \sqrt{\frac{\hbar\omega}{2}} \left(\alpha x \pm \frac{1}{\alpha} \frac{\partial}{\partial x} \right) \tag{2.116}$$

in which the constant $\alpha \equiv \sqrt{m\omega/\hbar}$ has been defined, corresponding to the inverse of the typical length scale of the system. That allows us to rewrite (2.115) in the simpler form:

$$X_+ X_- - X_- X_+ = \hbar\omega . \tag{2.117}$$

If, extending the operator formalism, we define

$$H \equiv -\frac{\hbar^2}{2m} \partial_x^2 + \frac{m\omega^2}{2} x^2 , \tag{2.118}$$

we can rewrite (2.114) as:

$$X_- X_+ = H - \frac{\hbar\omega}{2} , \tag{2.119}$$

then obtaining from (2.117):

$$X_+ X_- = H + \frac{\hbar\omega}{2} . \tag{2.120}$$

The Schrödinger equation can be finally written as:

$$H\psi_E(x) = E\psi_E(x) . \tag{2.121}$$

The operator formalism permits to get quite rapidly a series of results.

(a) The wave function which is solution of the equation

$$X_+ \psi_0(x) = \sqrt{\frac{m\omega^2}{2}} x\psi_0(x) + \frac{\hbar}{\sqrt{2m}} \partial_x \psi_0(x) = 0 , \tag{2.122}$$

is also a solution of (2.121) with $E = \hbar\omega/2$. In order to compute it we can rewrite (2.122) as:

$$\frac{\partial_x \psi_0(x)}{\psi_0(x)} = -\alpha^2 x ,$$

hence, integrating both members:

$$\ln \psi_0(x) = c - \frac{\alpha^2}{2} x^2 ,$$

from which it follows that

$$\psi_0(x) = e^c \, e^{-\alpha^2 x^2/2} ,$$

where the constant c can be fixed by the normalization condition given in (2.44), leading finally to

$$\psi_0(x) = \left(\frac{m\omega}{\pi\hbar}\right)^{\frac{1}{4}} e^{-m\omega x^2/(2\hbar)} . \tag{2.123}$$

We would like to remind the need for restricting the analysis to the so-called *square integrable* functions, which can be normalized according to (2.44). This is understood in the following.

(b) What we have found is the lowest energy solution, usually called the *ground state* of the system, as can be proved by observing that, for every normalized solution $\psi_E(x)$, the following relations hold:

$$\int_{-\infty}^{\infty} dx \, \psi_E(x)^* \left(\sqrt{\frac{m\omega^2}{2}}x - \frac{\hbar}{\sqrt{2m}}\partial_x\right)\left(\sqrt{\frac{m\omega^2}{2}}x + \frac{\hbar}{\sqrt{2m}}\partial_x\right)\psi_E(x)$$

$$= \int_{-\infty}^{\infty} dx\, |X_+\psi_E(x)|^2 = \int_{-\infty}^{\infty} dx\, \psi_E(x)^* \left(E - \frac{\hbar\omega}{2}\right)\psi_E(x)$$

$$= E - \frac{\hbar\omega}{2} \geq 0 , \tag{2.124}$$

where the derivative in X_- has been integrated by parts, exploiting the vanishing of the wave function at $x = \pm\infty$. Last inequality follows from the fact that the integral of the squared modulus of any function cannot be negative. Moreover it must be noticed that if the integral vanishes, i.e. if $E = \hbar\omega/2$, then necessarily $X_+\psi_E = 0$, so that ψ_E is proportional to ψ_0. That proves that the ground state is unique.

(c) If ψ_E satisfies (2.112) then $X_\pm\psi_E$ satisfies the same equation with E replaced by $E \mp \hbar\omega$, i.e. we have

$$HX_\pm\psi_E = (E \mp \hbar\omega)X_\pm\psi_E . \tag{2.125}$$

Notice that $X_+\psi_E$ vanishes if and only if $\psi_E = \psi_0$ while $X_-\psi_E$ never vanishes: one can prove this by verifying that if $X_-\psi_E = 0$ then ψ_E behaves as ψ_0 but with the sign $+$ in the exponent, hence it is not square integrable. In order to prove Eq. (2.125), from (2.119) and (2.120) we infer, for instance:

$$X_+X_-X_+\psi_E = X_+\left(H - \frac{\hbar\omega}{2}\right)\psi_E(x) = \left(H + \frac{\hbar\omega}{2}\right)X_+\psi_E(x)$$

$$= X_+\left(E - \frac{\hbar\omega}{2}\right)\psi_E(x) = \left(E - \frac{\hbar\omega}{2}\right)X_+\psi_E(x) \tag{2.126}$$

from which (2.125) follows in the $+$ case.

Last computations show again that operators combine in a fashion which resembles usual multiplication, however their product is strictly dependent on the order in which they appear. We say that the product is non-commutative; that is also evident from (2.117), which expresses what is usually known as the *commutator* of two operators.

Exchanging X_- and X_+ in previous equations we have:

$$X_- X_+ X_- \psi_E = X_- \left(H + \frac{\hbar\omega}{2} \right) \psi_E(x) = \left(H - \frac{\hbar\omega}{2} \right) X_- \psi_E(x)$$

$$= X_- \left(E + \frac{\hbar\omega}{2} \right) \psi_E(x) = \left(E + \frac{\hbar\omega}{2} \right) X_- \psi_E(x) \quad (2.127)$$

which completes the proof of (2.125).

(d) Finally, combining points (a–c), we can show that the only possible energy levels are:

$$E_n = \left(n + \frac{1}{2} \right) \hbar\omega . \quad (2.128)$$

In order to prove that, let us suppose instead that (2.112) admits the level $E = (m + 1/2)\hbar\omega + \delta$, where $0 < \delta < \hbar\omega$, and then repeatedly apply X_+ to ψ_E up to $m + 1$ times. If $X_+^k \psi_E = 0$ with $k \leq m + 1$ and $X_+^{k-1} \psi_E \neq 0$, then we would have $X_+(X_+^{k-1} \psi_E) = 0$ which, as we have already seen, is equivalent to $X_+^{k-1} \psi_E \sim \psi_0$, hence to $H X_+^{k-1} \psi_E = \hbar\omega/2 \psi_E$. However Eq. (2.125) implies $H X_+^{k-1} \psi_E = (E - (k - 1)\hbar\omega)\psi_E$, hence $E = (k - 1/2)\hbar\omega$, which is in contrast with the starting hypothesis ($\delta \neq 0$). On the other hand $X_+^k \psi_E \neq 0$ even for $k = m + 1$ would imply the presence of a solution with energy less that $\hbar\omega/2$, in contrast with (2.124). We have instead no contradiction if $\delta = 0$ and $k = m + 1$.

We have therefore shown that the spectrum of the harmonic oscillator consists of the energy levels $E_n = (n + 1/2) \hbar\omega$. We also know from (2.125) that $\sim X_-^n \psi_0$ is a possible solution with $E = E_n$: we will now show that this is actually the only possible solution.

(e) Any wave function corresponding to the n-th energy level is necessarily proportional to $X_-^n \psi_0$:

$$\psi_{E_n} \sim X_-^n \psi_0 . \quad (2.129)$$

We already know that this is true for $n = 0$ (ground state). Now let us suppose the same to be true for $n = k$ and we shall prove it for $n = k + 1$, thus concluding our argument by induction. Let $\psi_{E_{k+1}}$ be a solution corresponding to E_{k+1}, then, by (2.125) and by the uniqueness of ψ_{E_k}, we have

$$X_+ \psi_{E_{k+1}} = a \psi_{E_k} \quad (2.130)$$

for some constant $a \neq 0$, with $\psi_{E_k} \propto X_-^k \psi_0$. By applying X_- to both sides of last equation we obtain

$$
\begin{aligned}
X_- X_+ \psi_{E_{k+1}} &= \left(H - \frac{\hbar\omega}{2} \right) \psi_{E_{k+1}} = (k+1)\,\hbar\omega\,\psi_{E_{k+1}} \\
&= X_- a \psi_{E_k} \propto X_- X_-^k \psi_0 = X_-^{k+1} \psi_0 ,
\end{aligned}
\tag{2.131}
$$

which proves that also $\psi_{E_{k+1}}$ is proportional to $X_-^{k+1} \psi_0$.

In order to find the correct normalization factor, let us first find it for $\psi_{E_{k+1}}$, assuming that ψ_{E_k} is already correctly normalized. We notice that

$$
\begin{aligned}
\int_{-\infty}^{\infty} dx\, |X_- \psi_{E_k}|^2 &= \int_{-\infty}^{\infty} dx\, \psi_{E_k}^* X_+ X_- \psi_{E_k} = \int_{-\infty}^{\infty} dx\, \psi_{E_k}^* \left(H + \frac{\hbar\omega}{2} \right) \psi_{E_k} \\
&= \hbar\omega(k+1) \int_{-\infty}^{\infty} dx\, |\psi_{E_k}|^2 = \hbar\omega(k+1) ,
\end{aligned}
\tag{2.132}
$$

where in the first equality one of the X_- operators has been integrated by parts and in the second equality Eq. (2.120) has been used. We conclude that $\psi_{E_{k+1}} = (\hbar\omega(k+1))^{-1/2} X_- \psi_{E_k}$, hence, setting for simplicity $\psi_n \equiv \psi_{E_n}$:

$$
\psi_n = \sqrt{\frac{1}{n!}} \left(\frac{X_-}{\sqrt{\hbar\omega}} \right)^n \psi_0 \equiv \sqrt{\frac{1}{n!}} (A^\dagger)^n \psi_0
\tag{2.133}
$$

where one defines $A^\dagger \equiv X_-/\sqrt{\hbar\omega}$.

That concludes our analysis of the one-dimensional harmonic oscillator, which, based on an algebraic approach, has led us to finding both the possible energy levels, given in (2.128), and the corresponding wave functions, described by (2.123) and (2.129). In particular, confirming a general property of bound states in one dimension, we have found that the energy levels are non-degenerate. The operators X_+ and X_- permit us to transform a given solution into a different one, in particular by rising (X_-) or lowering (X_+) the energy level by one quantum $\hbar\omega$.

Also in this case, as for the square well, solutions have definite transformation properties under axis reflection, $x \to -x$, which follow from the symmetry of the potential, $V(-x) = V(x)$. In particular they are divided into even and odd functions according to the value of n, $\psi_n(-x) = (-1)^n \psi_n(x)$, as can be proved by noticing that ψ_0 is an even function and that the operator X_- transforms an even (odd) function into an odd (even) one.

Moreover we notice that, according to (2.129), (2.123) and to the expression for X_- given in (2.116), all wave functions are real. This is also a general property of bound states in one dimension, which can be easily proved and has a simple interpretation. Indeed, suppose ψ_E be the solution of the stationary Schrödinger equation (2.66) corresponding to a discrete energy level E; since obviously both E and the potential energy $V(x)$ are real, it follows, by taking the complex conjugate of both sides of (2.66), that also ψ_E^* is a good solution corresponding to the same energy. However the non-degeneracy of bound states in one dimension implies that ψ_E must be unique. The only possibility is $\psi_E^* \propto \psi_E$, hence $\psi_E^* = e^{i\phi}\psi_E$, so that, leaving aside an irrelevant overall phase factor, ψ_E is a real function.

On the other hand, recalling the definition of the probability current density J given in (2.43), it can be easily proved that the wave function is real if and only if the current density vanishes everywhere. Since we are considering a stationary problem, the probability density is constant in time by definition and the conservation equation (2.27) implies, in one dimension, that the current density J is a constant in space (the same is not true in more than one dimension, where that translates in J being a vector field with vanishing divergence, see Problem 2.47). On the other hand, for a bound state J must surely vanish as $|x| \to \infty$, hence it must vanish everywhere, implying a real wave function: in a one-dimensional bound state there is no current flow at all.

Our results admit various generalizations of great physical interest. First of all, let us consider their extension to the isotropic three-dimensional harmonic oscillator corresponding to the following Schrödinger equation:

$$-\frac{\hbar^2}{2m}\left(\partial_x^2 + \partial_y^2 + \partial_z^2\right)\psi_E + \frac{m\omega^2}{2}\left(x^2 + y^2 + z^2\right)\psi_E = E\,\psi_E\,, \quad (2.134)$$

where $\psi_E = \psi_E(x, y, z)$ is the three-dimensional wave function. This is the typical example of a *separable* Schrödinger equation: if we look for a particular class of solutions, written as the product of three functions depending separately on x, y and z, then Eq. (2.134) becomes equivalent[4] to three independent equations for three one-dimensional oscillators along x, y and z.

Therefore we conclude that the quantized energy levels are in this case:

$$E_{n_x,n_y,n_z} = \hbar\omega\left(n_x + n_y + n_z + \frac{3}{2}\right)\,, \quad (2.135)$$

and that the corresponding wave functions are

$$\psi_{n_x,n_y,n_z}(x, y, z) = \psi_{n_x}(x)\psi_{n_y}(y)\psi_{n_z}(z)\,. \quad (2.136)$$

Notice that, according to (2.135) and (2.136), in three dimensions several degenerate solutions can be found having the same energy, corresponding to all possible integers n_x, n_y, n_z such that $n_x + n_y + n_z = n$ where n is a non-negative integer. The number of such solutions is $(n + 1)(n + 2)/2$.

Since we have looked for particular solutions, having the dependence on x, y and z factorized, it is natural to ask if in this way we have exhausted the possible solutions of equation (2.134). In some sense this is not true: since the Schrödinger

---

[4]This is clear if we divide both sides of Eq. (2.134) by ψ_E: the resulting equation requires that the sum of three functions depending separately on x, y and z be a constant, implying that each function must be constant separately.

equation is linear, we can make linear combinations (with complex coefficients) of the $(n+1)(n+2)/2$ degenerate solutions described above, obtaining new solutions having the same energy $E_n = (n+3/2)\hbar\omega$ but not writable, in general, as the product of three functions of x, y and z. However we have exhausted all the possible solutions in some other sense: indeed it is possible to demonstrate that no further solution can be found beyond all the possible linear combinations of the particular solutions in equation (2.136). In other words, all the possible solutions of equation (2.134), which are found for $E = (n+3/2)\,\hbar\omega$, form a linear space of dimension $(n+1)(n+2)/2$, having the particular solutions in equation (2.136) as an orthonormal basis. We have thus found a possible *complete set* of solutions of equation (2.134): we shall find a different complete set (i.e. a different basis) for the same problem in Sect. 2.9 (see also Problem 2.47).

A further generalization is that regarding small oscillations around equilibrium for a system with N degrees of freedom, whose energy can be separated into the sum of the contributions from N one-dimensional oscillators having, in general, different proper frequencies (ν_i, $i = 1, \ldots, N$). In this case the quantization formula reads

$$E_{(n_1,\ldots,n_N)} = \sum_{i=1}^{N} \hbar\omega_i \left(n_i + \frac{1}{2} \right), \qquad (2.137)$$

and the corresponding wave function can be written as the product of the wave functions associated with every single oscillator.

Let us now take a short detour by recalling the analysis of the electromagnetic field resonating in one dimension. It can be shown that, from a dynamical point of view, the electromagnetic field can be described as an ensemble of harmonic oscillators, i.e. mechanical systems with definite frequencies. Applying the result of this Section we confirm Einstein's assumption that the electromagnetic field can only exchange quanta of energy equal to $\hbar\omega = h\nu$. That justifies the concept of a photon as a particle carrying an energy equal to $h\nu$. At the quantum level, the possible states of an electromagnetic field oscillating in a cavity can thus be seen as those of a system of photons, corresponding in number to the total quanta of energy present in the cavity, which bounce elastically between the walls.

2.8 Periodic Potentials and Band Spectra

In previous Sections we have encountered and discussed situations in which the energy spectrum is continuous, as for particles free to move far to infinity with or without potential barriers, and other cases presenting a discrete spectrum, like that of bound particles. We will now show that other different interesting situations exist, in particular those characterized by a band spectrum. That is the case for a particle in a periodic potential, like an electron in the atomic lattice of a solid.

An example, which can be treated in a relatively simple way, is that in which the potential energy can be written as the sum of an infinite number of thin barriers (Kronig-Penney model), each proportional to the Dirac delta function, placed at a constant distance a from each other:

$$V(x) = \sum_{n=-\infty}^{\infty} \mathcal{V}\delta(x - na).$$

(2.138)

It is clear that:

$$V(x + a) = V(x),$$

(2.139)

so that we are dealing with a periodic potential. Our analysis will be limited to the case of barriers, i.e. $\mathcal{V} > 0$.

Equation (2.139) expresses a symmetry property of the Schrödinger equation, which is completely analogous to the symmetry under axis reflection discussed for the square well and valid also in the case of the harmonic oscillator. With an argument similar to that used in the square well case, it can be shown that for periodic potentials, i.e. invariant under translations by a, if $\psi_E(x)$ is a solution of the stationary Schrödinger equation then $\psi_E(x + a)$ is a solution too, corresponding to the same energy, so that, by suitable linear combinations, the analysis can be limited to a particular class of functions which are not changed by the symmetry transformation but for an overall multiplicative constant. In the case of reflections that constant must be ± 1, since a double reflection must bring back to the original configuration. Instead, in the case of translations $x \to x + a$, solutions can be chosen so as to satisfy the following relation:

$$\psi_E(x + a) = \alpha \, \psi_E(x),$$

where α is in general a complex number. Clearly such functions, like plane waves, are not normalizable, so that we have to make reference to the collective physical interpretation, as in the case of the potential barrier. In this case probability densities which do not vanish in the limit $|x| \to \infty$ are acceptable, but those diverging in the same limit must be discarded anyway. That constrains α to be a pure phase factor, $\alpha = e^{i\phi}$, so that

$$\psi_E(x + a) = e^{i\phi} \, \psi_E(x).$$

(2.140)

This is therefore another application of the symmetry principle enunciated in Sect. 2.6.

The wave function $\psi_E(x)$ must satisfy both (2.140) and the free Schrödinger equation in each interval $(n - 1)a < x < na$:

$$-\frac{\hbar^2}{2m}\partial_x^2\psi_E(x) = E \, \psi_E(x),$$

which has the general solution

$$\psi_E(x) = a_n e^{i\sqrt{2mE}\,x/\hbar} + b_n e^{-i\sqrt{2mE}\,x/\hbar}.$$

Finally, at the position of each delta function, the wave function must be continuous while its first derivative must be discontinuous with a gap equal to $(2m\mathcal{V}/\hbar^2)\psi_E(na)$. Since, according to (2.140), the wave function is pseudo-periodic, these conditions will be satisfied in every point $x = na$ if they are satisfied in the origin.

The continuity (discontinuity) conditions in the origin can be written as

$$a_0 + b_0 = a_1 + b_1,$$
$$i\frac{\sqrt{2mE}}{\hbar}(a_1 - b_1 - a_0 + b_0) = \frac{2m\mathcal{V}}{\hbar^2}(a_0 + b_0), \qquad (2.141)$$

while (2.140), in the interval $-a < x < 0$, is equivalent to:

$$a_1 e^{i\sqrt{2mE}(x+a)/\hbar} + b_1 e^{-i\sqrt{2mE}(x+a)/\hbar} = e^{i\phi}\left(a_0 e^{i\sqrt{2mE}x/\hbar} + b_0 e^{-i\sqrt{2mE}x/\hbar}\right). \qquad (2.142)$$

Last equation implies:

$$a_1 = e^{i\left(\phi - \sqrt{2mE}a/\hbar\right)} a_0, \qquad b_1 = e^{i\left(\phi + \sqrt{2mE}a/\hbar\right)} b_0,$$

which replaced in (2.141) leads to a system of two homogeneous linear equation in two unknown quantities:

$$\left(1 - e^{i\left(\phi - \frac{\sqrt{2mE}}{\hbar}a\right)} - i\sqrt{\frac{2m}{E}}\frac{\mathcal{V}}{\hbar}\right)a_0 - \left(1 - e^{i\left(\phi + \frac{\sqrt{2mE}}{\hbar}a\right)} + i\sqrt{\frac{2m}{E}}\frac{\mathcal{V}}{\hbar}\right)b_0 = 0$$
$$\left(1 - e^{i\left(\phi - \frac{\sqrt{2mE}}{\hbar}a\right)}\right)a_0 + \left(1 - e^{i\left(\phi + \frac{\sqrt{2mE}}{\hbar}a\right)}\right)b_0 = 0. \qquad (2.143)$$

The system admits non-trivial solutions ($a_0, b_0 \neq 0$) if and only if the determinant of the coefficient matrix does vanish; that is equivalent to a second order equation for $e^{i\phi}$:

$$\left(1 - e^{i\left(\phi - \frac{\sqrt{2mE}}{\hbar}a\right)} - i\sqrt{\frac{2m}{E}}\frac{\mathcal{V}}{\hbar}\right)\left(1 - e^{i\left(\phi + \frac{\sqrt{2mE}}{\hbar}a\right)}\right)$$
$$+ \left(1 - e^{i\left(\phi + \frac{\sqrt{2mE}}{\hbar}a\right)} + i\sqrt{\frac{2m}{E}}\frac{\mathcal{V}}{\hbar}\right)\left(1 - e^{i\left(\phi - \frac{\sqrt{2mE}}{\hbar}a\right)}\right)$$
$$= 2e^{2i\phi} - \left(\left(2 - i\sqrt{\frac{2m}{E}}\frac{\mathcal{V}}{\hbar}\right)e^{i\frac{\sqrt{2mE}}{\hbar}a} + \left(2 + i\sqrt{\frac{2m}{E}}\frac{\mathcal{V}}{\hbar}\right)e^{-i\frac{\sqrt{2mE}}{\hbar}a}\right)e^{i\phi}$$
$$+ 2 = 0, \qquad (2.144)$$

which can be rewritten in the form:

$$e^{2i\phi} - \left(2\cos\left(\frac{\sqrt{2mE}}{\hbar}a\right) + \sqrt{\frac{2m}{E}}\frac{V}{\hbar}\sin\left(\frac{\sqrt{2mE}}{\hbar}a\right)\right)e^{i\phi} + 1$$

$$\equiv e^{2i\phi} - 2Ae^{i\phi} + 1 = 0\,. \tag{2.145}$$

Equation (2.145) can be solved by a real ϕ if and only if $A^2 < 1$, as can be immediately verified by using the resolutive formula for second degree equations.

We have therefore an inequality, involving the energy E together with the amplitude V and the period a of the potential, which is a necessary and sufficient condition for the existence of physically acceptable solutions of the Schrödinger equation:

$$\left(\cos\left(\frac{\sqrt{2mE}}{\hbar}a\right) + \sqrt{\frac{m}{2E}}\frac{V}{\hbar}\sin\left(\frac{\sqrt{2mE}}{\hbar}a\right)\right)^2 < 1\,, \tag{2.146}$$

hence

$$\cos^2\left(\frac{\sqrt{2mE}}{\hbar}a\right) + \frac{m}{2E}\frac{V^2}{\hbar^2}\sin^2\left(\frac{\sqrt{2mE}}{\hbar}a\right)$$

$$+ 2\sqrt{\frac{m}{2E}}\frac{V}{\hbar}\sin\left(\frac{\sqrt{2mE}}{\hbar}a\right)\cos\left(\frac{\sqrt{2mE}}{\hbar}a\right) < 1 \tag{2.147}$$

and therefore

$$1 - \cos^2\left(\frac{\sqrt{2mE}}{\hbar}a\right) - \frac{m}{2E}\frac{V^2}{\hbar^2}\sin^2\left(\frac{\sqrt{2mE}}{\hbar}a\right)$$

$$- 2\sqrt{\frac{m}{2E}}\frac{V}{\hbar}\sin\left(\frac{\sqrt{2mE}}{\hbar}a\right)\cos\left(\frac{\sqrt{2mE}}{\hbar}a\right)$$

$$= \left(1 - \frac{m}{2E}\frac{V^2}{\hbar^2}\right)\sin^2\left(\frac{\sqrt{2mE}}{\hbar}a\right)$$

$$- 2\sqrt{\frac{m}{2E}}\frac{V}{\hbar}\sin\left(\frac{\sqrt{2mE}}{\hbar}a\right)\cos\left(\frac{\sqrt{2mE}}{\hbar}a\right) < 0\,, \tag{2.148}$$

leading finally to:

$$\cot\left(\frac{\sqrt{2mE}}{\hbar}a\right) < \frac{1}{2}\left(\sqrt{\frac{2E}{m}}\frac{\hbar}{V} - \sqrt{\frac{m}{2E}}\frac{V}{\hbar}\right)\,. \tag{2.149}$$

Fig. 2.5 The plot of inequality (2.149) identifying the first three bands of allowed values of $\sqrt{2mE}\,a/\hbar$ for the choice of the Kronig–Penney parameter $\gamma = \hbar^2/(ma\mathcal{V}) = 1/2$

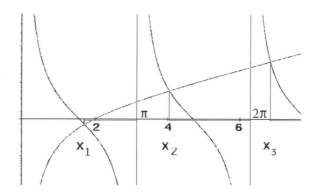

Both sides of last inequality are plotted in Fig. 2.5 for a particular choice of the parameter $\gamma = \hbar^2/(ma\mathcal{V}) = 1/2$. The variable used in the figure is $x = \sqrt{2mE}a/\hbar$, so that the two plotted functions are $f_1 = \cot x$ and $f_2 = (\gamma x - 1/(\gamma x))/2$. The intervals where the inequality (2.149) is satisfied are those enclosed between x_1 and π, x_2 and 2π, x_3 and 3π and so on. Indeed in these regions the uniformly increasing function f_2 is greater than the oscillating function f_1. The result shows therefore that the permitted energies correspond to a series of intervals $(x_n, n\pi)$, which are called *bands*, separated by a series of forbidden *gaps*.

As we shall discuss in the next chapter, electrons in a solid, which are compelled by the *Pauli exclusion principle* to occupy each a different energy level, may fill completely a certain number of bands, so that they can only absorb energies greater than a given minimum quantity, corresponding to the gap with the next free band: in such situation electrons behave as bound particles. Alternatively, if the electrons fill partially a given band, they can absorb arbitrarily small energies, thus behaving as free particles. In the first case the solid is an *insulator*, in the second it is a *conductor*.

Having determined the phase $\phi(E)$ from (2.145) and taking into account (2.140), it can be seen that, by a simple transformation of the wave function:

$$\psi_E(x) \equiv e^{i\,\phi(E)\,x/a}\hat{\psi}_E(x) \equiv e^{\pm i\,p(E)\,x/\hbar}\hat{\psi}_E(x)\,, \qquad (2.150)$$

equation (2.140) can be translated into a periodicity constraint:

$$\hat{\psi}_E(x + a) = \hat{\psi}_E(x)\,.$$

Therefore wave functions in a periodic potential can be written as in (2.150), i.e. like plane waves, which are called *Bloch waves*, modulated by periodic functions $\hat{\psi}_E(x)$.

It must be noticed that the momentum associated with Bloch waves, $p(E) = (\hbar/a)\phi(E)$, cannot take all possible real values, as in the case of free particles, but is limited to the interval $(-\hbar\pi/a,\ \hbar\pi/a)$, which is known as the *first Brillouin zone*. This limitation can be seen as the mathematical reason underlying the presence of bands.

On the other hand, the relation which in a given band gives the electron energy as a function of the Bloch momentum (dispersion relation) is very intricate from the analytical point of view. It is indeed the inverse function of $p(E) = (\hbar/a) \arccos(A(E))$ with $A(E)$ defined by (2.145). For that reason we limit ourselves to some qualitative remarks.

By noticing that in the lower ends of the bands, x_n, $n = 1, 2, \ldots$, the parameter A in (2.145) is equal to

$$\cos x_n + \frac{\sin x_n}{\gamma \, x_n} = \frac{\sin x_n}{2}\left(\gamma \, x_n + \frac{1}{\gamma \, x_n}\right) = (-1)^{n+1}, \qquad (2.151)$$

we have: $e^{i\phi}|_{x_n} = (-1)^{n+1}$. Hence $\phi(x_n) = 0$ for odd n and $\phi(x_n) = \pm\pi$ for even n. Instead in the upper ends, $x = n\pi$, we have $A = \cos n\pi = (-1)^n$ hence $\phi(n\pi) = 0$ for even n and $\phi(n\pi) = \pm\pi$ for odd n. Moreover, for a generic A between -1 and 1, there are two solutions: $A \pm i\sqrt{1 - A^2}$ corresponding to opposite phases ($\phi(E) = \pm\arctan\left(\sqrt{1 - A^2}/A\right)$) interpolating between 0 and $\pm\pi$.

Therefore, based on Fig. 2.5, we come to the conclusion that in odd bands the minimum energy corresponds to states with $p = 0$, while states at the border of the Brillouin zone have the maximum possible energy. The opposite happens instead for even bands. Finally we observe that the derivative dE/dp vanishes at the border of the Brillouin zone, where $A^2 = 1$ and A has a non-vanishing derivative, indeed we have

$$\frac{dE}{dp} = \frac{a}{\hbar}\left(\frac{d(\arccos A(E))}{dE}\right)^{-1} = \pm\frac{a}{\hbar}\frac{\sqrt{1 - A^2}}{d A(E)/dE}. \qquad (2.152)$$

2.9 The Schrödinger Equation in a Central Potential

In the case of a particle moving in three dimensions under the influence of a central force field, the symmetry properties of the problem play a dominant role.

Indeed, already at the classical level, the invariance of the Hamiltonian, $H = p^2/(2M) + V(r)$, under rotations around the center, identified with the origin, implies conservation of the angular momentum $L = r \wedge p$. Once L is specified, the motion must be planar on a plane orthogonal to it. The absolute value of the angular momentum L identifies the areal velocity $L/(2M) = r^2\dot{\theta}/2$. It follows that the kinetic energy on the plane, which is given by $M/2[(\dot{r})^2 + r^2(\dot{\theta})^2]$, is equal to $m(\dot{r})^2/2 + L^2/(2Mr^2)$ and hence the energy in a central potential is given by:

$$E = \frac{M(\dot{r})^2}{2} + \frac{L^2}{2Mr^2} + V(r) = \frac{p_r^2}{2M} + \frac{L^2}{2Mr^2} + V(r). \qquad (2.153)$$

Thus, if the angular momentum is specified, the energy appears as a function of the radius and of its time derivative and the equations of motions separate into an equation for the radial motion and another equation for the angular motion. One has separation of variables.

In the framework of quantum mechanics this simple approach to the motion in a central potential does not work, because of the uncertainty principle, which forbids a complete determination of the angular momentum vector. This vector corresponds to the vector valued operator $-i\hbar \mathbf{r} \wedge \nabla$. Indeed, considering the relation shown in Sect. 2.4 between uncertainty in the distribution of pairs of observables and lack of commutativity of the corresponding operators, the angular momentum uncertainty follows from the fact that the operators corresponding to its components do not commute. For example, the x-component of the particle angular momentum around the center is given by $L_x = i\hbar(z\partial_y - y\partial_z)$ and the y-component is $L_y = i\hbar(x\partial_z - z\partial_x)$. These operators do not commute. Indeed

$$[L_x, L_y] \equiv L_x L_y - L_y L_x = \hbar^2(y\partial_x - x\partial_y) \equiv i\hbar L_z . \tag{2.154}$$

In much the same way we find

$$[L_y, L_z] = i\hbar L_x, \quad [L_z, L_x] = i\hbar L_y . \tag{2.155}$$

We can still exploit the consequences of the rotation invariance in the analysis of solutions of the stationary Schrödinger equation:

$$-\frac{\hbar^2}{2M}\nabla^2\psi_E(\mathbf{r}) + V(r)\psi_E(\mathbf{r}) = E\psi_E(\mathbf{r}) . \tag{2.156}$$

The standard method is based on Group Theory, but we do not assume our readers to be Group Theory experts, hence we adopt a different approach. The only Group Theory result that we exploit, as we have already done a few times in the preceding part of this text, is what we have called *symmetry principle*; if the Schrödinger equation is left invariant by a coordinate transformation, we can always find a complete set of solutions which do not change, but for a phase factor, under the transformation. This set is complete when all square integrable, or locally square integrable, solutions of physical interest can be written as linear combinations of elements of the set.

We start considering, among all possible rotations, those around one particular axis, for instance the z axis. These rotations transform $x \rightarrow x' = x\cos\phi - y\sin\phi$ and $y \rightarrow y' = y\cos\phi + x\sin\phi$, while z is left unchanged. An equivalent and simpler way of representing these rotations, making use of complex combinations of coordinates, is:

$$x'_\pm \equiv x' \pm iy' = e^{\pm i\phi}x_\pm, \quad \text{and} \quad z' = z . \tag{2.157}$$

We can represent the same rotations in spherical coordinates (r, θ, φ), defined by

$$x_\pm = r\sin\theta\exp(\pm i\varphi), \quad \text{and} \quad z = r\cos\theta , \tag{2.158}$$

in which they are equivalent to the translations $\varphi \to \varphi' = \varphi + \phi$. According to the symmetry principle, we consider solutions of equation (2.156) transforming under the above rotations as $\psi_E \longrightarrow e^{i\Phi}\psi_E$.

The phase Φ is necessarily a linear function of ϕ, as it appears by observing that for two subsequent rotations around the same axis, with angles ϕ and ϕ', we have $\Phi(\phi) + \Phi(\phi') = \Phi(\phi + \phi')$. Then asking that for $\phi = 2\pi$ the wave function is left unchanged, i.e. that $\Phi(2\pi) = 2\pi m$ with m any relative integer, we obtain that, in spherical coordinates, we have a complete set of solutions of equation (2.156) of the form:

$$\psi_{E,m}(\mathbf{r}) \equiv \psi_{E,m}(r, \theta, \varphi) = \hat{\psi}_{E,m}(r, \theta)e^{im\varphi} . \tag{2.159}$$

It is an easy exercise to verify that:

$$L_z \psi_{E,m}(\mathbf{r}) = -i\hbar(x\partial_y - y\partial_x)\psi_{E,m}(\mathbf{r}) = -i\hbar\frac{\partial}{\partial\varphi}\psi_{E,m}(\mathbf{r}) = m\hbar\psi_{E,m}(\mathbf{r}) , \tag{2.160}$$

which shows that the wave function satisfies Bohr's quantization rule for L_z.

The operator iL_z, being proportional to the φ-derivative, appears as the generator of rotations around the z-axis, in much the same way as the x-derivative generates translations of functions of the x variable. For isotropy reasons, this property of generating rotations extends to the other components of the angular momentum.

If operators, wave functions and numbers are related by equations analogous to (2.160), the wave function is called *eigenfunction* of the operator, and the coefficient in the right-hand side *eigenvalue*. Bohr's quantization rule is thus interpreted as an equation for the eigenvalues of L_z.

Now we must see how the other components of the angular momentum operator act on the solutions $\psi_{E,m}(\mathbf{r})$. We first note that any component of the angular momentum operator commutes with both the Laplacian operator and the distance from the center r. For example, the Leibniz rule gives:

$$\nabla^2 L_x = i\hbar(\partial_x^2 + \partial_y^2 + \partial_z^2)(z\partial_y - y\partial_z)$$
$$= i\hbar(z\partial_y - y\partial_z)(\partial_x^2 + \partial_y^2 + \partial_z^2) + 2i\hbar(\partial_z\partial_y - \partial_y\partial_z) = L_x\nabla^2 . \tag{2.161}$$

It is easy to verify that one gets analogous results replacing ∇^2 with r^2 and/or L_x with any other component of the angular momentum. Therefore we have, e.g., for the x component

$$L_x\left[-\frac{\hbar^2}{2M}\nabla^2 + V(r)\right]\psi_E(\mathbf{r}) = \left[-\frac{\hbar^2}{2M}\nabla^2 + V(r)\right]L_x\psi_E(\mathbf{r}) = EL_x\psi_E(\mathbf{r}) . \tag{2.162}$$

It means that the action of any component of the angular momentum on a solution of equation (2.157) gives either a solution or zero.

This result extends to the square of the angular momentum corresponding to the operator

$$L^2 \equiv L_x^2 + L_y^2 + L_z^2 , \qquad (2.163)$$

because L^2 commutes with both the Laplacian and $V(r)$. If A and B are operators commuting with C, then also A^m and B^n commute with C^k and, e.g., $A^2 + B^2$ commutes with C. Here A and B correspond to the components of L, while C is either the Laplacian, or $V(r)$.

It also is an easy exercise to verify, taking into account Eqs. (2.154) and (2.155), that L^2 commutes with all the components of the angular momentum. Thus, using equations analogous to (2.162), we can show that L^2 transforms any solution $\psi_{E,m}$ of (2.157) into another solution, possibly proportional to the first one, with the same E and m. If there is a single solution, with given E and m, it is obvious that $L^2 \psi_{E,m} \sim \psi_{E,m}$ and hence $\psi_{E,m}$ is an eigenfunction of L^2. But in general there are many such solutions. Let us denote the mentioned solutions by $\psi_{E,m,i}$. We have:

$$L^2 \psi_{E,m,i} = \sum_{j=1}^{N_m} l_{E,m,i,j} \psi_{E,m,j} . \qquad (2.164)$$

This means that the operator L^2 acts as the multiplication by the matrix $l_{E,m,i,j}$ on the wave function space spanned by the $\psi_{E,m,i}$'s, which is the set of linear combinations of the $\psi_{E,m,i}$'s with fixed m.

Identifying eigenfunctions and eigenvalues of L^2 is a crucial step in the analysis of the solutions to equation (2.156). For this it is convenient to choose the complex coordinates (2.157), introducing the operators:

$$L_\pm \equiv L_x \pm i L_y = \pm \hbar (2z \partial_{x_\mp} - x_\pm \partial_z) , \qquad (2.165)$$

for which the commutation relations (2.155) are translated into[5]:

$$[L_z, L_\pm] = \pm \hbar L_\pm, \quad [L_+, L_-] = 2\hbar L_z . \qquad (2.166)$$

Thus we have:

$$L_z L_\pm \psi_{E,m,i} = \hbar (m \pm 1) L_\pm \psi_{E,m,i} , \qquad (2.167)$$

which means that either $L_\pm \psi_{E,m,i}$ vanishes, or it satisfies (2.160) with the quantum number m increased/decreased by one.

Now, for a given value of E, m cannot increase indefinitely. Indeed, from Eq. (2.153), we see that the energy of the particle is the sum of a purely radial part, $p_r^2/(2M) + V(r)$, and of two positive terms: $(L_x^2 + L_y^2)/(2Mr^2)$ and $L_z^2/(2Mr^2) = \hbar^2 m^2/(2Mr^2)$. Given a square integrable wave function which is a single particle

---

[5]In spherical coordinates one has $L_\pm = \hbar i \exp(\pm i\varphi)[\cot\theta \partial_\varphi \mp i\partial_\theta]$.

bound state solution of equation (2.156) with a given energy, the single above mentioned terms are affected by uncertainties. But we can compute their average values.

Since L_+ does not act on the radial variable, it is clear that the average value of the first radial term remains fixed when m increases.[6] In contrast the average values of $(L_x^2 + L_y^2)/(2Mr^2)$ and $L_z^2/(2Mr^2)$ change, the second one increasing by $(\hbar^2 m/2M)\langle 1/r^2 \rangle$. Here, given an operator X, $\langle X \rangle$ denotes its average value. The average value of positive quantities are necessarily positive and $\langle 1/r^2 \rangle$ does not change under the action of L_+ because it is a radial property. Therefore the variations of $\langle (L_x^2 + L_y^2)/(2Mr^2) \rangle$ cannot compensate the indefinite increase of $\hbar^2 m^2/(2M)\langle 1/r^2 \rangle$. Thus this must stop at a certain value of m. We denote by l_E the maximum, integer value of m which depends on E. The corresponding wave functions $\psi_{E,l_E,\alpha}$ satisfy $L_+ \psi_{E,l_E,\alpha} = 0$. We have:

$$L^2 \psi_{E,l_E,\alpha} = \hbar^2 l_E (l_E + 1) \psi_{E,l_E,\alpha} \ . \tag{2.168}$$

because, from the commutation rule in (2.154), we get:

$$L^2 = L_- L_+ + L_z^2 + \hbar L_z = L_+ L_- + L_z^2 - \hbar L_z \ . \tag{2.169}$$

Also, writing $\psi_{E,l_E,\alpha}$ as a function of the variables x_+, $x_- x_+$ and z, and using

$$L_z = \hbar(x_+ \partial_{x_+} - x_- \partial_{x_-}) \ , \tag{2.170}$$

we have, from $L_z \psi_{E,l_E,\alpha} = l_E \psi_{E,l_E,\alpha}$:

$$\psi_{E,l_E,\alpha} = x_+^{l_E} g_{E,l_E,\alpha}(x_- x_+, z) \tag{2.171}$$

and, from $L_+ \psi_{E,l_E,\alpha} = 0$, we have:

$$\psi_{E,l_E,\alpha} = x_+^{l_E} f_{E,l_E,\alpha}(r) \ . \tag{2.172}$$

Therefore we have separation of variables because this wave function is equal to the product of a purely angular factor, $(\sin\theta \exp(i\varphi))^{l_E}$ and a purely radial one, $r^{l_E} f_{E,l_E,\alpha}(r)$.

Starting from $\psi_{E,l_E,\alpha}$, we can build a sequence of solutions of equation (2.156), that we denote by $\psi_{E,l_E,m,\alpha}$, being understood the identification $\psi_{E,l_E,l_E,\alpha} \equiv \psi_{E,l_E,\alpha}$. These new wave functions are identified, up to a multiplicative constant, by:

$$\psi_{E,l_E,m,\alpha} \sim L_-^{l_E-m} \psi_{E,l_E,\alpha} \ , \tag{2.173}$$

---

[6]This argument assumes wave function factorization into radial and angular factors. An alternative and simpler argument is based on the assumption that $\sum_m N_m$ is finite. This is easily justified in the case of a finite range potential, since we know that the number of independent states with limited energy in a finite volume is finite.

This sequence must stop when m reaches $-l_E$ because, using the second identity in (2.169) and $L^2 \psi_{E,l_E,m,\alpha} = \hbar^2 l_E(l_E+1)\psi_{E,l_E,m,\alpha}$, we have:

$$L_+L_-\psi_{E,l_E,m,\alpha} = \hbar^2 (l_E(l_E+1) - m^2 + m)\psi_{E,l_E,m,\alpha} \,, \tag{2.174}$$

which implies that $L_-\psi_{E,l_E,-l_E,\alpha} = 0$. Then from (2.165) we find that

$$\psi_{E,l_E,-l_E,\alpha} \sim x_-^{l_E} f_{E,l_E,\alpha}(r) \,. \tag{2.175}$$

So, for each α, we have found a multiplet of $2l_E + 1$ eigenfunctions of L^2 which are built starting from (2.172) and repeatedly acting on it by L_-. All the eigenfunctions of the multiplet have the same radial dependence. One might wonder if the alternate action of L_+ and L_- might produce more solutions with the same m belonging to the same multiplet but not proportional to each other. The negative answer follows directly from Eq. (2.174) which shows that this alternate action changes the wave function by a multiplicative constant. Hence the $2l_E + 1$ wave functions $\psi_{E,l_E,m,\alpha}$ span, for each value of α, an independent linear space invariant under the action of the angular momentum components.

The above analysis, which has begun from the solutions of equation (2.156) with maximum m ($m = l_E$), can be repeated considering the remaining solutions of equation (2.156) which are linearly independent of those belonging to the identified multiplets. We start considering the independent solutions of the set $\psi_{E,l_E-1,i}$. If their number exceeds, by p, that of the already identified multiplets, we can select p linear combinations of the $\psi_{E,l_E-1,i}$'s that, using the same notation introduced above, we denote by $\psi_{E,l_E-1,l_E-1,\beta}$, which are annihilated by L_+, that is, such that $L_+\psi_{E,l_E-1,l_E-1,\beta} = 0$. Indeed, if $L_+\psi_{E,l_E-1,l_E-1,\beta}$ does not vanish, it must be linearly dependent on the $L_+\psi_{E,l_E,l_E-1,\alpha}$'s, since these span the linear space of the solutions with maximum m. That is, we must have:

$$L_+\left[\psi_{E,l_E-1,l_E-1,\beta} - \sum_\alpha c_{\alpha\beta}\psi_{E,l_E,l_E-1,\alpha}\right] = 0 \,. \tag{2.176}$$

But the wave function in brackets does not vanish because $\psi_{E,l_E-1,l_E-1,\beta}$ is chosen linearly independent of the members of the already built multiplets, therefore

$$\hat{\psi}_{E,l_E-1,l_E-1,\beta} \equiv \psi_{E,l_E-1,l_E-1,\beta} - \sum_\alpha c_{\alpha\beta}\psi_{E,l_E,l_E-1,\alpha} \,, \tag{2.177}$$

is a solution with $m = l_E - 1$ which is annihilated by L_+. In this way we can finally build the chosen set of p independent solutions annihilated by L_+.

From these solutions, repeatedly acting with L_-, we can build p new multiplets of eigenfunctions of L^2. Continuing this procedure we show that, among the $N = \sum_i N_i$ independent solutions of equation (2.156), we can select N independent linear combinations belonging to multiplets of eigenfunctions of L^2 and, of course L_z, with eigenvalues $\hbar^2 l(l+1)$ ($l_E \geq l \geq 0$) and $\hbar l \geq \hbar m \geq -\hbar l$, respectively.

We have seen that the angular dependence of the wave functions is identified once the quantum number l and m are given, while the radial dependence is identified by the multiplet. In much the same way as in (2.172) and (2.175), the wave functions are given as the product of radial functions, left invariant by the action of $L_\pm$, by homogeneous polynomials, called *harmonic polynomials*, that we denote by $\mathcal{Y}_{l,m}(\mathbf{r})$.

$$\psi_{E,l,m,\alpha}(\mathbf{r}) = \mathcal{Y}_{l,m}(\mathbf{r}) \, f_{E,l,\alpha}(r) \,. \tag{2.178}$$

We can also deduce the transformation properties of the wave functions given in (2.178) under spatial reflection $\mathbf{r} \to -\mathbf{r}$, usually called *parity transformation*, using the fact that the $\mathcal{Y}_{l,m}(\mathbf{r})$'s are homogeneous polynomials of degree l. The result is $\psi_{E,l,m,\alpha}(-\mathbf{r}) = (-1)^l \psi_{E,l,m,\alpha}(\mathbf{r})$. The harmonic polynomials, which are built by repeatedly acting on $x_\pm^l$ by $L_\mp$, are called harmonic since they satisfy[7]

$$\nabla^2 \mathcal{Y}_{l,m}(\mathbf{r}) = 0, \quad \mathbf{r} \cdot \nabla \mathcal{Y}_{l,m}(\mathbf{r}) = l \mathcal{Y}_{l,m}(\mathbf{r}) = 0 \,, \tag{2.179}$$

the second equation being equivalent to homogeneity. Indeed both Eq. (2.179) hold true for $x_\pm^l$ and the homogeneous operators $L_\pm$ commute with the Laplacian. For the time being we do not specify any normalization prescription for these harmonic polynomials except that implied by the equation

$$\mathcal{Y}_{l,-m} = (-1)^m \mathcal{Y}_{l,m}^* \,. \tag{2.180}$$

In order to make the construction clear it is convenient to give a few simple examples of harmonic polynomials:

$$\mathcal{Y}_{0,0}(\mathbf{r}) \sim 1, \quad \mathcal{Y}_{1,\pm}(\mathbf{r}) \sim \mp x_\pm, \quad \mathcal{Y}_{1,0}(\mathbf{r}) \sim z \tag{2.181}$$
$$\mathcal{Y}_{2,\pm 2}(\mathbf{r}) \sim x_\pm^2, \quad \mathcal{Y}_{2,\pm 1}(\mathbf{r}) \sim \mp z x_\pm, \quad \mathcal{Y}_{2,0}(\mathbf{r}) \sim 2z^2 - x_+ x_- \,,$$

where the normalization is left free, but the sign is fixed assuming positive sign of $\mathcal{Y}_{1,0}$ on the positive z-axis and considering the sign induced by the action of $L_\pm$ defined in (2.165).

It is now possible to consider that the basic purpose of the presented construction was to insert into the Schrödinger equation in a central potential (2.156) the maximum possible information about the angular momentum of the particle. This is equivalent to choosing the solutions of the form in (2.178) and reducing (2.156) to an equation for the radial wave function $f_{E,l,\alpha}(r)$, which is deduced computing the action of the Laplacian on a generic product $\mathcal{Y}_{l,m}(\mathbf{r}) f(r)$. We have:

---

[7] In complex coordinates the Laplacian operator is given by: $4\partial_{x_+}\partial_{x_-} + \partial_z^2$.

$$\nabla^2(\mathcal{Y}_{l,m}(\mathbf{r})f(r)) = f(r)\nabla^2\mathcal{Y}_{l,m}(\mathbf{r}) + \mathcal{Y}_{l,m}(\mathbf{r})\nabla^2 f(r)$$

$$+ 2(\nabla\mathcal{Y}_{l,m}(\mathbf{r})) \cdot \nabla f(r) = \mathcal{Y}_{l,m}(\mathbf{r})\nabla \cdot (\mathbf{r}\frac{f'(r)}{r}) + 2(\mathbf{r} \cdot \nabla\mathcal{Y}_{l,m}(\mathbf{r}))\frac{f'(r)}{r}$$

$$= \mathcal{Y}_{l,m}(\mathbf{r})\left[2(l+1)\frac{f'(r)}{r} + f''(r)\right], \tag{2.182}$$

therefore the Schrödinger equation for $f_{E,l,\alpha}(r)$ becomes:

$$-\frac{\hbar^2}{2M}\left(f''_{E,l,\alpha}(r) + 2(l+1)\frac{f'_{E,l,\alpha}(r)}{r}\right) + V(r)f_{E,l,\alpha}(r) = E\,f_{E,l,\alpha}(r). \tag{2.183}$$

In spherical coordinates the harmonic polynomials appear as polynomials in $\cos\theta$ multiplied by $r^l \sin^m \theta \exp(\pm im\varphi)$. In order to complete the separations of the r dependence in the solutions of (2.156) we introduce the *spherical harmonics*:

$$Y_{l,m}(\theta, \varphi) = \frac{\mathcal{Y}_{l,m}(\mathbf{r})}{r^l}, \tag{2.184}$$

now specifying the normalization conditions:

$$\int_0^{2\pi} d\varphi \int_{-1}^{1} d\cos\theta\, Y^*_{l,m}(\theta, \varphi)Y_{l',m'}(\theta, \varphi) = \delta_{l,l'}\delta_{m,m'}. \tag{2.185}$$

This is an orthonormalization condition for the spherical harmonics, which is particularly convenient for normalizing the wave functions. For reader's convenience, we give the explicit form of spherical harmonics up to $l = 2$:

$$Y_{0,0} = \sqrt{\frac{1}{4\pi}}; \quad Y_{1,0} = \sqrt{\frac{3}{4\pi}}\cos\theta; \quad Y_{1,\pm1} = \mp\sqrt{\frac{3}{8\pi}}\sin\theta e^{\pm i\varphi};$$

$$Y_{2,0} = \sqrt{\frac{5}{16\pi}}(3\cos^2\theta - 1); \quad Y_{2,\pm1} = \mp\sqrt{\frac{15}{8\pi}}\sin\theta\cos\theta e^{\pm i\varphi};$$

$$Y_{2,\pm2} = \sqrt{\frac{15}{32\pi}}\sin^2\theta e^{\pm 2i\varphi}. \tag{2.186}$$

The solutions of the Schrödinger equation are written in terms of the spherical harmonics in the form:

$$\psi_{E,l,m,\alpha}(\mathbf{r}) = Y_{l,m}(\theta, \varphi)\frac{\chi_{E,l,\alpha}(r)}{r}, \tag{2.187}$$

where, if the wave function is normalized, $|\chi_{E,l,\alpha}(r)|^2$ is the probability density of finding the particle at a distance r from the center. The equation for $\chi_{E,l,\alpha}(r)$ is obtained replacing in (2.183) $f_{E,l,\alpha}(r)$ by $r^{-(l+1)}\chi_{E,l,\alpha}(r)$. The radial Schrödinger equation becomes:

$$-\frac{\hbar^2}{2M}\chi''_{E,l,\alpha}(r)+\frac{\hbar^2 l(l+1)}{2Mr^2}\chi_{E,l,\alpha}(r)+V(r)g\chi_{E,l,\alpha}(r) = E\,\chi_{E,l,\alpha}(r)\,, \quad (2.188)$$

as expected from Eq. (2.153). Indeed this is the one-dimensional stationary
Schrödinger equation corresponding to the energy associated with the radial motion
given in (2.153).

Let us discuss this point in few more details. Consider the energy appearing
in (2.153) as an operator, whose action on the wave function specifies the right-
hand side of the stationary Schrödinger equation (2.156). The term proportional to
the Laplacian in (2.156), written in spherical coordinates, apparently corresponds
to the sum of the first two terms in (2.153). The first term, which is proportional
to the square of the radial momentum p_r, corresponds to the first term in (2.188),
because the radial momentum, which is the variable conjugate to r, corresponds
to the operator $-i\hbar\partial_r$. The second term is proportional to L^2. Having written the
solutions to equation (2.156) as products of radial functions and of eigenfunctions
of the operator L^2, whose eigenvalues are $\hbar^2 l(l+1)$ for non-negative integer l, it
is clear that the second term in (2.188) corresponds to the term proportional to L^2
in (2.153).

This proves that we have obtained the quantum mechanical equivalent of the
classical separation of variables described at the beginning of this section. In the
following subsections we shall study the simplest solutions to the radial equation in
a few cases with simple potentials.

2.9.1 A Piecewise Constant Potential and the Free Particle Case

The strategy for the solution to the Schrödinger equation in the case of a piecewise
constant potential $V(r)$ is essentially the same as in the one dimensional case, we have
only to pay special attention to the additional constraint that the wave function must
vanish in $r=0$, otherwise the related three dimensional probability density would be
divergent in a position where the potential is flat. In particular, in the S wave case (that
being the usual way of indicating the case $l=0$) Eq. (2.188) coincides with the one
dimensional Schrödinger equation, therefore we can obtain its solutions as a linear
combination of the functions $\sin(\sqrt{2M(E-V)}r/\hbar)$ and $\cos(\sqrt{2M(E-V)}r/\hbar)$ for
$E>V$ and $\exp(\pm\sqrt{2M(E-V)}r/\hbar)$ in the opposite case. In the case of a spherical
potential well:

$$V(r) = -V_0\Theta(R-r)\,, \quad (2.189)$$

where $V_0>0$ and $\Theta(x)$ is the step function ($\Theta(x)=1$ for $x>0$ and $\Theta(x)=0$ for $x<0$), the (S wave) radial equation coincides with the one-dimensional
Schrödinger equation discussed in Sect. 2.6 for the parity odd wave functions in a
square well with width $L=2R$. Thus one can find the equation for the binding
energy in (2.106).

For $l > 0$ Eq. (2.188) can be written in the form:

$$\chi_{E,l}''(r) + \left[\sigma q^2 - \frac{l(l+1)}{r^2}\right]\chi_{E,l}(r) = 0 .\tag{2.190}$$

with

$$q = \sqrt{\frac{2M}{\hbar^2}|E - V|} \quad \text{and} \quad \sigma = \frac{E - V}{|E - V|} ,\tag{2.191}$$

this implies that sine, cosine and exponentials must be replaced by new special functions (which are called *spherical Bessel functions*), which can be explicitly constructed using the recursive equation:

$$q\,\chi_{E,l+1}(r) = \frac{l+1}{r}\chi_{E,l}(r) - \chi_{E,l}'(r) .\tag{2.192}$$

This recursive equation is proved as follows. Assuming Eq. (2.192) and using (2.190) we obtain

$$q\chi_{E,l+1}' = \sigma q^2 \chi_{E,l} - q(l+1)\frac{\chi_{E,l+1}}{r} .\tag{2.193}$$

Using again Eqs. (2.192) and (2.193),

$$\begin{aligned}
\chi_{E,l+1}'' &= \sigma q\chi_{E,l}' - \frac{(l+1)}{r}[\chi_{E,l+1}' - \frac{\chi_{E,l+1}}{r}] \\
&= -\sigma q^2\chi_{E,l+1} + \frac{l+1}{r}[\sigma q\chi_{E,l} - \chi_{E,l+1}'] + (l+1)\frac{\chi_{E,l+1}}{r^2} \\
&= -\sigma q^2\chi_{E,l+1} + \frac{(l+2)(l+1)}{r^2}\chi_{E,l+1} .
\end{aligned}\tag{2.194}$$

This shows that, by (2.192), we can obtain, from a solution of equation (2.190), another solution of the same equation with l increased by one.[8] For the internal case, $\sigma = 1$, we have in particular:

$$\chi_{E,0}(r) = \sin(qr), \quad \chi_{E,1}(r) = \frac{\sin(qr)}{qr} - \cos(qr) .\tag{2.195}$$

while analogous solutions (with real exponentials replacing trigonometric functions, as usual) are found in the external region for $E < 0$, i.e. for bound states. The asymptotic property:

$$\chi_{E,l}(r) \xrightarrow[qr\to\infty]{} \sin(qr - \frac{l\pi}{2}) ,\tag{2.196}$$

is a direct consequence of Eq. (2.192).

---

[8]The spherical Bessel functions, $j_l(qr)$ are identified with $\chi_{E,l+1}(r)/(qr)$ normalized according to (2.192) and $\chi_{E,0}(qr) = \sin(qr)$.

If, for instance, we want to study the possible bound states in P wave (i.e. $l = 1$) in the above potential well, setting the energy to $-B$ and defining $q_i = \sqrt{2MB/\hbar^2}$ and $q_e = \sqrt{2M(V_0 - B)/\hbar^2}$ we must continuously connect the internal solution $\sin(q_i r)/q_i r - \cos(q_i r)$, which vanishes in $r = 0$, with the external solution which vanishes as $r \to \infty$, i.e. $a\, e^{-q_e(r-R)}(1/q_e r + 1)$. This leads to the system:

$$\frac{\sin(q_i R) - q_i R \cos(q_i R)}{q_i} = a \frac{1 + q_e R}{q_e}$$

$$\frac{\sin(q_i R)(1 - q_i^2 R^2) - q_i R \cos(q_i R)}{q_i} = a \frac{1 + q_e R + q_e^2 R^2}{q_e}.$$

Therefore, setting $y = \sqrt{2MV_0}R/\hbar$, $x = q_i R$ and hence $q_e R = \sqrt{y^2 - x^2}$, we have the transcendental equation:

$$\tan x = x \frac{y^2 - x^2}{y^2 + x^2\sqrt{y^2 - x^2}}.$$

Considering the bound state condition: $0 \leq x \leq y$, the above equation requires $0 \leq \tan x/x \leq 1$ and $\tan y/y = 0$. This implies the absence of $l = 1$ bound states for $y < \pi$ while we have seen, comparing with the one dimensional case, that the first $l = 0$ bound state appears for $y = \pi/2$.

We explicitly notice that, for $V = 0$, solutions to equation (2.190) provide the wave functions for the free particle problem. Let us discuss in particular the case $l = 0$: from (2.186), (2.187) and (2.195) we deduce that solutions with zero angular momentum and $E > 0$ are

$$\psi_{E,l=0}(\boldsymbol{r}) \propto \frac{\sin kr}{r}, \tag{2.197}$$

where $\hbar k = \sqrt{2ME}$. If we insert time dependence explicitly, such a solution can be rewritten in the form

$$\psi_{E,l=0}(\boldsymbol{r}, t) \propto e^{-iEt/\hbar} \frac{\sin kr}{r} \propto \frac{e^{i(pr-Et)/\hbar}}{r} - \frac{e^{-i(pr+Et)/\hbar}}{r} \tag{2.198}$$

where $p = k\hbar$. From that it is clear that the solution is the sum of two spherical waves, the first propagating outwards, the second inwards. This can be confirmed also by an explicit computation of the probability current density, which will show that the two waves lead to the same probability flux across every spherical surface centered around the origin, with a different sign for the inward and outward solution.

Analogous considerations hold for solutions with $l > 0$. In this way one finds wave functions associated with particles with fixed energy and angular momentum, which are alternative to the standard plane wave solutions, corresponding to fixed energy and momentum. We will go back to free particle solutions when we discuss the scattering problem.

2.9.2 The Coulomb Potential

It is obviously of great interest to study bound states in a Coulomb potential, which permits an analysis of the energy levels of the hydrogen atom. To this purpose, let us consider the motion of a particle of mass m in a central potential $V(r) = -e^2/(4\pi\epsilon_0 r)$, where ϵ_0 is the vacuum dielectric constant and e is (minus) the charge of the (electron) proton in MKS units; M is actually the reduced mass of the proton-electron system, $M = m_e m_p/(m_e + m_p)$, which is equal to the electron mass within a good approximation. In this case it is convenient to start from Eq. (2.183), which we rewrite as

$$f''_{B,l}(r) + 2(l+1)\frac{f'_{B,l}(r)}{r} + \frac{2Me^2}{4\pi\epsilon_0\hbar^2 r}f_{B,l}(r) = \frac{2MB}{\hbar^2}f_{B,l}(r), \qquad (2.199)$$

where $B \equiv -E$ is the binding energy. Before proceeding further, let us perform a change of variables specifying the relevant parameters: we introduce Bohr's radius $a_0 = 4\pi\epsilon_0\hbar^2/(Me^2) \simeq 0.52 \times 10^{-10}$ m and Rydberg's energy constant $E_R \equiv hR \equiv Me^4/(2\hbar^2(4\pi\epsilon_0)^2) \simeq 13.6\,\mathrm{eV}$, which are the typical length and energy scales which can be constructed in terms of the physical constants involved in the problem. Equation (2.199) can be rewritten in terms of the dimensionless radial variable $\rho \equiv r/a_0$ and of the dimensionless binding energy $B/E_R \equiv \lambda^2$ (with $\lambda \geq 0$), as follows:

$$f''_{\lambda,l}(\rho) + 2(l+1)\frac{f'_{\lambda,l}(\rho)}{\rho} + \frac{2}{\rho}f_{\lambda,l}(\rho) = \lambda^2 f_{\lambda,l}(\rho). \qquad (2.200)$$

Let us first consider the asymptotic behavior of the solution as $\rho \to \infty$: in this limit the second and the third term on the left hand side can be neglected, so that the solution of equation (2.200) is asymptotically also solution of $f''_{\lambda,l}(\rho) = \lambda^2 f_{\lambda,l}(\rho)$, i.e. $f_{\lambda,l}(\rho) \sim e^{\pm\lambda\rho}$ for $\rho \gg 1$. The asymptotically divergent behavior must obviously be rejected since we are looking for a solution corresponding to a normalizable, single particle, bound state. We shall therefore write our solution in the form $f_{\lambda,l}(\rho) = h_{\lambda,l}(\rho)e^{-\lambda\rho}$, where $h_{\lambda,l}(\rho)$ should not diverge too strongly for $\rho \to \infty$, thus overcoming the damping exponential factor. The differential equation satisfied by $h_{\lambda,l}(\rho)$ easily follows from (2.200):

$$h''_{\lambda,l} + \left(\frac{2(l+1)}{\rho} - 2\lambda\right)h'_{\lambda,l} + \frac{2}{\rho}(1 - \lambda(l+1))h_{\lambda,l} = 0. \qquad (2.201)$$

Because the coefficients of this linear differential equation are analytic for ρ finite and strictly positive, $h_{\lambda,l}(\rho)$ should also be an analytic function in this domain. Therefore we can expand $h_{\lambda,l}(\rho)$ in power series of ρ, finding a recursion relation for its coefficients. We shall then impose that the series stops at some finite order so as to keep the asymptotic behavior of $f_{\lambda,l}(\rho)$ as $\rho \to \infty$ unchanged.

In order to understand what is the first term ρ^s of the series that we must take into account, let us consider the behavior of Eq. (2.201) as $\rho \to 0$. In this limit, setting $h_{\lambda,l} \sim \rho^s$, it can be easily checked that (2.201) is satisfied at the leading order in ρ only if $s(s-1) = -2s(l+1)$, whose solutions are $s = 0$ and $s = -2l - 1$. Last possibility must be rejected, otherwise the probability density related to our solution would not be integrable around the origin. Hence we write:

$$h_{\lambda,l}(\rho) = c_0 + c_1\rho + c_2\rho^2 + \cdots + c_h\rho^h + \cdots = \sum_{h=0}^{\infty} c_h\rho^h, \qquad (2.202)$$

with $c_0 \neq 0$. Inserting the last expression into (2.201), we obtain the following recurrence relation for the coefficients c_h:

$$c_{h+1} = 2\frac{\lambda(h+l+1) - 1}{(h+1)(h+2(l+1))} c_h \qquad (2.203)$$

which, apart from an overall normalization constant fixing the starting coefficient c_0, completely determines our solution in terms of l and λ. However, if the recurrence relation never stops, it becomes asymptotically (i.e. for large h):

$$c_{h+1} \simeq \frac{2\lambda}{h} c_h$$

which can be easily checked to be the same relation relating the coefficients in the Taylor expansion of $\exp(2\lambda\rho)$. Therefore, if the series does not stop, the asymptotic behavior of $f_{\lambda,l}(\rho)$ is corrupted, bringing in fact back the unwanted divergent behavior $f_{\lambda,l}(\rho) \sim e^{\lambda\rho}$. The series stops if and only if the coefficient on the right hand side of (2.203) vanishes for some given value $h = k \geq 0$, hence

$$\lambda(k+l+1) - 1 = 0 \quad \Rightarrow \quad \lambda = \frac{1}{k+l+1}. \qquad (2.204)$$

In this case $h_{\lambda,l}(\rho)$ is simply a polynomial of degree k in ρ, which is completely determined (neglecting an overall normalization) as a function of l and k: these polynomials belong to a well know class of special functions and are usually called Laguerre's associated polynomials. We have so found that, for a given value of l, the admissible solutions with negative energy, i.e. the hydrogen bound states, can be enumerated according to a non-negative integer k and the energy levels are quantized according to (2.204).

If we replace k by a new, integer and strictly positive, quantum number n given by:

$$n = k + l + 1 = \lambda^{-1}, \qquad (2.205)$$

which is usually called the *principal quantum number*, then, according to (2.204) and to the definition of λ, the energy levels of the hydrogen atom are given by

$$E_n = -\frac{E_R}{n^2} = -\frac{Me^4}{8\epsilon_0^2 h^2 n^2},$$

in perfect agreement with the Balmer–Rydberg series for line spectra and with the qualitative result obtained in Sect. 2.2 using Bohr's quantization rule.

It is important to note that, in the general case of a motion in a central field, energy levels related to different values of the angular momentum l are expected to be different, since they are related to the solutions of different equations of the form given in (2.183). Stated otherwise, the only expected degeneracy is that related to the rotational symmetry of the problem, leading to degenerate wave function multiplets of dimension $2l + 1$, as discussed above. However, in the hydrogen atom case, we have found a quite different result: according to Eq. (2.205), for a fixed value of the integer $n > 0$, there are n different multiplets, corresponding to $l = 0, 1, \ldots, (n-1)$, having the same energy. The degeneracy is therefore

$$\sum_{l=0}^{n-1}(2l + 1) = n^2$$

instead of $2l + 1$. Unexpected additional degeneracies like this one are usually called "accidental", even if in this case the degeneracy is not so accidental. Indeed the motion in a Coulomb (or gravitational) field has a larger symmetry than simply the rotational one. We will not go into details, but just remind the reader of a particular integral of motion which is only present, among all possible central potentials, in the case of the Coulomb (gravitational) field: that is Lenz's vector, which completely fixes the orientation of classical orbits. Another central potential leading to a similar "accidental" degeneracy is that corresponding to the three-dimensional isotropic harmonic oscillator. Actually, the Coulomb potential and the harmonic oscillator are joined in Classical Mechanics by Bertrand's theorem, which states that they are the only central potentials whose classical orbits are always closed.

Let us finish by giving the explicit form of the hydrogen wave functions in a few cases. Writing them in a form similar to that given in (2.187), and in particular as

$$\psi_{n,l,m}(r, \theta, \phi) = R_{n,l}(r)Y_{l,m}(\theta, \varphi),$$

we have

$$R_{1,0}(r) = 2(a_0)^{-3/2}\exp(-r/a_0),$$
$$R_{2,0}(r) = 2(2a_0)^{-3/2}\left(1 - \frac{r}{2a_0}\right)\exp(-r/2a_0),$$
$$R_{2,1}(r) = (2a_0)^{-3/2}\frac{r}{\sqrt{3}a_0}\exp(-r/2a_0).$$

2.9.3 The Isotropic Harmonic Oscillator

We go on with our introduction to the Schrödinger equation with central potentials reconsidering the case of the isotropic harmonic oscillator, that we shall discuss in a moment. We briefly recall the main results. The Schrödinger equation:

$$\left[-\frac{\hbar^2}{2m}\nabla^2 + \frac{k}{2}r^2\right]\psi_E = E\psi_E \,, \tag{2.206}$$

written in the form (2.134), appears separable in Cartesian coordinates and it is possible to find solutions written as the product of one-dimensional solutions $\psi_{n_x,n_y,n_z}(x, y, z) = \psi_{n_x}(x)\psi_{n_y}(y)\psi_{n_z}(z)$, and the corresponding energy is the sum of one-dimensional energies,

$$E_{n_x,n_y,n_z} = \hbar\omega(n_x + n_y + n_z + 3/2) = \hbar\omega(n + 3/2)\,,$$

where $n = n_x + n_y + n_z$ and $\omega = \sqrt{k/m}$. In particular the ground state wave function is $\psi_0(r) = (\alpha^2/\pi)^{3/4}\exp(-\alpha^2r^2/2)$, where $\alpha = \sqrt{m\omega/\hbar}$ is the inverse of the typical length scale of the system introduced in (2.116). Using the operator formalism we introduce three raising operators

$$A_x^\dagger = \frac{1}{\sqrt{\hbar\omega}}X_- = \frac{1}{\sqrt{2}}\left(\alpha x - \frac{1}{\alpha}\partial_x\right)$$

$$A_y^\dagger = \frac{1}{\sqrt{\hbar\omega}}Y_- = \frac{1}{\sqrt{2}}\left(\alpha y - \frac{1}{\alpha}\partial_y\right)$$

$$A_z^\dagger = \frac{1}{\sqrt{\hbar\omega}}Z_- = \frac{1}{\sqrt{2}}\left(\alpha z - \frac{1}{\alpha}\partial_z\right)\,, \tag{2.207}$$

and we write the generic solution shown above in the form:

$$\psi_{n_x,n_y,n_z}(x, y, z) = \frac{(A_x^\dagger)^{n_x}(A_y^\dagger)^{n_y}(A_z^\dagger)^{n_z}}{\sqrt{n_x!n_y!n_z!}}\psi_0(r)\,, \tag{2.208}$$

where the square root in the denominator is the normalization factor (see Eq. (2.133)). As we have shown in Sect. 2.7, these solutions are degenerate, in the sense that there are $(n+1)(n+2)/2$ solutions corresponding to the same energy $E_n = \hbar\omega(n+3/2)$ if $n = n_x + n_y + n_z$, and form a complete set. They also have the same transformation properties under reflection of all coordinate axes (parity transformation): indeed, since the ground state is parity even and each raising operator is parity odd, it is apparent that the solution corresponding to n_x, n_y, n_z has parity $(-1)^{n_x+n_y+n_z} = (-1)^n$. However they have no well defined angular momentum property, their form does not correspond to that shown in (2.187). Our purpose is to identify the solutions with well defined angular momentum quantum numbers, that is l and m: they will form an alternative complete set, i.e. a different orthonormal basis for the linear space of solutions corresponding to each energy level.

With this purpose it is useful to study the commutation rules of the angular momentum components with the raising operators and to take into account that the ground state is rotation invariant, that is $L_i \psi_0(r) = 0$, for $i = \pm, z$. In order to simplify the commutation rules we adapt our raising operators to the choice of complex coordinates $x_\pm, z$ introducing:

$$A_\pm^\dagger = \frac{1}{\sqrt{2}} \left(\alpha x_\pm - \frac{2}{\alpha} \partial_{x_\mp} \right) = A_x^\dagger \pm i A_y^\dagger . \tag{2.209}$$

They satisfy the commutation rules:

$$[L_z, A_z^\dagger] = 0, \quad [L_z, A_\pm^\dagger] = \pm \hbar A_\pm^\dagger, \quad [L_\pm, A_z^\dagger] = \mp \hbar A_\pm^\dagger$$

$$[L_\pm, A_\mp^\dagger] = \pm 2\hbar A_z^\dagger, \quad [L_\pm, A_\pm^\dagger] = 0, \tag{2.210}$$

which can be easily verified to coincide with those between angular momentum components and coordinates, after the substitution $A_z^\dagger \leftrightarrow z$ and $A_\pm^\dagger \leftrightarrow x_\pm$.

Due to the same correspondence and to the rotation invariance of ψ_0, given a polynomial $P(x_+, x_-, z)$ in the coordinates and the wave function $P(x_+, x_-, z)\psi_0(r)$, together with another wave function written in the operator formalism as $P(A_+^\dagger, A_-^\dagger, A_z^\dagger)\psi_0(r)$, we can state that if

$$L_i P(x_+, x_-, z)\psi_0(r) = Q_i(x_+, x_-, z)\psi_0(r) , \tag{2.211}$$

then

$$L_i P(A_+^\dagger, A_-^\dagger, A_z^\dagger)\psi_0(r) = Q_i(A_+^\dagger, A_-^\dagger, A_z^\dagger)\psi_0(r) . \tag{2.212}$$

Notice that the $A^\dagger$ operators commute among themselves, thus $P(A_+^\dagger, A_-^\dagger, A_z^\dagger)$ is a well defined differential operator and $P(A_+^\dagger, A_-^\dagger, A_z^\dagger)\psi_0(r)$ is a well defined wave function. The left-hand sides of the above equations are computed by repeatedly commuting L_i with the coordinates x_+, x_-, z, in the first equation, and with the raising operators $A_+^\dagger, A_-^\dagger, A_z^\dagger$ in the second one, until L_i reaches and annihilates ψ_0. The strict correspondence of the commutation rules guarantees the validity of the above equations.[9]

---

[9]This one-to-one correspondence between the action of the generators of rotations on the coordinates and on the raising operators can be generalized to other linear, in fact unitary, transformations of the coordinates, transforming homogeneous polynomials into homogeneous polynomials of the same degree. These transformations act within degenerate multiplets of solutions of the Schrödinger equation and clarify the origin of the additional degeneracy which is found for the central harmonic potential.

Hence in particular, considering the harmonic homogeneous polynomials, introduced in (2.178), and recalling that:

$$L^2 \mathcal{Y}_{l,m}(x_+, x_-, z)\psi_0(r) = \hbar^2 l(l+1)\mathcal{Y}_{l,m}(x_+, x_-, z)\psi_0(r) \qquad (2.213)$$

and that

$$L_z \mathcal{Y}_{l,m}(x_+, x_-, z)\psi_0(r) = \hbar m \mathcal{Y}_{l,m}(x_+, x_-, z)\psi_0(r)], \qquad (2.214)$$

we have:

$$L^2 \mathcal{Y}_{l,m}(A_+^\dagger, A_-^\dagger, A_z^\dagger)\psi_0(r) = \hbar^2 l(l+1)\mathcal{Y}_{l,m}(A_+^\dagger, A_-^\dagger, A_z^\dagger)\psi_0(r) , \qquad (2.215)$$

and

$$L_z \mathcal{Y}_{l,m}(A_+^\dagger, A_-^\dagger, A_z^\dagger)\psi_0(r) = \hbar m \mathcal{Y}_{l,m}(A_+^\dagger, A_-^\dagger, A_z^\dagger)\psi_0(r) . \qquad (2.216)$$

In this way we have identified a degenerate set of solutions of the Schrödinger equation corresponding to the energy $E_l = \hbar\omega(l + 3/2)E$ and with the angular momentum given above. However this does not exhaust the solutions with the same energy. Indeed for any positive integer $k \leq [l/2]$, considering that

$$L^2 \mathcal{Y}_{l-2k,m}(x_+, x_-, z)(r^2)^k \psi_0(r)$$
$$= \hbar^2 (l - 2k)(l - 2k + 1)\mathcal{Y}_{l-2k,m}(x_+, x_-, z)(r^2)^k \psi_0(r) \qquad (2.217)$$

and that

$$L_z \mathcal{Y}_{l-2k,m}(x_+, x_-, z)(r^2)^k \psi_0(r) = \hbar m \mathcal{Y}_{l-2k,m}(x_+, x_-, z)(r^2)^k \psi_0(r) \qquad (2.218)$$

we have

$$L^2 \mathcal{Y}_{l-2k,m}(A_+^\dagger, A_-^\dagger, A_z^\dagger)(A_+^\dagger A_-^\dagger + (A_z^\dagger)^2)^k \psi_0(r) \qquad (2.219)$$
$$= \hbar^2 (l - 2k)(l - 2k + 1)\mathcal{Y}_{l-2k,m}(A_+^\dagger, A_-^\dagger, A_z^\dagger)(A_+^\dagger A_-^\dagger + (A_z^\dagger)^2)^k \psi_0(r)$$

and

$$L_z \mathcal{Y}_{l-2k,m}(A_+^\dagger, A_-^\dagger, A_z^\dagger)(A_+^\dagger A_-^\dagger + (A_z^\dagger)^2)^k \psi_0(r)$$
$$= \hbar m \mathcal{Y}_{l-2k,m}(A_+^\dagger, A_-^\dagger, A_z^\dagger)(A_+^\dagger A_-^\dagger + (A_z^\dagger)^2)^k \psi_0(r). \qquad (2.220)$$

Therefore we have $2l - 4k + 1$ further solutions with the same energy E_l and angular momentum $l - 2k$. Notice that we have not considered the problem of normalizing the above wave functions. In this way, for any l, we have identified

$$(2l + 1) + (2l - 4 + 1) + \cdots = (l + 1)(l + 2)/2$$

independent solutions with energy E_l. These solutions form a complete degenerate set, i.e. a new basis, alternative to that described by Eq. (2.136), for the linear space of solutions of energy E_l. Their angular momenta correspond to all possible non-negative integers ranging from l down to zero (or one), but keeping the same parity of l. This last property can be easily understood, recalling that all solutions belonging to the same energy level of the isotropic harmonic oscillator have the same transformation properties under parity (they are even or odd depending on the parity of l, i.e. they have parity $(-1)^l$), and that, on the other hand, the solutions with fixed angular momentum $\bar{l}$ have parity $(-1)^{\bar{l}}$.

2.9.4 The Scattering Solutions

In the previous examples we have considered the determination of bound state solutions for some simple cases of central potentials. Now, to conclude, we consider a scattering experiment: a beam of particles of given energy is launched towards a fixed target. One is interested in determining the distribution of products after the collision, with the assumption that far away from the target, both before and after the collisions, a free particle approximation holds true. We limit ourselves to the case of elastic scattering in which the final scattered particles have the same nature as the beam ones and their energy loss is only due to the recoil of the scatterer. The experiment is based on the measure of the angular distribution of particles scattered at a certain angle with respect to the initial beam, which is assumed to be as monochromatic and parallel as possible. In principle, there should be a single scatterer which should initially be at rest and should interact with a single beam particle. However in practice the effect of a single scatterer, at the atomic level, would be too tiny to be observable. Therefore, in most cases, one uses many scatterers which are contained, e.g., in a piece of matter, a target, or else in a second beam crossing the first one.

A typical and renown example is Rutherford's experiment in which an almost parallel beam of alpha particles emitted by some radioactive material (Radium) and selected by a suitable diaphragm, crosses a thin golden target.[10] Since the alpha particles are heavy, their interaction with the atomic electrons is negligible and, if the target is thin enough, the scattered alpha particles have interacted with a single atomic nucleus of gold.

The scattered particles are detected by observing the sparks they produce impinging on a phosphorescent screen. The physical goal is to compare the intensity of particles scattered at small angles with that at large angles. Nuclei with radii of the same order of magnitude as the atomic radii should not scatter alpha particles at large angles, while this should happen for much smaller nuclei, because the electric forces generated by large nuclei are much smaller than those due to almost point-like ones. Rutherford was able to show that the nuclear radii are smaller than 10^{-14} m, and

---

[10]As a matter of fact the experiment was suggested by Rutherford, but performed by Geiger and Madsen.

therefore he suggested his atomic model which has been discussed in the first pages of this chapter.

In order to translate these qualitative considerations into quantitative ones, one usually counts the number of scattered particles (sparks) detected per unit time in a small solid angle around a given direction. This number should be proportional to the number of particles carried by the initial beam per unit time, that is, that of the particles crossing the beam section per unit time, and to the number of gold atoms which are present in the beam-target intersection, this is proportional to the area of the beam section. The coefficient of proportionality should thus have the dimensions of an area divided by a solid angle; it is usually written in differential form as $d\sigma/d\Omega(E, \theta)$ and is called the differential cross section. $d\Omega$ is the infinitesimal solid angle, θ is the polar angle taken with respect to the beam direction. The quantity σ, which is obtained after integration of the previous quantity over $d\Omega$, is an area called *cross section*, because classically it coincides with the cross section of the scatterer as seen by the beam. It is evident that θ is the only relevant angular variable here: we have assumed isotropy of the scatterer and invariance of the beam under rotations around the z axis.

Even if in many cases the mass of the scatterer M_s is much larger than that of the beam particles, m_p, and therefore practically the scatterer does not recoil, the process can always be described in relative coordinates. Then the recoil effects are taken into account replacing m_p by the reduced mass $m = M_s m_p/(M_s + m_p)$. Once this is done the above described elastic scattering process is represented assuming a wave function $\psi_k(r, \theta)$ which, at large distances, tends to

$$\psi_k(r, \theta) \underset{r \to \infty}{\to} e^{ikr\cos\theta} + \frac{f(\cos\theta)}{r} e^{ikr} . \tag{2.221}$$

We note that the collective interpretation of this wave function corresponds to a flux, parallel to the z axis, of ingoing particles with flux density (the number of particles per unit time crossing a unitary area orthogonal to the beam axis) $J_z = \hbar k/m$, and a radially scattered outgoing angular flux. The number of particles scattered per unit time and unit solid angle is equal to $J_r = |f(\cos\theta)|^2 \hbar k/m$. The ratio of these two quantities which can be computed from Eq. (2.221) is just the *differential cross section*:

$$\frac{d\sigma}{d\Omega}(\cos\theta) = |f(\cos\theta)|^2 , \tag{2.222}$$

and can be directly measured as described above. Therefore the physically interesting question is how one can compute $f(\cos\theta)$.

To start with, let us consider how a plane wave proceeding along the z axis, $\exp(ikz) = \exp(ikr\cos\theta)$, i.e. like the ingoing wave in (2.221), can be expanded in a series of spherical harmonics multiplied by functions of r, i.e. in a series of free particle solutions with fixed angular momentum. As we have seen, if the particle is free, the radial functions are solutions of equation (2.190), i.e. they coincide with spherical Bessel functions $j_l(kr) = \chi_{E,l}(r)/(kr)$. We note that the above plane wave is left invariant by the rotations around the z axis, thus only the $Y_{l,0}$'s should appear among the spherical harmonics appearing in the expansion. Indeed we have:

$$\exp(ikz) = \sum_{l=0}^{\infty} i^l \sqrt{4\pi(2l+1)}\, j_l(kr) Y_{l,0}(\cos\theta) \, . \tag{2.223}$$

The values of the coefficients in this expansion follow from the orthonormality of the spherical harmonics, Eq. (2.185), and from the equation

$$2\pi \int_{-1}^{1} dx\, e^{iyx} Y_{l,0}(x) \xrightarrow[y\to\infty]{} i^l \sqrt{4\pi(2l+1)} \frac{\sin(y - \frac{l\pi}{2})}{y} \, , \tag{2.224}$$

which can be proved taking into account that the $Y_{l,0}(z)$'s are polynomials in z with parity $(-1)^l$, that the following two equations hold:

$$Y_{l,0}(1) = \sqrt{\frac{2l+1}{4\pi}} \, , \tag{2.225}$$

$$\int_{-1}^{1} dy\, e^{ixy} x^n \xrightarrow[y\to\infty]{} \frac{e^{iy} - (-1)^n e^{-iy}}{iy} = i^n \frac{2}{y} \sin(y - \frac{n\pi}{2}) \, , \tag{2.226}$$

and, finally, comparing (2.226) with the asymptotic behavior of the spherical Bessel functions given in (2.196).

The physical interesting question is how the plane wave is deformed in the presence of a central potential $V(r)$ which we assume vanishing at large r faster than r^{-1}. It is clear that, because of the presence of a non-trivial potential $V(r)$, the asymptotic behavior of the component with angular momentum l of the radial wave functions does not coincide with the free one, that is, with the spherical Bessel function $\sim \sin(kr - l\pi/2))/r$. In contrast we have an asymptotic radial wave function $\sim \sin(kr - l\pi/2 + \delta_l))/r$, where δ_l is called *phase shift*. Here we show how, given the whole sequence δ_l, for $l = 0, \dots, \infty$, we can compute $f(\cos\theta)$.

We note, first of all, that the asymptotic form of the plane wave in (2.223) is:

$$e^{ikz} \xrightarrow[kr\to\infty]{} \sum_{l=0}^{\infty} i^l \sqrt{4\pi(2l+1)} Y_{l,0}(\cos\theta) \frac{\sin(kr - \frac{l\pi}{2})}{kr}$$

$$= \sum_{l=0}^{\infty} \frac{\sqrt{4\pi(2l+1)}}{2ikr} Y_{l,0}(\cos\theta) \left[e^{ikr} - (-1)^l e^{-ikr} \right] \, . \tag{2.227}$$

Assuming for $f(\cos\theta)$ the decomposition

$$f(\cos\theta) = \sum_{l=0}^{\infty} \sqrt{4\pi(2l+1)} f_l Y_{l,0}(\cos\theta) \, , \tag{2.228}$$

we translate equation (2.221) into:

$$\psi_k(r,\theta) \underset{r\to\infty}{\to} \sum_{l=0}^{\infty} \frac{\sqrt{4\pi(2l+1)}}{2ikr} Y_{l,0}(\cos\theta)\left[(1+2ikf_l)e^{ikr} - (-1)^l e^{-ikr}\right].$$

(2.229)

Here we must insert the above given information about the asymptotic behavior of the radial wave functions related to the phase shifts. This is possible if, and only if, the *unitarity constraint*

$$2ikf_l = e^{2i\delta_l} - 1,$$

(2.230)

is fulfilled, and one has

$$\psi_k(r,\theta) \underset{r\to\infty}{\to} \sum_{l=0}^{\infty} \frac{i^l\sqrt{4\pi(2l+1)}}{kr} e^{i\delta_l} Y_{l,0}(\cos\theta)\sin(kr - \frac{l\pi}{2} + \delta_l),$$

(2.231)

and

$$f(\cos\theta) = \sum_{l=0}^{\infty} \frac{\sqrt{4\pi(2l+1)}}{k} e^{i\delta_l}\sin\delta_l \, Y_{l,0}(\cos\theta),$$

(2.232)

from which one can compute the differential and the total cross section defined as

$$\sigma = 2\pi \int_{-1}^{1} d\cos\theta |f(\cos\theta)|^2 = \frac{4\pi}{k^2}\sum_{l=0}^{\infty}(2l+1)\sin^2\delta_l$$

$$= 4\pi \sum_{l=0}^{\infty} \frac{2l+1}{k^2 + k^2\cot^2\delta_l}.$$

(2.233)

Taking into account (2.225) it is easy to verify that:

$$\sigma = 4\pi \frac{\mathrm{Im}\,f(1)}{k}.$$

(2.234)

This is a very general property of scattering, which relates the total cross section to the differential cross section in the forward direction ($\cos\theta = 1$), which is usually called *the optical theorem*.

Few examples of phase shift calculations are given in the problems (see in particular 2.41, 2.52 and 2.53), here we consider the example of the scattering by a rigid ball of radius R, in which the radial wave function satisfies the free particle equation with the vanishing condition at $r = R$. For $l = 0$ we have $\sin(kR - \delta_0) = 0$ and hence $\delta_0 = kR$. For $l = 1$ the free wave function is $\frac{\sin(kR-\delta_1)}{kR} - \cos(kR - \delta_1)$ and hence we have $\delta_1 = kR - \tan^{-1}(kR)$, which for small energies vanishes as $(kR)^3$. Going further, it is possible to verify that for small energies the phase shifts $\delta_l \sim (kR)^{2l+1}$. This property holds true in general, if one identifies R with the (finite) range of the potential. The coefficient of k in the first term of the low energy expansion of δ_0 is called *scattering length*.

In the case at hand, for small energies, we have $\sigma \simeq 4\pi R^2(1 + O((kR)^2))$, that is, four times the geometric section of the ball, which is equal to the classical cross section. This proves the existence of diffractive contributions to the cross section. In the general expression of the low energy cross section the scattering length replaces R in the above formula.

For finite range potentials, at high energy, $\sin \delta_l$ must vanish for all l since an infinite number of l's contribute to the cross section, which must approach the classical value, thus remaining finite. Therefore $\lim_{k\to\infty} \delta_l = n_l\pi$ with n_l integer.

Suggestions for Supplementary Readings

- E. H. Wichman: *Quantum Physics - Berkeley Physics Course*, volume 4 (Mcgraw-Hill Book Company, New York 1971)
- L. D. Landau, E. M. Lifchitz: *Quantum Mechanics - Non-relativistic Theory*, Course of Theoretical Physics, volume 3 (Pergamon Press, London 1958)
- L. I. Schiff: *Quantum Mechanics* 3d edn (Mcgraw-Hill Book Company, Singapore 1968)
- J. J. Sakurai: *Modern Quantum Mechanics* (The Benjamin-Cummings Publishing Company Inc., Menlo Park 1985)
- E. Persico: *Fundamentals of Quantum Mechanics* (Prentice - Hall Inc., Englewood Cliffs 1950).

Problems

2.1 A diatomic molecule can be simply described as two point-like objects of mass $m - 10^{-26}$ kg placed at a fixed distance $d - 10^{-9}$ m. Describe what are the possible values of the molecule energy according to Bohr's quantization rule. Compute the energy of the photons which are emitted when the system decays from the $(n+1)$-th to the n-th energy level.

Answer: $E_{n+1} - E_n = (\hbar^2/2I)(n+1)^2 - (\hbar^2/2I)n^2 = (2n+1)\hbar^2/(md^2) \simeq 1.1 \times 10^{-24}(2n+1)$ J. Notice that in Sommerfeld's perfected theory, mentioned in Sect. 2.2, the energy of a rotator is given by $E_n = \hbar^2 n(n+1)/2I$, so that the factor $2n+1$ in the solution must be replaced by $2n+2$.

2.2 An artificial satellite of mass $m = 1$ kg rotates around the Earth on a circular orbit of radius practically equal to that of the Earth itself, i.e. roughly 6370 km. If the satellite orbits are quantized according to Bohr's rule, what is the radius variation when going from one quantized level to the next (i.e. from n to $n+1$)?

Answer: If g indicates the gravitational acceleration at the Earth surface, the radius of the n-th orbit is given by $r_n = n^2\hbar^2/(m^2R^2g)$. Therefore, if $r_n = R$, $\delta r_n \equiv r_{n+1} - r_n \simeq 2\hbar/(m\sqrt{Rg}) \simeq 2.6 \times 10^{-38}$ m .

2.3 An electron is accelerated through a potential difference $\Delta V = 10^8$ V, what is its wavelength according to de Broglie?

Answer: The energy gained by the electron is much greater than mc^2, therefore it is ultra-relativistic and its momentum is $p \simeq E/c$. Hence $\lambda \simeq hc/e\Delta V \simeq 12.4 \times 10^{-15}$ m. The exact formula is instead $\lambda = hc/\sqrt{(e\Delta V + mc^2)^2 - m^2c^4}$.

2.4 An electron is constrained to bounce between two reflecting walls placed at a distance $d = 10^{-9}$ m from each other. Assuming that, as in the case of a stationary electromagnetic wave confined between two parallel mirrors, the distance d be equal to n half wavelengths, determine the possible values of the electron energy as a function of n.

Answer: $E_n = \hbar^2 \pi^2 n^2/(2md^2) \simeq n^2\, 6.03 \times 10^{-20}$ J .

2.5 An electron of kinetic energy 1 eV is moving upwards under the action of its weight. Can it reach an altitude of 1 km? If yes, what is the variation of its de Broglie wavelength?

Answer: The maximum altitude reachable by the electron in a constant gravitational field would be $h = T/mg \simeq 1.6 \times 10^{10}$ m. After one kilometer the kinetic energy changes by $\delta T/T \simeq 5.6 \times 10^{-8}$, hence $\delta\lambda/\lambda \simeq 2.8 \times 10^{-8}$. Since the starting wavelength is $\lambda = h/\sqrt{2mT} \simeq 1.2 \times 10^{-9}$ m, the variation is $\delta\lambda \simeq 3.4 \times 10^{-17}$ m.

2.6 Ozone (O_3) is a triatomic molecule made up of three atoms of mass $m \simeq 2.67 \times 10^{-26}$ kg placed at the vertices of an equilateral triangle with sides of length l. The molecule can rotate around an axis P going through its center of mass and orthogonal to the triangle plane, or around another axis L which passes through the center of mass as well, but is orthogonal to the first axis. Making use of Bohr's quantization rule and setting $l = 10^{-10}$ m, compare the possible rotational energies in the two different cases of rotations around P or L.

Answer: The moments of inertia are $I_P = ml^2 = 2.67 \times 10^{-46}$ kg m^2 and $I_L = ml^2/2 = 1.34 \times 10^{-46}$ kg m^2. The rotational energies are then $E_{L,n} = 2E_{P,n} = \hbar^2 n^2/2I_L \simeq n^2\, 4.2 \times 10^{-23}$ J.

2.7 A table salt crystal is irradiated with an X-ray beam of wavelength $\lambda = 2.5 \times 10^{-10}$ m, the first diffraction peak ($d \sin \theta = \lambda$) is observed at an angle equal to 26.3°. What is the interatomic distance of salt?

Answer: $d = \lambda/\sin \theta \simeq 5.6 \times 10^{-10}$ m.

2.8 In β decay a nucleus, with a radius of the order of $R = 10^{-14}$ m, emits an electron with a kinetic energy of the order of 1 MeV $= 10^6$ eV. Compare this value with that which according to the uncertainty principle is typical of an electron initially localized inside the nucleus (thus having a momentum $p \sim \hbar/R$).

Answer: The order of magnitude of the momentum of the particle is $p \sim \hbar/R \sim 10^{-20}$ N s, thus $pc \simeq 3 \times 10^{-12}$ J, which is much larger that the electron rest energy

$m_e c^2 \simeq 8 \times 10^{-14}$ J. Therefore the kinetic energy of the electron in the nucleus is about $pc = 3.15 \times 10^{-12}$ J $\simeq 20$ MeV.

2.9 An electron is placed in a constant electric field $\mathcal{E} = 1000$ V/m directed along the x axis and going out of a plane surface orthogonal to the same axis. The surface also acts on the electron as a reflecting plane where the electron potential energy $V(x)$ goes to infinity. The behavior of $V(x)$ is therefore as illustrated in the following figure.

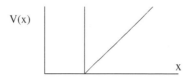

Evaluate the order of magnitude of the minimal electron energy according to Heisenberg's Uncertainty Principle.

Answer: The total energy is given by $\epsilon = p^2/2m + V(x) = p^2/2m + e\mathcal{E}x$, with the constraint $x > 0$. From a classical point of view, the minimal energy would be realized for a particle at rest ($p = 0$) in the minimum of the potential. The uncertainty principle states instead that $\delta p\, \delta x \sim \hbar$, where δx is the size of a region around the potential minimum where the electron is localized. Therefore the minimal total energy compatible with the uncertainty principle can be written as a function of δx as $E(\delta x) \equiv \hbar^2/(2m\delta x^2) + e\mathcal{E}\delta x$ ($\delta x > 0$) and has a minimum

$$\epsilon_{\min} \sim \frac{3}{2}\left(\frac{\hbar^2 e^2 \mathcal{E}^2}{m}\right)^{1/3} \sim 0.6 \times 10^{-4}\ \text{eV}.$$

2.10 An atom of mass $M = 10^{-26}$ kg is attracted towards a fixed point by an elastic force of constant $k = 1$ N/m; the atom is moving along a circular orbit in a plane orthogonal to the x axis. Determine the energy levels of the system by making use of Bohr's quantization rule for the angular momentum computed with respect to the fixed point.

Answer: Let ω be the angular velocity and r the orbital radius. The centripetal force is equal to the elastic one, hence $\omega = \sqrt{k/M}$. The total energy is given by $E = (1/2)M\omega^2 r^2 + (1/2)kr^2 = M\omega^2 r^2 = L\omega$, where $L = M\omega r^2$ is the angular momentum. Since $L = n\hbar$, we finally infer $E_n = n\hbar\omega \simeq n\, 1.05 \times 10^{-21}$ J $\simeq n\, 0.66 \times 10^{-2}$ eV.

2.11 Compute the number of photons emitted in one second by a lamp of power 10 W, if the photon wavelength is 0.5×10^{-6} m.

Answer: The energy of a single photon is $E = h\nu$ and $\nu = c/\lambda = 6 \times 10^{14}$ Hz, hence $E \simeq 4 \times 10^{-19}$ J. Therefore the number photons emitted in one second is 2.5×10^{19}.

2.12 A particle of mass $m = 10^{-28}$ kg is moving along the x axis under the influence of a potential energy given by $V(x) = v\sqrt{|x|}$, where $v = 10^{-15}$ J m$^{-1/2}$. Determine what is the order of magnitude of the minimal particle energy according to the uncertainty principle.

Answer: The total energy of the particle is given by

$$E = \frac{p^2}{2m} + v\sqrt{|x|} \,.$$

If the particle is localized in a region of size δx around the minimum of the potential ($x = 0$), according to the uncertainty principle it has a momentum at least of the order of $\delta p = \hbar/\delta x$. It is therefore necessary to minimize the quantity

$$E = \frac{\hbar^2}{2m\delta x^2} + v\sqrt{\delta x}$$

with respect to δx, finally finding that

$$E_{min} \simeq \left(\frac{\hbar^2 v^4}{m}\right)^{1/5} \left(2^{1/5} + 2^{-9/5}\right) \simeq 0.092 \text{ eV}.$$

It is important to notice that our result, apart from a numerical factor, could be also predicted on the basis of simple dimensional remarks. Indeed, it can be easily checked that $(\hbar^2 v^4/m)^{1/5}$ is the only possible quantity having the dimensions of an energy and constructed in terms of m, v and $\hbar$, which are the only physical constants involved in the problem. In the analogous classical problem $\hbar$ is missing, and v and m are not enough to build a quantity with the dimensions of energy, hence the classical problem lacks the typical energy scale appearing at the quantum level.

2.13 An electron beam with kinetic energy equal to 10 eV is split into two parallel beams placed at different altitudes in the terrestrial gravitational field. If the altitude gap is $d = 10$ cm and if the beams recombine after a path of length L, say for which values of L the phases of the recombining beams are opposite (destructive interference). Assume that the upper beam maintains its kinetic energy, that the total energy is conserved for both beams and that the total phase difference accumulated during the splitting and the recombination of the beams is negligible.

Answer: De Broglie's wave describing the initial electron beam is proportional to $\exp(\mathrm{i}\, px/\hbar - \mathrm{i}\, Et/\hbar)$, where $p = \sqrt{2mE_k}$ is the momentum corresponding to a kinetic energy E_k and $E = E_k + mgh$ is the total energy. The beam is split into two beams, the first travels at the same altitude and is described by the same wave function, the second travels 10 cm lower and is described by a wave function $\propto \exp(\mathrm{i}\, p'x/\hbar - \mathrm{i}\, Et/\hbar)$ where $p' = \sqrt{2mE_k'} = \sqrt{2m(E_k + mgd)}$ (obviously the total energy E stays unchanged). The values of L for which the two beams recombine with opposite phases are solutions of $(p' - p)L/\hbar = (2n + 1)\pi$ where n is an integer. The smallest value of L is $L = \pi\hbar/(p' - p)$. Notice that $mgd \simeq 10^{-30}\ J \ll 10\ \mathrm{eV} \simeq 1.6 \times 10^{-18}\ \mathrm{J}$ hence $p' - p \simeq \sqrt{2mE_k}(mgd/2E_k)$ and $L \simeq 2\pi\hbar E_k/(mgd\sqrt{2mE_k}) \simeq 696$ m. This experiment, which clearly demonstrates the wavelike behavior of material particles, has been really performed using neutrons in place of electrons: that has various advantages, among which that of leading to smaller values of L because of the much heavier mass, as it is clear from the solution. That makes the setting of the experimental apparatus simpler.

2.14 An electron is moving in the $x - y$ plane under the influence of a magnetic induction field parallel to the z-axis. What are the possible energy levels according to Bohr's quantization rule?

Answer: The electron is subject to the force $e\boldsymbol{v} \wedge \boldsymbol{B}$ where $\boldsymbol{v}$ is its velocity. Classically the particle, being constrained in the $x - y$ plane, would move on circular orbits with constant angular velocity $\omega = eB/m$, energy $E = 1/2\, m\omega^2 r^2$ and any radius r. Bohr's quantization instead limits the possible values of r by $m\omega r^2 = n\hbar$. Finally one finds $E = 1/2\, n\hbar\omega = n\hbar eB/(2m)$. This is an approximation of the exact solution for the quantum problem of an electron in a magnetic field (Landau's levels).

2.15 The positron is a particle identical to the electron but carrying an opposite electric charge. It forms a bound state with the electron, which is similar to the hydrogen atom but with the positron in place of the proton: that is called *positronium*. What are its energy levels according to Bohr's rule?

Answer: The computation goes exactly along the same lines as for the proton–electron system, but the reduced mass $\mu = m^2/(m + m) - m/2$ has to be used in place of the electron mass. Energy levels are thus

$$E_n = -\frac{me^4}{16\epsilon_0^2 h^2 n^2}\ .$$

2.16 A particle of mass $M = 10^{-29}$ kg is moving in two dimensions under the influence of a central potential

$$V = \sigma r\ ,$$

where $\sigma = 10^5$ N. Considering only circular orbits, what are the possible values of the energy according to Bohr's quantization rule?

Answer: Combining the equation for the centripetal force necessary to sustain the circular motion, $m\omega^2 r = \sigma$, with the quantization of angular momentum, $m\omega r^2 = n\hbar$, we obtain for the total energy, $E = 1/2\, m\omega^2 r^2 + \sigma r$, the following quantized values

$$E_n = \frac{3}{2}\left(\frac{\hbar^2\sigma^2}{m}\right)^{1/3} n^{2/3} \simeq 2\,n^{2/3}\;\mathrm{GeV}\;.$$

Notice that the only possible combination of the physical parameters available in the problem with energy dimensions is $(\hbar^2\sigma^2/m)^{1/3}$. The potential proposed in this problem is similar to that believed to act among quarks, which are the elementary constituents of hadrons (a wide family of particles including protons, neutrons, mesons ...); σ is usually known as the *string tension*. Notice that the lowest energy coincides, identifying $\sigma = e\mathcal{E}$, with that obtained in Problem 2.9 using the uncertainty principle for the one-dimensional problem.

2.17 The momentum probability distribution for a particle with wave function $\psi(x)$ is given by

$$\left|\int_{-\infty}^{\infty} dx \frac{1}{\sqrt{h}} e^{-i\,px/\hbar}\psi(x)\right|^2 = |\tilde{\psi}(p)|^2\;.$$

Compute the distribution for the following wave function $\psi(x) = e^{-a|x|/2}\sqrt{a/2}$ (a is real and positive) and verify the validity of the uncertainty principle in this case.

Answer: $\tilde{\psi}(p) = (\hbar a)^{3/2}/(\sqrt{4\pi}(p^2 + \hbar^2 a^2/4))$ hence

$$\Delta_x^2 = \frac{a}{2}\int_{-\infty}^{\infty} dx\; x^2 e^{-a|x|} = \frac{2}{a^2}\;,$$

$$\Delta_p^2 = \frac{(\hbar a)^3}{4\pi}\int_{-\infty}^{\infty} dp\; \frac{p^2}{(p^2 + \frac{a^2\hbar^2}{4})^2} = \frac{a^2\hbar^2}{4}\;,$$

so that $\Delta_x^2\Delta_p^2 = \hbar^2/2 > \hbar^2/4$.

2.18 Show that a wave packet described by a real wave function has always zero average momentum. Compute the probability current for such packet.

Answer: From the relation

$$\tilde{\psi}(p) = \int_{-\infty}^{\infty} dx \frac{1}{\sqrt{h}} e^{-i\,px/\hbar}\psi(x)$$

and $\psi^*(x) = \psi(x)$ we infer

$$\tilde{\psi}^*(p) = \int_{-\infty}^{\infty} dx \frac{1}{\sqrt{h}} e^{i\,px/\hbar}\psi(x) = \tilde{\psi}(-p)$$

hence $|\tilde{\psi}(p)|^2 = |\tilde{\psi}(-p)|^2$. The probability distribution function is even in momentum space, so that the average momentum is zero. The probability current is zero as well, in agreement with the average zero momentum, i.e. with the fact that there is not net matter transportation associated to this packet. Notice that the result does not change if $\psi(x)$ is multiplied by a constant complex factor $e^{i\phi}$.

2.19 The wave function of a free particle is

$$\psi(x) = \frac{1}{\sqrt{2P\hbar}} \int_{-P}^{P} dq \, e^{\frac{iqx}{\hbar}}$$

at time $t = 0$. What is the corresponding probability density $\rho(x)$ of locating the particle at a given point x? What is the probability distribution function in momentum space? Give an integral representation of the wave function at a generic time t, assuming that the particle mass is m.

Answer: The probability density is $\rho(x) = |\psi(x)|^2 = \hbar/(\pi P x^2) \sin^2 (Px/\hbar)$ while $\psi(x,t) = (1/\sqrt{2P\hbar}) \int_{-P}^{P} dq e^{i(qx - q^2 t/2m)/\hbar}$. The distribution in momentum in instead given by $|\tilde{\psi}(p)|^2 = \Theta(P^2 - p^2)/2P$, where Θ is the step function, $\Theta(y) = 0$ for $y < 0$ and $\Theta(y) = 1$ for $y \geq 0$. Notice that for the given distribution we have $\Delta_x^2 = \infty$. The divergent variance is strictly related to the sharp, step-like distribution in momentum space; indeed Δ_x^2 becomes finite as soon as the step is smoothed.

2.20 An electron beam hits the potential step sketched in the figure, coming from the right. In particular, the potential energy of the electrons is 0 for $x < 0$ and $-V = -300$ eV for $x > 0$, while their kinetic energy in the original beam (thus for $x > 0$) is $E_k = 400$ eV. What is the reflection coefficient?

Answer: The wave function can be written, leaving aside an overall normalization coefficient which is not relevant for computing the reflection coefficient, as

$$\psi(x) = be^{-ik'x} \text{ for } x < 0, \qquad \psi(x) = e^{-ikx} + ae^{ikx} \text{ for } x > 0$$

where $k = \sqrt{2mE_k}/\hbar = \sqrt{2m(E + V)}/\hbar$ and $k' = \sqrt{2m(E_k - V)}/\hbar = \sqrt{2mE}/\hbar$; m is the electron mass and $E = E_k - V$ is the total energy of the electrons. The continuity conditions at the position of the step read

$$b = 1 + a, \qquad bk' = (1 - a)k,$$

hence

$$b = \frac{2}{1 + k'/k}, \qquad a = \frac{1 - k'/k}{1 + k'/k},$$

and

$$R = |a|^2 = \left(\frac{k - k'}{k + k'}\right)^2 = \frac{2E + V - 2\sqrt{E(E + V)}}{2E + V + 2\sqrt{E(E + V)}} = \frac{1}{9}.$$

2.21 An electron beam hits the same potential step considered in Problem 2.20, this time coming from the left with a kinetic energy $E = 100$ eV. What is the reflection coefficient in this case?

Answer: In this case we write:

$$\psi(x) = e^{ik'x} + be^{-ik'x} \quad \text{for} \ x < 0, \qquad \psi(x) = ae^{ikx} \quad \text{for} \ x > 0,$$

where again $k = \sqrt{2m(E + V)}/\hbar$ and $k' = \sqrt{2mE}/\hbar$ with $E = E_k$ being the total energy. By solving the continuity conditions we find:

$$b = \frac{k'/k - 1}{1 + k'/k}; \qquad R = |b|^2 = \left(\frac{k' - k}{k + k'}\right)^2 = \frac{2E + V - 2\sqrt{E(E + V)}}{2E + V + 2\sqrt{E(E + V)}} = \frac{1}{9}.$$

We would like to stress that the reflection coefficient coincides with that obtained in Problem 2.20: electron beams hitting the potential step from the right or from the left are reflected in exactly the same way, if their total energy E is the same, as it is in the present case. In fact this is a general result which is valid for every kind of potential barrier and derives directly from the invariance of the Schrödinger equation under time reversal: the complex conjugate of a solution is also a solution. It may seem a striking result, but it should not be so striking for those familiar with reflection of electromagnetic signals in the presence of unmatching impedances.

Notice also that there is actually a difference between the two cases, consisting in a different sign for the reflected wave. That is irrelevant for computing R but significant for considering interference effects involving the incident and the reflected waves. In the present case interference is destructive, hence the probability density is suppressed close to the step, while in Problem 2.20 the opposite happens. To better appreciate this fact consider the analogy with an oscillating rope made up of two different ropes having different densities (which is a system in some sense similar to ours), and try to imagine the different behaviors observed if you enforce oscillations shaking the rope from the heavier (right-hand in our case) or from the lighter (left-hand in our case) side. As an extreme and easier case, you could think of a single rope with a free end (one of the densities goes to zero) or with a fixed end (one of the densities goes to infinity): the shape of the rope at the considered endpoint is cosine-like in the first case and sine-like in the second case, exactly as for the cases of respectively the previous and the present problem in the limit $V \to \infty$.

2.22 An electron beam hits, coming from the right, a potential step similar to that considered in Problem 2.20. However this time $-V = -10$ eV and the kinetic energy of the incoming electrons is $E_k = 9$ eV. If the incident current is equal to $J = 10^{-3}$ A, compute how many electrons can be found, at a given time, along the negative x axis, i.e. how many electrons penetrate the step barrier reaching positions which would be classically forbidden.

Answer: The solution of the Schrödinger equation can be written as

$$\psi(x) = c\, e^{p'x/\hbar} \ \ \text{for} \ \ x < 0 , \qquad \psi(x) = a\, e^{i\,px/\hbar} + b\, e^{-i\,px/\hbar} \ \ \text{for} \ \ x > 0$$

where $p = \sqrt{2mE}$ and $p' = \sqrt{2m(V-E)}$. Imposing continuity in $x = 0$ for both $\psi(x)$ and its derivative, we obtain $c = 2a/(1 + i\,p'/p)$ and $b = a(1 - i\,p'/p)/(1 + i\,p'/p)$. It is evident that $|b|^2 = |a|^2$, hence the reflection coefficient is one. Indeed the probability current $J(x) = -i\,\hbar/(2m)(\psi^*\partial_x\psi - \psi\partial_x\psi^*)$ vanishes on the left, where we have an evanescent wave function, hence no transmission. Nevertheless the probability distribution is non-vanishing for $x < 0$ and, on the basis of the collective interpretation, the total number of electrons on the left is given by

$$N = \int_{-\infty}^{0} |\psi(x)|^2 dx = |c|^2 \hbar/(2p') = \frac{2|a|^2 \hbar p^2}{p'(p^2 + p'^2)}.$$

The coefficient a can be computed by asking that the incident current $J_{el} = eJ = e|a|^2 p/m \equiv 10^{-3}$ A. The final result is $N \simeq 1.2$.

2.23 An electron is confined inside a cubic box with reflecting walls and an edge of length $L = 2 \times 10^{-9}$ m. How many stationary states can be found with energy less than 1 eV? Take into account the spin degree of freedom, which in practice doubles the number of available levels.

Answer: Energy levels in a cubic box are $E_{n_x,n_y,n_z} = \pi^2\hbar^2(n_x^2 + n_y^2 + n_z^2)/(2mL^2)$, where $m = 0.911 \times 10^{-30}$ kg and n_x, n_y, n_z are positive integers. The constraint $E < 1$ eV implies $n_x^2 + n_y^2 + n_z^2 < 10.7$, which is satisfied by 7 different combinations ((1,1,1), (2,1,1), (2,2,1) plus all possible different permutations). Taking spin into account, the number of available levels is 14.

2.24 When a particle beam hits a potential barrier and is partially transmitted, a forward going wave is present on the other side of the barrier which, besides having a reduced amplitude with respect to the incident wave, has also acquired a phase factor which can be inferred by the ratio of the transmitted wave coefficient to that of the incident one. Assuming a thin barrier, describable as

$$V(x) = v\, \delta(x) ,$$

and that the particles are electrons of energy $E = 10$ eV, compute the value of v for which the phase difference is $-\pi/4$.

Suppose now that we have two beams of equal amplitude and phase and that one beam goes through the barrier while the other goes free. The two beams recombine after paths of equal length. What is the ratio of the recombined beam intensity to that one would have without the barrier?

Answer: On one side of the barrier the wave function is $e^{i\sqrt{2mEx}/\hbar} + a\,e^{-i\sqrt{2mEx}/\hbar}$, while it is $b\,e^{i\sqrt{2mEx}/\hbar}$ on the other side. Continuity and discontinuity constraints read $1+a = b$ and $b-1+a = \sqrt{2m/E}\,vb/\hbar$, from which $b = (1+i\sqrt{m/2E}\,v/\hbar)^{-1}$ can be easily derived. Requiring that the phase of b be $-\pi/4$ is equivalent to $\sqrt{m/2E}\,v/\hbar = 1$, hence $v \simeq 2.0 \times 10^{-28}$ J m.

With this choice of v the recombined beam is $[1+1/(1+i)]e^{i\sqrt{2mEx}/\hbar}$. The ratio of the intensity of the recombined beam to that one would have without the barrier is $|[1 + 1/(1+i)]/2|^2 = 5/8$.

2.25 If a potential well in one dimension is so thin to be describable by a Dirac delta function:

$$V(x) = -VL\delta(x)$$

where V is the depth and L the width of the well, then it is possible to compute the bound state energies by recalling that for a potential energy of that kind the wave function is continuous while its first derivative has the following discontinuity:

$$\lim_{\epsilon\to 0}(\partial_x\psi(x+\epsilon) - \partial_x\psi(x-\epsilon)) = -\frac{2m}{\hbar^2}VL\psi(0)\,.$$

What are the possible energy levels?

Answer: The bound state wave function is

$$a\,e^{-\sqrt{2mB}x/\hbar} \quad \text{for } x>0 \quad \text{and} \quad a\,e^{\sqrt{2mB}x/\hbar} \quad \text{for } x<0$$

where the continuity condition for the wave function has been already imposed. $B = |E|$ is the absolute value of the energy (which is negative for a bound state). The discontinuity condition on the first derivative leads to an equation for B which has only one solution, $B = mV^2L^2/(2\hbar^2)$, thus indicating the existence of a single bound state.

2.26 A particle of mass m moves in the following one-dimensional potential:

$$V(x) = v(\alpha\delta(x-L) + \alpha\delta(x+L) - \frac{1}{L}\Theta(L^2-x^2))\,,$$

where Θ is the step function, $\Theta(y) = 0$ for $y<0$ and $\Theta(y) = 1$ for $y>0$. Constants are such that

$$\frac{2mvL}{\hbar^2} = \left(\frac{\pi}{4}\right)^2\,.$$

Find the conditions on $\alpha > 0$ for the existence of bound states.

Answer: The potential is such that $V(-x) = V(x)$: in this case the lowest energy level, if any, corresponds to an even wave function. We can thus limit the discussion to the region $x > 0$, where we have $\psi(x) = \cos kx$ for $x < L$ and $\psi(x) = ae^{-\beta x}$ for $x > L$, with the constraint $0 < k < \sqrt{2mv/(\hbar^2 L)} = \pi/(4L)$, since $\beta = \sqrt{2mv/(\hbar^2 L) - k^2}$ must be real in order to have a bound state, hence $kL < \pi/4$. Continuity and discontinuity constraints, respectively on $\psi(x)$ and $\psi'(x)$ in $x = L$, give: $\tan kL = (\beta + 2mv\alpha/\hbar^2)/k$. Setting $y \equiv kL$, we have

$$\tan y = \frac{\sqrt{\frac{\pi^2}{16} - y^2} + \frac{\pi^2}{16}\alpha}{y}.$$

The function on the left hand side grows from 0 to 1 in the interval $0 < y < \pi/4$, while the function on the right decreases from ∞ to $\alpha\pi/4$ in the same interval. Therefore an intersection (hence a bound state) exists only if $\alpha < 4/\pi$.

2.27 An electron moves in a one-dimensional potential corresponding to a square well of depth $V = 0.1$ eV and width $L = 3 \times 10^{-10}$ m. Show that in these conditions there is only one bound state and compute its energy in the thin well approximation, discussing also the validity of that limit.

Answer: There is one only bound state if the first odd state is absent. That is true if $y = \sqrt{2mVL}/(2\hbar) < \pi/2$, which is verified in our case since, using $m = 0.911 \times 10^{-30}$ kg, one obtains $y \simeq 0.243 < \pi/2$.

Setting $B \equiv -E$, where E is the negative energy of the bound state, B is obtained as a solution of

$$\tan\left(\frac{L\sqrt{2m(V - B)}}{2\hbar}\right) = \sqrt{\frac{B}{V - B}}.$$

The thin well limit corresponds to $V \to \infty$ and $L \to 0$ as the product VL is kept constant. Neglecting B with respect to V we can write

$$\tan\left(\frac{\sqrt{2mVL^2}}{2\hbar}\right) = \sqrt{\frac{B}{V}}.$$

In the thin well limit $VL^2 \to$ constant $\cdot L \to 0$, hence we can replace the tangent by its argument, obtaining finally $B = mV^2L^2/(2\hbar^2) \simeq 0.59 \times 10^{-2}$ eV, which coincides with the result obtained in Problem 2.25. In this case the argument of the tangent is $y \sim 0.24$ and we have $\tan 0.24 \simeq 0.245$; therefore the exact result differs from that obtained in the thin well approximation by roughly 4%.

2.28 An electron moves in one dimension and is subject to forces corresponding to a potential energy:

$$V(x) = \mathcal{V}[-\delta(x) + \delta(x - L)].$$

What are the conditions for the existence of a bound state and what is its energy if $L = 10^{-9}$ m and $\mathcal{V} = 2 \times 10^{-29}$ J m?

Answer: A solution of the Schrödinger equation corresponding to a binding energy $B \equiv -E$ can be written as

$$\psi(x) = e^{\sqrt{2mB}x/\hbar} \quad \text{for} \quad x < 0,$$

$$\psi(x) = ae^{\sqrt{2mB}x/\hbar} + be^{-\sqrt{2mB}x/\hbar} \quad \text{for} \quad 0 < x < L,$$

$$\psi(x) = ce^{-\sqrt{2mB}x/\hbar} \quad \text{for} \quad L < x.$$

Continuity and discontinuity constraints, respectively for the wave function and for its derivative in $x = 0$ and $x = L$, give: $a + b = 1$, $a - b - 1 = -\sqrt{2m/B}\mathcal{V}/\hbar$, $ae^{\sqrt{8mBL}/\hbar} + b = c$, $ae^{\sqrt{8mBL}/\hbar} - b + c = -c\sqrt{2m/B}\mathcal{V}/\hbar$.

The four equations are compatible if $e^{\sqrt{8mBL}/\hbar} = (1 - \frac{2B\hbar^2}{m\mathcal{V}^2})^{-1}$, which has a non-trivial solution $B \neq 0$ for any $L > 0$. Setting $y = \sqrt{2B/m}\hbar/\mathcal{V}$ the compatibility condition reads $e^{2m\mathcal{V}Ly/\hbar^2} = 1/(1 - y^2)$. Using the values of L and $\mathcal{V}$ given in the text one obtains $e^{2m\mathcal{V}L/\hbar^2} \gg 1$, hence $2B\hbar^2/(m\mathcal{V}^2) \simeq 1$ within a good approximation, i.e. $B \simeq m\mathcal{V}^2/(2\hbar^2)$, which coincides with the result obtained in the presence of a single thin well. This approximation is indeed equivalent to the limit of a large distance L (hence $e^{2m\mathcal{V}L/\hbar^2} \gg 1$) between the two Dirac delta functions; it can be easily verified that in the same limit one has $b \simeq 1$ and $a \simeq 0$, so that, in practice, the state is localized around the attractive delta function in $x = 0$, which is the binding part of the potential, and does not feel the presence of the other term in the potential which is very far away.

As L is decreased, the binding energy lowers and the wave function amplitude, hence the probability density, gets asymmetrically shifted on the left, until the binding energy vanishes in the limit $L \to 0$. In practice, the positive delta function in $x = L$ acts as a repulsive term which asymptotically extracts, as $L \to 0$, the particle from its thin well.

2.29 A particle of mass $M = 10^{-26}$ kg moves along the x axis under the influence of an elastic force of constant $k = 10^{-6}$ N/m. The particle is in its ground state: compute its wave function and the mean value of x^2, given by

$$\langle x^2 \rangle = \frac{\int_{-\infty}^{\infty} dx\, x^2 |\psi(x)|^2}{\int_{-\infty}^{\infty} dx\, |\psi(x)|^2}.$$

Answer:

$$\psi(x) = \left(\frac{kM}{\pi^2\hbar^2}\right)^{1/8} e^{-\sqrt{kM}x^2/2\hbar}; \qquad \langle x^2 \rangle = \frac{1}{2}\frac{\hbar}{\sqrt{kM}} \simeq 5\,10^{-19}\,\text{m}^2.$$

2.30 A particle of mass $M = 10^{-25}$ kg moves in a 3-dimensional isotropic harmonic potential of elastic constant $k = 10$ N/m. How many states have energy less than 2×10^{-2} eV?

Answer: $E_{n_x,n_y,n_z} = \hbar\sqrt{k/M}(3/2 + n_x + n_y + n_z)$. Therefore $E_{n_x,n_y,n_z} < 2 \times 10^{-2}$ eV is equivalent to $n_x + n_y + n_z < 1.54$, corresponding to 4 possible states, $(n_x, n_y, n_z) = (0,0,0), (1,0,0), (0,1,0), (0,0,1)$.

2.31 A particle of mass $M = 10^{-26}$ kg moves in one dimension under the influence of an elastic force of constant $k = 10^{-6}$ N/m and of a constant force $F = 10^{-15}$ N acting in the positive x direction. Compute the wave function of the ground state and the corresponding mean value of the coordinate x, given by

$$\langle x \rangle = \frac{\int_{-\infty}^{\infty} dx\, x |\psi(x)|^2}{\int_{-\infty}^{\infty} dx |\psi(x)|^2} .$$

Answer: As in the analogous classical case, the problem can be brought back to a simple harmonic oscillator with the same mass and elastic constant by a simple change of variable, $y = x - F/K$, which is equivalent to shifting the equilibrium position of the oscillator. Hence the energy levels are spaced as for the harmonic oscillator and the wave function of the ground state is

$$\psi(x) = \left(\frac{kM}{\pi^2\hbar^2}\right)^{1/8} e^{-\frac{\sqrt{kM}}{2\hbar}(x-F/k)^2} ,$$

while $\langle x \rangle = F/k = 10^{-9}$ m.

2.32 A particle of mass $m = 10^{-30}$ kg and kinetic energy equal to 50 eV hits a square potential well of width $L = 2 \times 10^{-10}$ m and depth $V = 1$ eV. What is the reflection coefficient computed up to the first non-vanishing order in $\frac{V}{2E}$?

Answer: Let us choose the square well endpoints in $x = 0$ and $x = L$ and fix the potential to zero outside the well. Let ψ_s, ψ_c and ψ_d be respectively the wave functions for $x < 0$, $0 < x < L$ and $x > L$. If the particle comes from the left, then $\psi_s = e^{i\sqrt{2mEx}/\hbar} + a\, e^{-i\sqrt{2mEx}/\hbar}$, $\psi_c = b\, e^{i\sqrt{2m(E+V)}x/\hbar} + c\, e^{-i\sqrt{2m(E+V)}x/\hbar}$ and $\psi_d = d\, e^{i\sqrt{2mEx}/\hbar}$ where a and c are necessarily of order $V/2E$ while b and d are equal to 1 minus corrections of the same order. Indeed, as $V \to 0$ the solution must tend to a single plane wave. By applying the continuity constraints we obtain:

$$1 + a = b + c ,$$

$$1 - a \simeq b - c + \frac{V}{2E} ,$$

$$be^{\frac{i\sqrt{2m(E+V)}L}{\hbar}} + ce^{-i\sqrt{2m(E+V)}L/\hbar} \simeq (b + \frac{V}{2E})e^{i\sqrt{2m(E+V)}L\hbar} - ce^{-i\sqrt{2m(E+V)}L/\hbar},$$

which are solved by $a \simeq \frac{V}{4E}(e^{2i\sqrt{2m(E+V)}L/\hbar} - 1)$ and

$$R = \frac{V^2}{4E^2}\sin^2\left(\frac{\sqrt{2m(E+V)}L}{\hbar}\right) \simeq 0.96 \times 10^{-4}.$$

2.33 A particle of mass $m = 10^{-30}$ kg is confined inside a line segment of length $L = 10^{-9}$ m with reflecting endpoints, which is centered around the origin. In the middle of the line segment a thin repulsive potential barrier, describable as $V(x) = W\delta(x)$, acts on the particle, with $W = 2 \times 10^{-28}$ J m. Compare the ground state of the particle with what found in absence of the barrier.

Answer: Let us consider how the solutions of the Schrödinger equation in a line segment are influenced by the presence of the barrier. Odd solutions, contrary to even ones, do not change since they vanish right in the middle of the segment, so that the particle never feels the presence of the barrier. In order to discuss even solutions, let us notice that they can be written, shifting the origin in the left end of the segment, as $\psi_s \sim \sin(\sqrt{2mE}x/\hbar)$ for $x < L/2$ and $\psi_d \sim \sin(\sqrt{2mE}(L-x)/\hbar)$ for $x > L/2$. Setting $z \equiv \sqrt{2mE}L/(2\hbar)$ the discontinuity in the wave function derivative in the middle of the segment gives $\tan z = -z2\hbar^2/(mLW) \simeq -10^{-1}z$. Hence we obtain, for the ground state, $E \simeq \frac{2\hbar^2}{mL^2}\pi^2(1 - 2 \times 10^{-1}) \simeq 2.75 \times 10^{-19}$ J, slightly below the first excited level.

Notice that, increasing the intensity of the repulsive barrier W from 0 to ∞, the ground level grows from $\pi^2\hbar^2/(2mL^2)$ to $2\pi^2\hbar^2/(mL^2)$, i.e. it is degenerate with the first excited level in the $W \to \infty$ limit. There is no contradiction with the expected non-degeneracy since, in that limit, the barrier acts as a perfectly reflecting partition wall which separates the original line segment into two non-communicating segments: the two degenerate lowest states (as well as all the other excited ones) can thus be seen as two different superpositions (symmetric and antisymmetric) of the ground states of each segment.

2.34 An electron beam corresponding to an electric current $I = 10^{-12}$ A hits, coming from the right, the potential step sketched in the figure. The potential energy diverges for $x < 0$ while it is $-V = -10$ eV for $0 < x < L$ and 0 for $x > L$, with $L = 10^{-11}$ m. The kinetic energy of the electrons is $E_k = 0.01$ eV for $x > L$. Compute the electric charge density as a function of x.

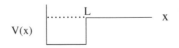

Answer: There is complete reflection in $x = 0$, hence the current density is zero along the whole axis and we can consider a real wave function. In particular we set $\psi(x) = a \sin(\sqrt{2mE}(x-L)/\hbar+\phi)$ for $x > L$ and $\psi(x) = b \sin(\sqrt{2m(E + V)}x/\hbar)$ for $0 < x < L$. Continuity conditions read $b \sin(\sqrt{2m(E + V)}L/\hbar) = a \sin\phi \simeq b \sin(\sqrt{2mV}L/\hbar)$, $b\cos(\sqrt{2m(E + V)}L/\hbar) = \sqrt{E/(E+V)}a \cos\phi \simeq \sqrt{E/V}$ $a \cos\phi \simeq b\cos(\sqrt{2mV}L/\hbar)$ (notice that $\sqrt{2mV}L/\hbar \simeq 0.57$ rad hence $\cos(\sqrt{2mV}L/\hbar) \simeq 0.85$). That fixes $\sqrt{E/V} \tan\left(\sqrt{2mV}L/\hbar\right) \simeq \tan\phi \simeq \phi$ and $b = a\sqrt{E/V}/\cos(\sqrt{2mV}L/\hbar)$, while the incident current fixes the value of a, $I = ea^2\sqrt{2E/m}$. Finally we can write, for the charge density,

$$e\rho = I\sqrt{\frac{m}{2E}} \sin^2\left(\frac{\sqrt{2mE}}{\hbar}(x - L) + \phi\right)$$

for $x > L$ and

$$e\rho \sim I\sqrt{\frac{mE}{2V^2}} \sin^2\left(\frac{\sqrt{2mE}}{\hbar}x\right)$$

for $0 < x < L$.

2.35 Referring to the potential energy given in Problem 2.34, determine the values of V for which there is one single bound state.

Answer: It can be easily realized that any possible bound state of the potential well considered in the problem will coincide with one of the odd bound states of the square well having the same depth and extending from $-L$ to L. The condition for the existence of a single bound state is therefore $\pi/2 < \sqrt{2mV}L/\hbar < 3\pi/2$.

2.36 A ball of mass $m = 0.05$ kg moves at a speed of 3 m/s and without rolling towards a smooth barrier of thickness $D = 10$ cm and height $H = 1$ m. Using the formula for the tunnel effect, give a rough estimate about the probability of the ball getting through the barrier.

Answer: The transmission coefficient is roughly

$$T \sim \exp\left(-\frac{2D}{\hbar}\sqrt{2m(mgH - \frac{mv^2}{2})}\right) \simeq 10^{-1.3 \times 10^{32}}.$$

2.37 What is the quantum of energy for a simple pendulum of length $l = 1$ m making small oscillations?

Answer: In the limit of small oscillations the pendulum can be described as a harmonic oscillator of frequency $\nu = 2\pi\omega = 2\pi\sqrt{g/l}$, where $g \simeq 9.81$ m/s is the gravitational acceleration on the Earth surface. The energy quantum is therefore $h\nu = 3.1 \times 10^{-34}$ J.

2.38 Compute the mean value of x^2 in the first excited state of a harmonic oscillator of elastic constant k and mass m.

Answer: The wave function of the first excited state is $\psi_1 \propto x\, e^{-x^2\sqrt{km/(4\hbar^2)}}$, hence

$$\langle x^2 \rangle = \frac{\int_{-\infty}^{\infty} x^4 e^{-x^2\sqrt{km/\hbar^2}}\,dx}{\int_{-\infty}^{\infty} x^2 e^{-x^2\sqrt{km/\hbar^2}}\,dx} = \frac{3}{2}\sqrt{\frac{\hbar^2}{km}}\,.$$

2.39 A particle of mass M moves in a line segment with reflecting endpoints placed at distance L. If the particle is in the first excited state ($n = 2$), what is the mean quadratic deviation of the particle position from its average value, i.e. $\sqrt{\langle x^2 \rangle - \langle x \rangle^2}$?

Answer: Setting the origin in the middle of the segment, the wave function is $\psi(x) = \sqrt{2/L}\,\sin(2\pi x/L)$ inside the segment and vanishes outside. Obviously $\langle x \rangle = 0$ by symmetry, while

$$\langle x^2 \rangle = \frac{2}{L}\int_{-\frac{L}{2}}^{\frac{L}{2}} x^2 \sin^2\left(\frac{2\pi x}{L}\right)dx = L^2\left(\frac{1}{12} - \frac{1}{8\pi^2}\right)$$

whose square root gives the requested mean quadratic deviation.

2.40 An electron beam of energy E hits, coming from the left, the following potential barrier: $V(x) = \mathcal{V}\delta(x)$ where $\mathcal{V}$ is tuned to $\hbar\sqrt{2E/m}$. Compute the probability density on both sides of the barrier.

Answer: The wave function can be set to $e^{ikx} + a\,e^{-ikx}$ for $x < 0$ and to $b\,e^{ikx}$ for $x > 0$, where $k = \sqrt{2mE}/\hbar$. Continuity and discontinuity constraints for ψ and ψ' in $x = 0$ lead to

$$a = \frac{1}{\frac{ik\hbar^2}{m\mathcal{V}} - 1} = \frac{1}{i-1}\,, \qquad b = \frac{\frac{ik\hbar^2}{m\mathcal{V}}}{\frac{ik\hbar^2}{m\mathcal{V}} - 1} = \frac{1}{i+1}\,.$$

The probability density is therefore $\rho = 1/2$ for $x > 0$, while for $x > 0$ it is $\rho = 3/2 - \sqrt{2}\sin(2kx + \pi/4)$.

2.41 A particle moves in one dimension under the influence of the potential given in Problem 2.34. Assuming that

$$\sqrt{2mV}\frac{L}{\hbar} = \frac{\pi}{2} + \delta\,, \quad \text{with} \quad \delta \ll 1\,,$$

show that, at the first non-vanishing order in δ, one has $B \simeq V\delta^2$, where $B = -E$ and E is the energy of the bound state. Compute the ratio of the probability of the particle being inside the well to that of being outside.

Answer: The depth of the potential is slightly above the minimum for having at least one bound state (see the solution of Problem 2.36), therefore we expect a small binding energy. In particular the equation for the bound state energy, which can be derived by imposing the continuity constraints, is $\cot\left(\sqrt{2m(V-B)}L/\hbar\right) = -\sqrt{B/(V-B)}$. The particular choice of parameters implies $B \ll V$, so that $\cot\left(\sqrt{2m(V-B)}L/\hbar\right) \simeq \cot(\pi/2(1-B/(2V))+\delta) \simeq -\delta + \pi B/(4V) \simeq -\sqrt{B/V}$, hence $B \simeq V\delta^2$. Therefore the wave function is well approximated by $k\sin(\pi x/2L)$ inside the well and by $ke^{-\sqrt{2mB}(x-L)/\hbar} \simeq ke^{-\pi\delta(x-L)/2L}$ outside, where k is a normalization constant. The ratio of probabilities is $\pi\delta/2$: the very small binding energy is reflected in the large probability of finding the particle outside the well.

2.42 A particle of mass $m = 10^{-30}$ kg and kinetic energy $E = 13.9$ eV hits a square potential barrier of width $L = 10^{-10}$ m and height $V = E$. Compute the reflection coefficient R.

Answer: Let us fix in $x = 0$ and in $x = L$ the edges of the square potential barrier, and suppose the particle comes from the left. The wave function is $\psi(x) = e^{ikx} + ae^{-ikx}$ for $x < 0$ and $\psi(x) = d\,e^{ikx}$ for $x > L$, where $k = \sqrt{2mE}/\hbar \simeq 2 \times 10^{10}$ m^{-1}. Instead for $0 \le x \le L$ the wave function satisfies the differential equation $\psi'' = 0$, which has the general integral $\psi(x) = bx + c$. The continuity conditions for ψ and ψ' in $x = 0$ and $x = L$ read

$$1 + a = c; \quad ik(1-a) = b; \quad bL + c = d\,e^{ikL}; \quad b = ik\,d\,e^{ikL}.$$

Dividing last two equations and substituting the first two we get

$$a = \frac{ikL}{ikL - 2}; \quad R = |a|^2 = \frac{k^2L^2}{4 + k^2L^2} \simeq \frac{1}{2}.$$

It is interesting to verify that the same result can be obtained by taking carefully the limit $E \to V$ in (2.80).

2.43 A particle whose wave function is, for asymptotically large negative times (that is $-t \gg m/(\hbar k_0 \Delta)$), a Gaussian wave packet

$$\psi(x,t) = \frac{1}{\sqrt{(2\pi)^{3/2}\Delta}} \int dk\, e^{i(kx - \hbar k^2 t/(2m))} e^{-(k-k_0)^2/(2\Delta)^2}$$

with $k_0/\Delta \gg 1$, interacts in the origin through the potential $V(x) = \mathcal{V}\delta(x)$ and its wave function splits into reflected and transmitted components. Considering values of the time for which the spreading of the packets can be neglected, that is $|t| \ll m/(\hbar\Delta^2)$ (see Sect. 2.4), compute the transmitted and reflected components of the wave packet.

Answer: The Gaussian wave packet has been studied in detail in Sect. 2.4, it is therefore straightforward to check that, for large negative times, the solution that we are seeking is a wave packet centered in $x = vt$, with $v = \hbar k_0/m$, i.e. a packet approaching the barrier from the left and hitting it at $t \simeq 0$.

It has been shown in Sect. 2.5.2 (see also Problem 2.24) that the generic solution of the time independent Schrödinger equation, obtained in the case of a single plane wave e^{ikx} hitting the barrier from the left, is

$$\psi_k(x) = \Theta(-x)[\exp(ikx) - i\,\kappa/(k + i\,\kappa)\exp(-i\,kx)] + \Theta(x)k/(k + i\,\kappa)\exp(i\,kx)$$

where Θ is the step function ($\Theta(x) = 0$ for $x < 0$ and $\Theta(x) = 1$ for $x \geq 0$) and $\kappa = mV/\hbar^2$.

The present problem consists in finding a solution of the time dependent Schrödinger equation which, for asymptotically large negative times and $x < 0$, must be a given superposition of progressive plane waves corresponding to the incoming wave packet. Given the linearity of the Schrödinger equation, the solution must be a linear superposition of the generic solutions given above, with the same coefficients of the incoming packet, i.e.

$$\psi(x, t) = \frac{1}{\sqrt{(2\pi)^{3/2}\Delta}} \int dk\, \psi_k(x)e^{-i\,\hbar k^2 t/(2m)}e^{-(k-k_0)^2/(2\Delta)^2} \,.$$

This decomposes into two components for $x < 0$ and a single transmitted component for $x > 0$.

The first, ingoing component on the negative semi-axis, which corresponds to $\psi_k(x) = \exp(i\,kx)$, is a standard Gaussian packet which, as discussed above, crosses the origin for $t \sim 0$ and hence disappears for larger times. On the contrary, as we shall show in a while, the second, reflected component describes a packet moving backward, which crosses the origin for $t \sim 0$, hence appears as a part of the solution for $x < 0$ for positive times (i.e. after reflection of the original packet), when it must be taken into account. In much the same way we shall compute the transmitted component, which is a packet moving forward and which appears on the positive semi-axis for positive times. Now we work out the details.

We represent the transmitted and reflected wave packets by

$$\frac{1}{\sqrt{(2\pi)^{3/2}\Delta}} \int_{-\infty}^{\infty} dt\, \exp(-F_{T/R}(k, x, t))$$

$$F_T(k, x, t) = (k - k_0)^2/(2\Delta)^2 - i\,(kx - \hbar k^2 t/(2m)) - \ln(k/(k + i\,\kappa))$$

$$F_R(k, x, t) = (k - k_0)^2/(2\Delta)^2 + i\,(kx + \hbar k^2 t/(2m)) - \ln(-i\,\kappa/(k + i\,\kappa)) \,.$$

Then we have the equations:

$$\partial_k F_T(k, x, t) = (k - k_0)/\Delta^2 - i\,(x - \hbar t k/m + \kappa/(k(k + i\,\kappa))) = 0$$

$$\partial_k F_R(k, x, t) = (k - k_0)/\Delta^2 + i\,(x + \hbar t k/m - i/(k + i\,\kappa)) = 0\,.$$

The equation for F_T has three solutions: $k_1 \sim k_0$, $k_2 \sim 0$ and $k_3 \sim -i\,\kappa$ up to corrections of order Δ^2. The first solution has a second derivative of order $1/\Delta^2$, to be compared with the second derivatives of the other two solutions, which are of order $1/\Delta^4$, hence it is the dominant solution, in the same sense discussed in Sect. 2.4, thus we concentrate on it. Setting again $v = \hbar k_0/m$ we have $k_1 = k_0 + i\,\Delta^2(x - vt + \kappa/(k_0(k_0 + i\,\kappa))) + O(\Delta^4)$ and

$$F_T(k_1, x, t) = F_T(k_0, x, t) - \partial_k^2 F_T(k_0, x, t)(k_1 - k_0)^2/2$$

$$= -i\,(k_0 x - \hbar k_0^2 t/(2m)) - \ln(k_0/(k_0 + i\,\kappa)) + [x - vt + \kappa/(k_0(k_0 + i\,\kappa))]^2 \Delta^2/2) + O(\Delta^4)\,.$$

Therefore we have a wave packet centered in $x = vt - \kappa/(k_0^2 + \kappa^2)$. An analogous analysis on the reflected packet gives:

$$F_R(k_1, x, t) = F_R(k_0, x, t) - \partial_k^2 F_R(k_0, x, t)(k_1 - k_0)^2/2$$

$$= i\,(k_0 x + \hbar k_0^2 t/(2m)) - \ln(-i\,\kappa/(k_0 + i\,\kappa)) + (x + vt - i/(k_0 + i\,\kappa))^2 \Delta^2/2) + O(\Delta^4)\,.$$

Now the packet is centered in $x = -vt + \kappa/(k_0^2 + \kappa^2)$.

The result is almost as anticipated, apart from the fact that the appearance of the transmitted and reflected wave packets is delayed (advanced) with respect to the time the incoming packet hits the potential barrier (well). For large, positive times the particle is in a superposition of reflected and transmitted state, the probability of finding it in one of the two states after a measurement of its position (i.e. the integral of the probability density over the corresponding packets) is given approximately by the reflection or transmission coefficients computed for $k \sim k_0$.

2.44 A particle of mass m moves in one dimension under the influence of the potential

$$V(x) = V_0 \Theta(x) - \mathcal{V}\delta(x)\,.$$

If $\mathcal{V} = 3 \times 10^{-29}$ J m and $m = 10^{-30}$ kg, identify the values of V_0 for which the particle has bound states. Assuming the existence of a bound state whose binding energy is $B \ll V_0$, compute the ratio of the probabilities for the particle to be found on the right and on the left-hand side of the origin.

Answer: The wave function of a bound state with energy $-B$ would be

$$\psi_B(x) = N[\Theta(-x)\exp(\sqrt{2mB}x/\hbar) + \Theta(x)\exp(-\sqrt{2m(B+V_0)}x/\hbar)]$$

where Θ is the step function, N is the normalization factor and the condition $\sqrt{B} + \sqrt{B+V_0} = \sqrt{2m\mathcal{V}}/\hbar$ must be satisfied. Since $\sqrt{V_0} \le \sqrt{B} + \sqrt{B+V_0} < \infty$ the above condition has a solution provided $\sqrt{2m\mathcal{V}}/\hbar \ge \sqrt{V_0}$, hence for $V_0 \le 2m\mathcal{V}^2/\hbar^2 \simeq 1.62 \times 10^{-19}$ J $\simeq 1$ eV. The probabilities for the particle to be found on the right and on the left of the origin are respectively $N^2\hbar/(2\sqrt{2m(B+V_0)})$ and $N^2\hbar/(2\sqrt{2mB})$, their ratio for small B is $\sqrt{B/V_0} \simeq \sqrt{2m/V_0}\mathcal{V}/\hbar - 1$.

2.45 A particle of mass m is bound between two spherical perfectly reflecting walls of radii R and $R + \Delta$. The potential energy between the walls is $V_0 = -\hbar^2\pi^2/(2m\Delta^2)$. If the total energy of the particle cannot exceed $E_M = 6\hbar^2/(2mR^2)$ compute, in the $\Delta \to 0$ limit in which the particle is bound on the sphere of radius R, the maximum possible value of its superficial probability density on the intersection point of the sphere with the positive z axis.

Answer: In the $\Delta \to 0$ limit, the radial Schrödinger equation tends to the one-dimensional Schrödinger equation of a particle between two reflecting walls with potential energy between the walls equal to $\bar{V} = -\hbar^2\pi^2/(2m\Delta^2) + l(l+1)\hbar^2/(2mR^2)$. Therefore the possible energy values are $E_{n,l} = (n^2-1)\hbar^2\pi^2/(2m\Delta^2) + l(l+1)\hbar^2/(2mR^2)$. Only the energies $E_{1,l} = l(l+1)\hbar^2/(2mR^2)$ remain finite as $\Delta \to 0$. In spherical coordinates, the corresponding wave functions are, in the $\Delta \to 0$ limit, $\Psi_{l,m} = \sqrt{2/(\Delta R^2)}\sin(\pi(r-R)/\Delta)Y_{l,m}(\theta, \phi)$.

The harmonic functions $Y_{l,m}$ with $m \neq 0$ are proportional to powers of $x_\pm$, hence they vanish on the z axis, therefore and on account of the energy bound, among the possible solutions, we only consider $\Psi_{l,0}$ for $0 \le l \le 2$. Forgetting the radial dependence which, in the $\Delta \to 0$ limit corresponds to a probability density equal to $\delta(r-R)$, these are $\Psi_{0,0} = 1/(R\sqrt{4\pi})$, $\Psi_{1,0} = \sqrt{3/4\pi}\cos\theta/R$ and $\Psi_{2,0} = \sqrt{5/16\pi}(3\cos^2\theta-1)/R$. The wave function of the particle with the above constraints is written as the linear combination $a_0\Psi_{0,0} + a_1\Psi_{1,0} + a_2\Psi_{2,0}$, with the normalization condition $|a_0|^2 + |a_1|^2 + |a_2|^2 = 1$. On the positive z axis the wave function is $1/\sqrt{4\pi}[a_0 + \sqrt{3}a_1 + \sqrt{5}a_2]/R^2$. It is fairly obvious that the maximum absolute value is reached when $a_0 = a_1 = 0$, hence the maximum superficial probability density of the particle is $5/(4\pi R^2)$. The result can be generalized to the case in which different values of the angular momentum can be reached, indeed it can be proved that $|\Psi_{l,0}(\theta = 0)|^2 = (2l+1)/(4\pi R^2)$.

2.46 A particle of mass m moves along the x axis under the influence of the double well potential:

$$V(x) = -\mathcal{V}[\delta(x+L) + \delta(x-L)],$$

with $\mathcal{V} > 0$. Study the solutions of the stationary Schrödinger equation. Since the potential is even under x reflection, the solutions are either even or odd. Show that in the even case there is a single solution for any value of L, discuss the range of values of the binding energy B and, in particular, how B depends on L for small L, i.e. when $\alpha(L) \equiv 2m\mathcal{V}L/\hbar^2 \ll 1$. Compute the "force" between the two wells in this limit. In the odd solution case, compute the range of $\alpha(L)$ for which there are bound solutions and compare the even with the odd binding energies.

Answer: Starting from the even case, we write the solution between the wells as $\psi_I(x) = \cosh kx$, with $B = \hbar^2 k^2/(2m)$, and the external solution as $\psi_E(x) = a \exp(-k(|x| - L))$. The continuity conditions on the wells give: $a = \cosh(kL)$ and $k(\sinh(kL) + a) = \alpha(L)\cosh(kL)/L$. Setting $kL = y$ we get the transcendental equation $\tanh y = \alpha(L)/y - 1$. This equation has a single solution $y(\alpha(L))$, corresponding to a single bound state, for any positive value of $\alpha(L)$. In particular for small $\alpha(L)$ also $y(\alpha(L))$ is small and the equation is approximated by $y - y^3/3 = \alpha(L)/y - 1$ which, up to the second order in $\alpha(L)$, has the positive y solution $y(\alpha(L)) = \alpha(L) - \alpha^2(L)$. The corresponding binding energy is computed noticing that $B = \hbar^2 y^2/(2mL^2)$, from which we have $B = 2m\mathcal{V}^2/\hbar^2 - 8m^2\mathcal{V}^3 L/\hbar^4 + O(\alpha^4(L))$. This implies that there is an attractive force between the two wells which, in the small $\alpha(L)$ limit, is equal to $F = 8m^2\mathcal{V}^3/\hbar^4$, furthermore $B \leq B_{max} = 2m\mathcal{V}^2/\hbar^2$; notice that B_{max} is the binding energy for a single well $-2\mathcal{V}\delta(x)$, which is indeed the limit of $V(x)$ as $L \to 0$. For large $\alpha(L)$ also $y(\alpha(L))$ is large and the transcendental equation is well approximated by $1 - \alpha(L)/y - 1$, which gives $y(\alpha(L)) = \alpha(L)/2$, so that the binding energy reaches its minimum value $B_{min} = m\mathcal{V}^2/(2\hbar^2)$, which coincides with the binding energy of a single well.

In the odd case the solution between the wells becomes $\psi_I(x) = \sinh kx$ while the external one does not change, therefore the transcendental equation becomes $\tanh y = y/(\alpha(L) - y)$. Here the right-hand side is concave downward and positive, while the left-hand side is concave upward and positive for $0 < y < \alpha(L)$, it has a singularity in $\alpha(L)$ and it is negative beyond the singularity. Therefore the equation has a solution for $0 < y < \alpha$ if, and only if, the left-hand side is steeper in the origin than the right-hand side, that is if $\alpha(L) > 1$. For large values of $\alpha(L)$, $y(\alpha(L))$ tends to $\alpha(L)/2$ from below; notice that in the even case the same limit is reached from above.

In conclusion, for any value of the distance between the two wells, there is an even solution, whose binding energy is larger than that of a single well; on the contrary an odd solution exists only if $L > \hbar^2/(2m\mathcal{V})$, with a binding energy lower than that of a single well. In the limit of large separation between the two wells, both the odd and the even level approach the energy of a single well, one from above and the other from below, i.e. we get asymptotically two degenerate levels. The presence of two slightly splitted levels (the even ground state and the odd first excited state) is a phenomenon common to other symmetric double well potentials; an example is given by the Ammonia molecule (NH_3), in which the Nitrogen atom has two symmetric equilibrium positions on both sides of the plane formed by the three Hydrogen atoms.

2.47 A particle of mass m is constrained to move on a plane surface where it is subject to an isotropic harmonic potential of angular frequency ω. Which are the stationary states which are found, for the first excited level, by separation of variables in Cartesian coordinates? Show that the probability current density for such states vanishes. Are there any stationary states belonging to the same level having a non-zero current density? Find those having the maximum possible current density and give a physical interpretation for them.

Answer: For the first excited level, $E_1 = 2\hbar\omega$, the two following stationary states are found in Cartesian coordinates (see Eqs. (2.136) and (2.133)):

$$\psi_{1,0} = \frac{\sqrt{2}\,\alpha^2 x}{\sqrt{\pi}}e^{-\alpha^2(x^2+y^2)/2}; \qquad \psi_{0,1} = \frac{\sqrt{2}\,\alpha^2 y}{\sqrt{\pi}}e^{-\alpha^2(x^2+y^2)/2}$$

where $\alpha = \sqrt{m\omega/\hbar}$. In both cases the current density

$$\boldsymbol{J} = -\frac{i\,\hbar}{2m}\left(\psi^*\boldsymbol{\nabla}\psi - \psi\boldsymbol{\nabla}\psi^*\right) = \frac{\hbar}{m}\mathrm{Im}\left(\psi^*\boldsymbol{\nabla}\psi\right)$$

vanishes, since the wave functions are real. However it is possible to find different stationary states, corresponding to linear combinations of the two states above, having a non-zero current. Indeed for the most general state, which up to an overall irrelevant phase factor can be written as

$$\psi = a\,\psi_{1,0} + \sqrt{1-a^2}e^{i\phi}\psi_{0,1}$$

where $a \in [0, 1]$ is a real parameter, the probability current density is

$$\boldsymbol{J} = \frac{2\hbar\alpha^4}{m\pi}e^{-\alpha^2(x^2+y^2)}\,a\sqrt{1-a^2}\sin\phi\,\boldsymbol{j}$$

where $(j_x, j_y) = (-y, x)$. The current field is independent, up to an overall factor, of the particular state chosen and describes a circular flow around the origin, hence in general a state with a non-zero average angular momentum; moreover $\boldsymbol{\nabla}\cdot\boldsymbol{J} = 0$, as expected for a stationary state. The current vanishes for $a = 0$, 1 or $\phi = 0$, π and is maximum for $a = 1/\sqrt{2}$ and $\phi = \pm\pi/2$, corresponding to the states:

$$\psi_\pm = \frac{1}{\sqrt{2}}\left(\psi_{1,0} \pm i\,\psi_{0,1}\right) = \frac{\alpha^2}{\sqrt{\pi}}e^{-\alpha^2(x^2+y^2)/2}(x \pm i\,y) = \frac{\alpha^2}{\sqrt{\pi}}e^{-\alpha^2 r^2/2}x_\pm$$

where $r^2 = x^2 + y^2$, which are easily recognized as the states having a well defined angular momentum $L = \pm\hbar$.

2.48 Starting from the definition of the corresponding harmonic polynomials given in (2.182) and computing the normalization constants, verify Eq. (2.186) for the first spherical harmonics.

Answer: For $l = 0$: $\mathcal{Y}_{0,0}$ is a constant, after normalization over the solid angle one finds $\mathcal{Y}_{0,0} = \sqrt{\frac{1}{4\pi}}$. For $l = 1$: $\mathcal{Y}_{1,1} \sim -x_+$. The corresponding normalization constant for $Y_{1,1} = -c_{1,1} \sin\theta e^{i\varphi}$ is given by the equation: $1/|c_{1,1}|^2 = 2\pi \int_{-1}^{1} dx(1 - x^2) = 8\pi/3$. This, together with (2.180), fixes $Y_{1,\pm 1}$. The normalization constant for $Y_{1,0} = c_{1,0} \cos\theta$ is given by the equation: $1/|c_{1,0}|^2 = 2\pi \int_{-1}^{1} dx x^2 = 4\pi/3$. The sign is fixed by the condition that $Y_{l,0}(1) > 0$. For $l = 2$: $\mathcal{Y}_{2,2} \sim x_+^2$. The corresponding normalization constant for $Y_{2,2} = c_{2,2} \sin^2\theta e^{2i\varphi}$ is given by the equation: $1/|c_{2,2}|^2 = 2\pi \int_{-1}^{1} dx(1 - x^2)^2 = 32\pi/15$. This, together with (2.180), fixes $Y_{2,\pm 2}$. $\mathcal{Y}_{2,1} \sim zx_+$. The corresponding normalization constant for $Y_{2,1} = c_{2,1} \cos\theta \sin\theta e^{i\varphi}$ is given by the equation: $1/|c_{2,1}|^2 = 2\pi \int_{-1}^{1} dx x^2 (1 - x^2) = 8\pi/15$. This, together with (2.180), fixes $Y_{2,\pm 1}$. The normalization constant for $Y_{2,0} = c_{2,0}(3 \cos^2\theta - 1)$ is given by the equation: $1/|c_{2,0}|^2 = 2\pi \int_{-1}^{1} dx(3x^2 - 1)^2 = 16\pi/5$. The sign is fixed by the condition that $Y_{l,0}(1) > 0$.

2.49 A particle of mass m moves in three dimensions under the influence of the central potential $V(r) = -\hbar^2 \alpha/(2mR) \, \delta(r - R)$, with α positive. Compute the values of α for which the particle has a bound state with non-zero angular momentum.

Answer: For zero angular momentum (S-wave), the solution to the differential equation for the radial wave function $\chi(r)$ defined in (2.187), satisfying the regularity conditions in the origin and at infinity, is equivalent to the odd solution for the one-dimensional double well potential given in Problem 2.46 (setting $L = R$), hence $\chi_< \propto \sinh(kr)$ for $r < R$ and $\chi_> \propto \exp(k(R - r))$ for $r > R$, the bound state energy being $E = -\hbar^2 k^2/(2m)$ with $k > 0$. We know, from Problem 2.46, that such solution exists only if $\alpha > 1$.

We consider now the P-wave case ($l = 1$). The solution satisfying the correct regularity conditions can be obtained from that written in the S-wave case by applying the recursive equation (2.192). It is, up to an overall normalization factor, $\chi_< = \sinh(kr)/(kr) - \cosh(kr)$ for $r < R$ and $\chi_> = a \exp(k(R - r))[1/(kr) + 1]$ for $r > R$. The continuity conditions at $r = R$, written in terms of $kR = x$, are given by $\sinh x/x - \cosh x = a(1 + x)/x$ and $\cosh x/x - \sinh x(1 + x^2)/x^2 + a(1 + 1/x + 1/x^2) = \alpha(\sinh x/x^2 - \cosh x/x)$, from which we have $\tanh x = x(1 + x - x^2/\alpha)/(1 + x + x^3/\alpha)$. We know that the graphs of both sides of this equation cross at most once for $x > 0$ since we have seen in the one-dimensional case, e.g. in Sect. 2.6, that a thin potential well has at most a single bound state. It remains to be verified if they cross. The graphs are tangent to each other in the origin and, for $x \to \infty$, the left-hand side tends to $+1$ and the right-hand side to -1, therefore if the left-hand side is steeper in the origin the graphs do not cross, otherwise they cross once and there is a bound state. Considering the Taylor expansions of both sides we have $\tanh x \simeq x - x^3/3$ and $x(1 + x - x^2/\alpha)/(1 + x + x^3/\alpha) \simeq x - x^3/\alpha$. The conclusion is that there is a bound state if $\alpha > 3$. It should be clear that if no bound state can be found for $l = 1$ (i.e. $\alpha < 3$), none will be found for $l > 1$ as well, because of the increased centrifugal potential.

2.50 Compute the S-wave scattering length for a particle with mass M in the potential well given in Eq. (2.189).

Answer: Introducing the dimensionless parameters $x = \sqrt{2MER}/\hbar$ and $y = \sqrt{2MV_0R}/\hbar$, one has, up to an over all normalization factor, the internal solution, for $r < R$, $\chi_< = \sin(\sqrt{x^2 + y^2}\,r/R)$ and, the external solution, for $r > R$, $\chi_> = a\sin(xr/R + \delta_0)$. The continuity conditions at $r = R$ give $x\tan\sqrt{x^2 + y^2} = \sqrt{x^2 + y^2}\tan(x + \delta_0)$, from which we get

$$x\cot\delta_0 = \frac{1 + x\tan x\tan\sqrt{x^2 + y^2}/\sqrt{x^2 + y^2}}{\tan\sqrt{x^2 + y^2}/\sqrt{x^2 + y^2} - \tan x/x}.$$

The scattering length is apparently equal to $y/(\tan y - y)$. It is positive if $y < \pi/2$, that is in the absence of bound states.

2.51 A particle with mass m moves in three dimensions under the influence of the central potential $V(r) = -\hbar^2\alpha/(2mR)\delta(r - R)$ with α positive. Compute which are the values of α for which the particle has a bound state in S and P waves.

Answer: We note that the S-wave equation gives a bound state if $\alpha > 1$ as we can see comparing with the one dimensional case with a reflecting wall in the origin. Then we consider the solution of the P-wave radial Schrödinger equation. Due to the regularity conditions in the origin and at infinity, the radial wave function defined in Eq. (2.195) is, up to an over all normalization factor, $\chi_< = \sinh(kr)/(kr) - \cosh(kr)$ for $r < R$ and $\chi_> = a\exp(k(R - r))[1/(kr) + 1]$ for $r > R$, the bound state energy being $-\hbar^2k^2/(2m)$ and $k > 0$. The continuity conditions at $r = R$, written in terms of $kr = x$, are given by $\sinh x/x - \cosh x = a(1 + x)/x$ and $\cosh x/x - \sinh x(1 + x^2)/x^2 + a(1 + 1/x + 1/x^2) = \alpha(\sinh x/x^2 - \cosh x/x)$, from which we have $\tanh x = x(1 + x - x^2/\alpha)/(1 + x + x^3/\alpha)$. The graphs of both sides of this equation are tangent to each other in the origin and, for $x \to \infty$, the left-hand side tends to $+1$ and the right-hand side to -1, therefore if the left-hand side is steeper in the origin the graphs do not cross, otherwise they cross once and there is a bound state. Considering the Taylor expansions of both sides we have $\tanh x \simeq x - x^3/3$ and $x(1 + x - x^2/\alpha)/(1 + x + x^3/\alpha) \simeq x - x^3/\alpha$. The conclusion is that there is a bound state if $\alpha > 3$.

2.52 A particle with mass m moves in three dimensions under the influence of the central potential $V(r) = -\hbar^2\alpha/(2mR)\delta(r - R)$. Compute the S-wave phase shift comparing the case of α positive with that of α negative.

Answer: Denoting the solutions as in Problem 2.51 one has, up to an over all normalization factor, $\chi_<(r) \sim \sin kr$, $\chi_>(r) \sim a\sin(kr + \delta_0)$. The (dis-)continuity relations give, for $x = kR$, $x\cot\delta_0 = (2x^2/\alpha - x\sin(2x))/2\sin^2 x$. The solution of physical interest corresponds to δ_0 vanishing with x, thus $\delta_0 = \alpha x/(1 - \alpha) + O(x^2)$. Therefore in the present case the scattering length is equal to $\alpha R/(1 - \alpha)$, which is negative either with $\alpha > 1$, or with α negative, otherwise the scattering length is positive. If the scattering length is positive the low energy phase shift increases with the energy, otherwise it decreases. It is worth recalling here that the potential has a bound state with $l = 0$ for $\alpha > 1$. Thus we see that, when there is a bound state, the scattering length is negative.

If the equation $2x/\alpha = \sin 2x$ has solutions, these correspond to energies for which $\delta_0 = \pi/2 \pm n\pi$ and hence $\sin^2 \delta_0$ reaches its maximum value. If $1/\alpha = 1 - \epsilon$ with ϵ and x small, from Eq. (2.233) we have that the S-wave contribution to the total cross section is

$$\sigma_S \simeq \frac{4\pi R^2}{x^2 + \epsilon^2} \, .$$

For $1 > \alpha > 0$ the phase shift starts increasing without reaching π and vanishes at high energy. If α is negative the phase shift starts decreasing but for large enough energy goes back to zero.

2.53 A particle with mass m moves in three dimensions under the influence of the central potential $V(r) = -\hbar^2 y^2/(2mR^2)\Theta(R - r) + \hbar^2\alpha/(2mR)\delta(r - R)$ with $y = \pi - \epsilon$, $\alpha = \pi/\epsilon - 1/2$ and $\epsilon \ll 1$. Compute the S-wave phase shift in the energy range $kR = O(\epsilon)$.

Answer: Denoting the internal and external solutions as in Problems 2.51 and 2.52, and introducing the variable $x = kR$ we have, up to an over all normalization factor, $\chi_<(r) \sim \sin(\sqrt{y^2 + x^2}r/R)$, $\chi_>(r) \sim a\sin(xr/R + \delta_0)$. The (dis-)continuity relations give $\sqrt{y^2 + x^2}\cot\sqrt{y^2 + x^2} = x\cot(x + \delta_0) - \alpha$. After short calculations we get:

$$x\cot\delta_0 = \frac{\sqrt{y^2 + x^2}\cot\sqrt{y^2 + x^2} + \alpha + x\tan x}{1 - \tan x(\sqrt{y^2 + x^2}\cot\sqrt{y^2 + x^2} + \alpha)/x} \, ,$$

and, with the given choice of the parameters, $\sqrt{y^2 + x^2}\cot\sqrt{y^2 + x^2} + \alpha = (1 - x^2/\epsilon^2)/2 + O(\epsilon^2)$. Thus $x\cot\delta_0 \simeq (1 - x^2/\epsilon^2)/(1 + x^2/\epsilon^2)$. The corresponding S-wave cross section is

$$\sigma_0(x) = 4\pi R^2 \frac{(1 + x^2/\epsilon^2)^2}{x^2(1 + x^2/\epsilon^2)^2 + (1 - x^2/\epsilon^2)^2} \, .$$

For $x = 0$ we have $\sigma_0(0) \simeq 4\pi R^2$ and the cross section has a sharp maximum for $x = \epsilon$, indeed $\sigma(\epsilon) = 4\pi R^2/\epsilon^2$. This is usually called a resonance. The phase shift grows and crosses $\pi/2$, while the cross section reaches its maximum possible value, λ^2/π.

2.54 A particle with mass m moves in three dimensions under the influence of the central potential $V(r) = -\hbar^2 y^2/(2mR^2)\Theta(R - r) + \hbar^2\alpha/(2mR)\delta(r - R)$ with $y = \pi - \epsilon$, $\alpha = \pi/\epsilon - 1/2$ and $\epsilon \ll 1$. Discuss the existence of bound states in S wave.

Answer: Denoting the internal and external solutions as in Problems 2.51 and 2.52, and introducing the parameter $x = \sqrt{2m\bar{B}R}/\hbar$ we have, up to an over all normalization factor, $\chi_<(r) \sim \sin(\sqrt{y^2 - x^2}r/R)$, $\chi_>(r) \sim a\exp(-xr/R)$. Then, the (dis-)continuity relations give: $\sqrt{y^2 - x^2}\cot\sqrt{y^2 - x^2} = -x - \alpha$. It is not difficult to verify that for large α and $y < \pi$ the above equation has no solution. Once again we find positive scattering length in the absence of bound states.

Chapter 3
Introduction to the Statistical Theory of Matter

In Chap. 2 we have discussed the existence and the order of magnitude of quantum effects, showing in particular their importance for microscopic physics. We have seen that quantum effects are relevant for electrons at energies of the order of the electron-volt, while for the dynamics of atoms in crystals, which have masses three or four order of magnitudes larger, significant effects appear at considerably lower energies, corresponding to low temperatures. Hence, in order to study these effects, a proper theoretical framework is needed for describing systems made up of particles at thermal equilibrium and for deducing their thermodynamic properties from the (quantum) nature of their states.

Boltzmann identified the thermal contact among systems as a series of shortly lasting and random interactions with limited energy exchange. These interactions can be considered as collisions among components of two different systems taking place at the surface of the systems themselves. Collisions generate sudden transitions among the possible states of motion for the parts involved. The sequence of collision processes is therefore analogous to a series of dice casts by which subsequent states of motion are chosen by drawing lots.

It is clear that in these conditions it is not sensible to study the time evolution of the system, since that is nothing but a random succession of states of motion. Instead it makes sense to study the *distribution* of states among those accessible to the system, i.e. the number of times a particular state occurs in **N** different observations. In case of completely random transitions among all the possible states, the above number is independent of the particular state considered and equal to the number of observations **N** divided by the total number of possible states. In place of the distribution of results of subsequent observations we can think of the *distribution of the probability* that the system be in a given state: under the same hypothesis of complete randomness, the probability distribution is independent of the state and equal to the inverse of the total number of accessible states.

However the problem is more complicated if we try to take energy conservation into account. Although the energy exchanged in a typical microscopic collision is very small, the global amount of energy (heat) transferred in a great number of

© Springer International Publishing Switzerland 2016
C.M. Becchi and M. D'Elia, *Introduction to the Basic Concepts
of Modern Physics*, Undergraduate Lecture Notes in Physics,
DOI 10.1007/978-3-319-20630-1_3

interactions can be macroscopically relevant; on the other hand, the total energy of all interacting macroscopic systems must be constant.

The american physicist J. Willard Gibbs proposed a method to evaluate the probability distribution for the various possible states in the case of thermal equilibrium.[1] His method is based on the following points.

(1) Thermal equilibrium is independent of the nature of the heat reservoir, which must be identified with a system having infinite thermal capacity (that is a possible enunciation of the so-called *zero-th principle of thermodynamics*). According to Gibbs, the heat reservoir is a set of $\mathcal{N}$ systems, identical to the one under consideration, which are placed in thermal contact. $\mathcal{N}$ is so great that each heat (energy) exchange between the system and the reservoir, being distributed among all different constituent systems, does not alter their average energy content, hence their thermodynamic state.

(2) Gibbs assumed transitions to be induced by completely random collisions. Instead of following the result of a long series of random transitions among states, thus extracting the probability distribution by averaging over time evolution histories, Gibbs proposed to consider a large number of simultaneous draws and to take the average over them. That is analogous to drawing a large number of dice simultaneously instead of a single die for a large number of times: time averages are substituted by *ensemble averages*. Since in Gibbs scheme the system–reservoir pair (*Macrosystem*) can be identified with the $\mathcal{N} + 1$ identical systems in thermal contact and in equilibrium, if at any time the distribution of the states occupied by the various systems is measured, one has automatically an average over the ensemble and the occupation probability for the possible states of a single system can be deduced. On the other hand, computing the ensemble distribution does not require the knowledge of the state of each single system, but instead that of the number of systems in each possible state.

(3) The macrosystem is isolated and internal collisions induce random changes of its state. However, all possible macrosystem states with the same total energy are assumed to be equally probable. That clearly implies that the probability associated to a given distribution of the $\mathcal{N} + 1$ systems is directly proportional to the number of states of the macrosystem realizing the given distribution: this number is usually called *multiplicity*. If i is the index distinguishing all possible system states, any distribution is fixed by a succession of integers $\{N_i\}$, where N_i is the number of systems occupying state i. It is easy to verify that the multiplicity $\mathcal{M}$ is given by

$$\mathcal{M}(\{N_i\}) = \frac{\mathcal{N}!}{\prod_i N_i!}, \tag{3.1}$$

---

[1] Notice that the states considered by Gibbs in the XIX century were small cells in the space of states of motion (the phase space) of the system, while we shall consider quantum states corresponding to independent solutions of the stationary Schrödinger equation for the system. This roughly corresponds to choosing the volume of Gibbs cells of the order of magnitude of $h^{\mathcal{N}}$, where $\mathcal{N}$ is the number of degrees of freedom of the system.

with the obvious constraint

$$\sum_i N_i = \mathcal{N}.$$ (3.2)

(4) The accessibility criterion for states is solely related to their energy which, due to the limited energy exchange in collision processes, is reduced to the sum of the energies of the constituent systems. Stated otherwise, if E_i is the energy of a single constituent system when it is in state i, disregarding the interaction energy between systems, the total energy of the macrosystem is identified with:

$$E_{\text{tot}} = \sum_i E_i N_i \equiv \mathcal{N}U.$$ (3.3)

Thus U can be identified with the average energy of the constituent systems: it characterizes the thermodynamic state of the reservoir and must therefore be related to its temperature in some way to be determined by computations.

(5) Gibbs identified the probability of the considered system being in state i with:

$$p_i = \frac{\bar{N}_i}{\mathcal{N}},$$ (3.4)

where the distribution $\{\bar{N}_i\}$ is the one having maximum multiplicity among all possible distributions:

$$\mathcal{M}(\{\bar{N}_i\}) \geq \mathcal{M}(\{N_i\}) \quad \forall \ \{N_i\},$$

i.e. that is realized by the largest number of states of the macrosystem. We call p_i the *occupation probability* of state i.

The identification made by Gibbs is justified by the fact that the multiplicity function has only one sharp peak in correspondence of its maximum, whose width $(\Delta\mathcal{M}/\mathcal{M})$ vanishes in the limit of an ideal reservoir, i.e. as $\mathcal{N} \to \infty$. Later on we shall discuss a very simple example, even if not very significant from the physical point of view, corresponding to a system with only three possible states, so that the multiplicity $\mathcal{M}$, given the two constraints in (3.2) and (3.3), will be a function of a single variable, thus allowing an easy computation of the width of the peak.

(6) The analysis of thermodynamic equilibrium described above can be extended to the case in which also the number of particles in each system is variable: not only energy transfer by collisions can take place at the surfaces of the systems, but also exchange of particles of various species (atoms, molecules, electrons, ions and so on). In this case the various possible states of the system are characterized not only by their energy but also by the number of particles of each considered species. We will indicate by $n_i^{(s)}$ the number of particles belonging to species s and present in state i; therefore, besides the energy E_i, there will be as many fixed quantities as the number of possible species characterizing each possible state of the system.

The distribution of states in the macrosystem, $\{N_i\}$, will then be subject to further constraints, besides (3.2) and (3.3), related to the conservation of the total number of particles for each species, namely

$$\sum_i n_i^{(s)} N_i = \bar{n}^{(s)} \mathcal{N} \tag{3.5}$$

for each s. The kind of thermodynamic equilibrium described in this case is very different from the previous one. While in the first case equilibrium corresponds to the system and reservoir having the same temperature ($T_1 = T_2$), in the second case also Gibbs potential $g^{(s)}$ will be equal, for each constrained single particle species separately. In place of $g^{(s)}$ it is usual to consider the quantity known as *chemical potential*, defined as $\mu^{(s)} \equiv g^{(s)}/N_A$, where $N_A = 6.02 \times 10^{23}$ is Avogadro's number.

The distributions corresponding to the two different kinds of equilibrium are named differently. For a purely thermal equilibrium we speak of *Canonical Distribution*, while when considering also particle number equilibrium we speak of *Grand Canonical Distribution*. We will start by studying simple systems by means of the Canonical Distribution and will then make use of the Grand Canonical Distribution for the case of perfect quantum gases.

As the simplest possible systems we shall consider in particular an isotropic three-dimensional harmonic oscillator (*Einstein's crystal*) and a particle confined in a box with reflecting walls. Let us briefly recall the nature of the states for the two systems.

Einstein's Crystal

In this model atoms do not exchange forces among themselves but in rare collisions, whose nature is not well specified and whose only role is that of assuring thermal equilibrium. Atoms are instead attracted by elastic forces towards fixed points corresponding to the vertices of a crystal lattice.

The attraction point for the generic atom is identified by the coordinates ($m_x a$, $m_y a, m_z a$), where m_x, m_y, m_z are relative integer numbers with $|m_i|a < L/2$: L is the linear size and a is the spacing of the crystal lattice, which is assumed to be cubic. To summarize, each atom corresponds to a vector $\boldsymbol{m}$ of components m_x, m_y, m_z.

Hence Einstein's crystal is equivalent to a large number of isotropic harmonic oscillators and can be identified with the macrosystem itself. According to the analysis of the harmonic oscillator made in previous chapter, the *microscopic* quantum state of the crystal is characterized by three non-negative integer numbers ($n_{x,\boldsymbol{m}}, n_{y,\boldsymbol{m}}, n_{z,\boldsymbol{m}}$) for every vertex ($\boldsymbol{m}$). The corresponding energy level is given by

$$E_{n_{x,\boldsymbol{m}}, n_{y,\boldsymbol{m}}, n_{z,\boldsymbol{m}}} = \sum_{\boldsymbol{m}} \hbar \omega \left(n_{x,\boldsymbol{m}} + n_{y,\boldsymbol{m}} + n_{z,\boldsymbol{m}} + \frac{3}{2} \right). \tag{3.6}$$

It is clear that several different states correspond to the same energy level: following the same notation as in Chap. 2, they are called *degenerate*.

In our analysis of the harmonic oscillator we have seen that states described as above correspond to solutions of the stationary Schrödinger equation, i.e. to wave functions depending on time through the phase factor $e^{-iEt/\hbar}$. Therefore the state of the macrosystem would not change in absence of further interactions among the various oscillators, and the statistical analysis would make no sense. If we instead admit the existence of rare random collisions among the oscillators leading to small energy exchanges, then the state of the macrosystem evolves while its total energy stays constant.

The Particle in a Box with Reflecting Walls

In this case the reservoir is made up of $\mathcal{N}$ different boxes, each containing one particle. Energy is transferred from one box to another by an unspecified collisional mechanism acting through the walls of the boxes. We have seen in previous chapter that the quantum states of a particle in a box are described by three positive integer numbers (k_x, k_y, k_z), which are related to the wave number components of the particle and correspond to an energy

$$E_k - \frac{\hbar^2 \pi^2}{2mL^2}[k_x^2 + k_y^2 + k_z^2]. \tag{3.7}$$

3.1 Thermal Equilibrium by Gibbs' Method

Following Gibbs' description given above, let us consider a system whose states are enumerated by an index i and have energy E_i. We are interested in the distribution which maximizes the multiplicity $\mathcal{M}$ defined in (3.1) when the constraints in (3.2) and (3.3) are taken into account. Since $\mathcal{M}$ is always positive, in place of it we can maximize its logarithm

$$\ln \mathcal{M}(\{N_i\}) = \ln \mathcal{N}! - \sum_i \ln N_i!. \tag{3.8}$$

If $\mathcal{N}$ is very large, thus approaching the so-called *Thermodynamic Limit* corresponding to an ideal reservoir, and if distributions corresponding to negligible multiplicity are excluded, we can assume that all N_i's get large as well. In these conditions we are allowed to replace factorials by Stirling formula:

$$\ln N! \simeq N (\ln N - 1). \tag{3.9}$$

If we set $N_i \equiv \mathcal{N} x_i$, then the logarithm of the multiplicity is approximately

$$\ln \mathcal{M}(\{N_i\}) \simeq -\mathcal{N} \sum_i x_i (\ln x_i - 1), \tag{3.10}$$

and the constraints in (3.2) and (3.3) are rewritten as:

$$\sum_i x_i = 1,$$

$$\sum_i E_i x_i = U. \tag{3.11}$$

In order to look for the maximum of expression (3.10) in the presence of the constraints (3.11), it is convenient to apply the method of Lagrange's multipliers.

Let us remind that the stationary points of the function $F(x_1, \ldots, x_n)$ in the presence of the constraints: $G_j(x_1, \ldots, x_n) = 0$, with $j = 1, \ldots, k$ and $k < n$, are solutions of the following system of equations:

$$\frac{\partial}{\partial x_i}\left[F(x_1, \ldots, x_n) + \sum_{j=1}^k \lambda_j G_j(x_1, \ldots, x_n) \right] = 0, \quad i = 1, \ldots, n,$$

and of course of the constraints themselves. Therefore we have $n + k$ equations in $n + k$ variables x_i, $i = 1, \ldots, n$, and λ_j, $j = 1, \ldots, k$. In the generic case, both the unknown variables x_i and the multipliers λ_j will be determined univocally. In our case the system reads:

$$\frac{\partial}{\partial x_i}\left[\ln \mathcal{M} - \beta \mathcal{N}(\sum_j E_j x_j - U) + \alpha \mathcal{N}(\sum_j x_j - 1) \right]$$

$$= -\mathcal{N}\frac{\partial}{\partial x_i}\sum_j \left[x_j(\ln x_j - 1) + \beta E_j x_j - \alpha x_j \right]$$

$$= -\mathcal{N}\left[\ln x_i + \beta E_i - \alpha \right] = 0 \tag{3.12}$$

where $-\beta$ and α are the Lagrange multipliers which can be computed by making use of (3.11).

Taking into account (3.4) and the discussion given in the introduction to this Chapter, we can identify the variables x_i solving system (3.12) with the occupation probabilities p_i in the Canonical Distribution, thus obtaining

$$\ln p_i + 1 + \beta E_i - \alpha = 0, \tag{3.13}$$

hence

$$p_i = e^{-1-\beta E_i+\alpha} \equiv k\, e^{-\beta E_i}, \tag{3.14}$$

where β must necessarily be positive in order that the sums in (3.11) be convergent. The constraints give:

$$p_i = \frac{e^{-\beta E_i}}{\sum_j e^{-\beta E_j}} = -\frac{1}{\beta} \frac{d}{dE_i} \ln \sum_j e^{-\beta E_j} \equiv -\frac{1}{\beta} \frac{d}{dE_i} \ln Z,$$

$$U = \sum_i E_i p_i = -\frac{d}{d\beta} \ln \sum_j e^{-\beta E_j} \equiv -\frac{d}{d\beta} \ln Z, \qquad (3.15)$$

where we have introduced the function

$$Z \equiv \sum_j e^{-\beta E_j}, \qquad (3.16)$$

which is known as the *partition function*.

The second of Eq. (3.15) expresses the relation between the Lagrange multiplier β and the average energy U, hence implicitly between β and the equilibrium temperature. It can be easily realized that in fact β is a universal function of the temperature, which is independent of the particular system under consideration.

To show that, let us consider the case in which each component system S can be actually described in terms of two independent systems s and s', whose possible states are indicated by the indices i and a corresponding to energies e_i and ϵ_a. The states of S are therefore described by the pair (a, i) corresponding to the energy:

$$E_{a,i} = \epsilon_a + e_i.$$

If we give the distribution in terms of the variables $x_{a,i} \equiv N_{a,i}/\mathcal{N}$ and we repeat previous analysis, we end up with searching for the maximum of

$$\ln \mathcal{M}(\{N_{a,i}\}) \simeq -\mathcal{N} \sum_{b,j} x_{b,j} (\ln x_{b,j} - 1), \qquad (3.17)$$

constrained by:

$$\sum_{b,j} x_{b,j} = 1,$$

$$\sum_{b,j} (\epsilon_b + e_j) x_{b,j} = U. \qquad (3.18)$$

Following previous analysis we finally find:

$$p_{a,i} = -\frac{1}{\beta} \frac{d}{dE_{a,i}} \ln Z,$$

$$U = -\frac{d}{d\beta} \ln Z, \qquad (3.19)$$

but Z is now given by:

$$Z = \sum_{b,j} e^{-\beta(\epsilon_b + e_j)} = \sum_b \sum_j e^{-\beta\epsilon_b} e^{-\beta e_j} = \sum_b e^{-\beta\epsilon_b} \sum_j e^{-\beta e_j} = Z_s Z_{s'} \quad (3.20)$$

so that the occupation probability factorizes as follows:

$$p_{a,i} = \frac{e^{-\beta(\epsilon_a + e_i)}}{Z} = \frac{e^{-\beta\epsilon_a}}{Z_s} \frac{e^{-\beta e_i}}{Z_{s'}} = p_a \, p_i.$$

We have therefore learned that the two systems have independent distributions, but corresponding to the same value of β: that is a direct consequence of having written a single constraint on the total energy (leading to a single Lagrange multiplier β linked to energy conservation), and on its turn this is an implicit way of stating that the two systems are in contact with the same heat reservoir, i.e. that they have the same temperature: hence we conclude that β is a universal function of the temperature, $\beta = \beta(T)$. We shall explicitly exploit the fact that $\beta(T)$ is independent of the system under consideration, by finding its exact form through the application of Gibbs method to systems as simple as possible.

3.1.1 Einstein's Crystal

Let us consider a little cube with an edge of length L and, following Gibbs method, let us put it in thermal contact with a great (infinite) number of similar little cubes, thus building an infinite crystal corresponding to the macrosystem, which is therefore imagined as divided into many little cubes. Actually, since by hypothesis single atoms do not interact with each other but in very rare thermalizing collisions, we can consider the little cube so small as to contain a single atom, which is then identified with an isotropic harmonic oscillator: we shall obtain the occupation probability of its microscopic states at equilibrium and evidently the properties of a larger cube can be deduced by combining those of the single atoms independently.

We recall that the microscopic states of the oscillator are associated with a vector n having integer non-negative components, the corresponding energy level being given in (2.135). We can then easily compute the partition function of the single oscillator:

$$Z_o = \sum_{n_x=0}^{\infty} \sum_{n_y=0}^{\infty} \sum_{n_z=0}^{\infty} e^{-\beta\hbar\omega(n_x+n_y+n_z+3/2)} = e^{-3\beta\hbar\omega/2} \left[\sum_{n=0}^{\infty} \left(e^{-\beta\hbar\omega} \right)^n \right]^3$$

$$= \left[\frac{e^{-\beta\hbar\omega/2}}{1 - e^{-\beta\hbar\omega}} \right]^3 = \left[\frac{e^{\beta\hbar\omega/2}}{e^{\beta\hbar\omega} - 1} \right]^3. \quad (3.21)$$

The average energy is then:

$$U = -\frac{\partial}{\partial \beta} \ln Z_o = -3 \frac{\partial}{\partial \beta} \ln \frac{e^{\beta \hbar \omega / 2}}{e^{\beta \hbar \omega} - 1} = \frac{3 \hbar \omega}{2} \left(\frac{e^{\beta \hbar \omega} + 1}{e^{\beta \hbar \omega} - 1} \right). \quad (3.22)$$

Approaching the classical limit, in which $\hbar \to 0$, this result gives direct information on β. Indeed in the classical limit Dulong–Petit's law must hold, stating that $U = 3kT$ where k is Boltzmann's constant. Instead, since $e^{\beta \hbar \omega} \simeq 1 + \beta \hbar \omega$ as $\hbar \to 0$, our formula tells us that in the classical limit we have $U \simeq 3/\beta$, which is also the result we would have obtained by directly applying Gibbs method to a classical harmonic oscillator (see Problem 3.12). Therefore we must set

$$\beta = \frac{1}{kT}.$$

This result will be confirmed later by studying the statistical thermodynamics of perfect gases. The specific heat, defined as

$$C = \frac{\partial U}{\partial T},$$

can then be computed through (3.22):

$$C = \frac{\partial \beta}{\partial T} \frac{\partial U}{\partial \beta} = -\frac{1}{kT^2} \frac{3}{2} \hbar^2 \omega^2 e^{\beta \hbar \omega} \left[\frac{1}{e^{\beta \hbar \omega} - 1} - \frac{e^{\beta \hbar \omega} + 1}{\left(e^{\beta \hbar \omega} - 1 \right)^2} \right]$$
$$= \frac{3 \hbar^2 \omega^2}{kT^2} \frac{e^{\hbar \omega / kT}}{\left(e^{\hbar \omega / kT} - 1 \right)^2}. \quad (3.23)$$

Setting $x = kT / \hbar \omega$, the behavior of the atomic specific heat is shown in Fig. 3.1. It is clearly visible that when $x \geq 1$ Dulong–Petit's law is reproduced within a very good approximation. The importance of Einstein's model consists in having given the first qualitative explanation of the violations of the Dulong–Petit law at low temperatures, in agreement with experimental measurements showing atomic specific heats systematically below $3k$. Einstein was the first showing that the specific heat vanishes at low temperatures, even if failing in predicting the exact behavior: that is cubic in T in insulators and linear in conductors (if the superconducting transition is not taken into account), while Einstein's model predicts C vanishing like $e^{-\hbar \omega / kT}$. The different behaviour can be explained, in the insulator case, through the fact that the hypothesis of single atoms being independent of each other, which is at the basis of Einstein's model, is far from being realistic. As a matter of fact, atoms move under the influence of forces exerted by nearby atoms, so that the crystal lattice itself is elastic and not rigid, as assumed in the model. A refinement to Einstein's model was proposed, a few years later, by P. Debye who, instead of considering non-interacting harmonic oscillators, decomposed the motion of the whole crystal in

Fig. 3.1 A plot of the atomic specific heat in Einstein's model in k units as a function of $x = \hbar\omega/kT$, showing the vanishing of the specific heat at low temperatures and its asymptotic agreement with Dulong–Petit's value $3k$

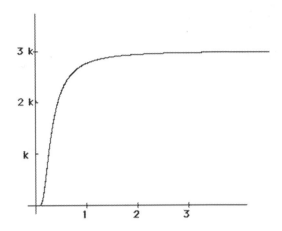

normal oscillation modes, corresponding to different frequencies ω_n, ranging from the fundamental frequency of the crystal (which depends on its macroscopic geometry) to the highest possible frequency (which depends on its internal geometry): in this way the various oscillation modes are excited sequentially as the temperature grows, and an explicit computation shows that in fact the specific heat grows like T^3. In the case of conductors, instead, the linear behavior is due to electrons in the conducting band.

Notwithstanding the wrong quantitative prediction, Einstein's model furnishes the correct qualitative interpretation for the vanishing of the specific heat as $T \rightarrow 0$. Since typical thermal energy exchanges are of the order of kT and since the system can only exchange quanta of energy equal to $\hbar\omega$, we infer that if $kT \ll \hbar\omega$ then quantum effects suppress the energy exchange between the system and the reservoir: the system cannot absorb any energy as the temperature is increased starting from zero, hence its specific heat vanishes. Notice also that quantum effects disappear as $kT \gg \hbar\omega$, when the typical energy exchange is much larger then the energy level spacing of the harmonic oscillator: energy quantization is not visible any more and the system behaves as if it were a classical oscillator.

Therefore we learn whether quantum effects are important or not from the comparison between the quantum energy scale of the system and the thermal energy scale, hence from the parameter $\hbar\omega/kT$.

3.1.2 The Particle in a Box with Reflecting Walls

In this case Gibbs reservoir is made up of $\mathcal{N}$ boxes of size L. The state of the particle in the box is specified by a vector with positive integer components k_x, k_y, k_z corresponding to the energy given in (3.7). The partition function of the system is therefore given by

$$Z = \sum_{k_x=1}^{\infty} \sum_{k_y=1}^{\infty} \sum_{k_z=1}^{\infty} e^{-\beta \frac{\hbar^2 \pi^2}{2mL^2}(k_x^2+k_y^2+k_z^2)} = \left(\sum_{k=1}^{\infty} e^{-\beta \frac{\hbar^2 \pi^2}{2mL^2}k^2} \right)^3. \tag{3.24}$$

For large values of β, hence for small temperatures, we have:

$$Z \simeq \left(e^{-\beta \frac{\hbar^2 \pi^2}{2mL^2}} \right)^3, \tag{3.25}$$

because the first term in the series on the right hand side of (3.24) dominates over the others. Instead for large temperatures, noticing that the quantity which is summed up in (3.24) changes very slowly as a function of k, we can replace the sum by an integral:

$$Z = \left(\sum_{k=1}^{\infty} e^{-\beta \frac{\hbar^2 \pi^2}{2mL^2}k^2} \right)^3 \simeq \left(\int_0^{\infty} dk \, e^{-\beta \frac{\hbar^2 \pi^2}{2mL^2}k^2} \right)^3$$

$$= \left(\frac{2mL^2}{\beta \pi^2 \hbar^2} \right)^{\frac{3}{2}} \left(\int_0^{\infty} dx e^{-x^2} \right)^3 = \left(\frac{m}{2\beta\pi} \right)^{\frac{3}{2}} \frac{L^3}{\hbar^3} = \left(\frac{m}{2\beta\pi\hbar^2} \right)^{\frac{3}{2}} V. \tag{3.26}$$

We conclude that while at low temperatures $(\beta \frac{\hbar^2 \pi^2}{2mL^2} \gg 1)$ the mean energy tends to

$$U \to -\frac{d}{d\beta} \left(-3\beta \frac{\hbar^2 \pi^2}{2mL^2} \right) = 3 \frac{\hbar^2 \pi^2}{2mL^2}, \tag{3.27}$$

i.e. to the energy of the ground state of the system, at high temperatures $(\beta \frac{\hbar^2 \pi^2}{2mL^2} \ll 1)$ we have

$$U \to -\frac{d}{d\beta} \left[\ln V + \frac{3}{2} \left(\ln m - \ln \beta - \ln(2\pi\hbar^2) \right) \right] = \frac{3}{2\beta} = \frac{3}{2}kT. \tag{3.28}$$

This confirms that $\beta = \frac{1}{kT}$, since the system under consideration corresponds to a perfect gas containing a single particle, whose mean energy in the classical limit is precisely $3kT/2$, if T is the absolute temperature.

3.2 The Pressure and the Equation of State

It is well known that the equation of state for a homogeneous and isotropic system fixes a relation among the pressure, the volume and the temperature of the same system when it is at thermal equilibrium. We shall discuss now how the pressure of the system can be computed once the distribution over its states is known.

178

3 Introduction to the Statistical Theory of Matter

The starting point for the computation of the pressure is a theorem which is valid both in classical and quantum mechanics and is known as the *adiabatic theorem*. In the quantum version the theorem makes reference to a system defined by parameters which change very slowly in time (where it is understood that a change in the parameters may also change the energy levels of the system) and asserts that in these conditions the system maintains its quantum numbers unchanged. What is still unspecified in the enunciation above is what is the meaning of *slow*, i.e. with respect to what time scale. We shall clarify that by an example.

Let us consider a particle of mass M moving in a one-dimensional segment of length L with reflecting endpoints. Suppose the particle is in the n-th quantum state corresponding to the energy $E_n = \hbar^2 n^2 \pi^2/(2ML^2)$ (see Eq. (2.107)) and that we slowly reduce the distance L, where slowly means that $|\delta L|/L \ll 1$ in a time interval of the order of $\hbar/\Delta E_n$, where ΔE_n is the energy difference between two successive levels. In this case the adiabatic theorem applies and states that the particle keeps staying in the n-th level as L is changed. That of course means that the energy of the particle increases as we bring the two endpoints closer to each other, $\delta E_n = (\partial E_n/\partial L)\delta L$, and we can interpret this energy variation as the work that must be done to move them. On the other hand, assuming the endpoints to be practically massless, in order to move them slowly we must exactly balance the force exerted on them by the presence of the particle inside, which is the analogous of the pressure in the one-dimensional case. Therefore we obtain the following force:

$$F(n, L) = -\frac{dE_n(L)}{dL} = \frac{\hbar^2 n^2 \pi^2}{ML^3}. \tag{3.29}$$

If we now consider the three-dimensional case of the particle in a box of volume $V = L^3$, for which, according to (2.111), we have:

$$E_{\boldsymbol{n}}(V) = \frac{\hbar^2 |\boldsymbol{n}|^2 \pi^2}{2ML^2} = \frac{\hbar^2 |\boldsymbol{n}|^2 \pi^2}{2MV^{\frac{2}{3}}}, \tag{3.30}$$

we can generalize (3.29) replacing the force by the pressure:

$$P(\boldsymbol{k}, V) = -\frac{dE_{\boldsymbol{k}}(V)}{dV} = -\frac{1}{3L^2}\frac{dE_{\boldsymbol{k}}}{dL} = \frac{2}{3}\frac{E_{\boldsymbol{k}}(V)}{V}. \tag{3.31}$$

Our choice to consider the pressure P instead of the force is dictated by our intention of treating the system without making explicit reference to the specific orientation of the cubic box. The force is equally exerted on all the walls of the box and is proportional to the area of each box, the pressure being the coefficient of proportionality.

Having learned how to compute the pressure when the system is in one particular quantum state, we notice that at thermal equilibrium, being the i-th state occupied with the probability p_i given in (3.15), the pressure can be computed by averaging that obtained for the single state over the Canonical Distribution, thus obtaining:

$$P = -\frac{1}{Z} \sum_i e^{-\beta E_i(V)} \frac{\partial E_i}{\partial V}. \tag{3.32}$$

In the case of a particle in a cubic box, using the result of (3.31), we can write:

$$P = \frac{2}{3V} \sum_{k_x,k_y,k_z=0}^{\infty} \frac{e^{-\beta E_k(V)}}{Z} E_k = \frac{2U}{3V}, \tag{3.33}$$

which represents the equation of state of our system. In the high temperature limit, taking into account (3.28), we easily obtain

$$P = \frac{kT}{V}, \tag{3.34}$$

which coincides with the equation of state of a classical perfect gas made up of a single atom in a volume V.

Starting from the definition of the partition function Z in (3.16), we can translate (3.32) into a formula of general validity:

$$P = \frac{1}{\beta} \frac{\partial \ln Z}{\partial V}. \tag{3.35}$$

3.3 A Three Level System

In order to further illustrate Gibbs method and in particular to verify what already stated about the behavior of the multiplicity function in the limit of an ideal reservoir, $\mathcal{N} \to \infty$ (i.e. that it has only one sharp peak in correspondence of its maximum whose width vanishes in that limit), let us consider a very simple system characterized by three energy levels $E_1 = 0$, $E_2 = \epsilon$ and $E_3 = 2\epsilon$, each corresponding to a single microscopic state. Let $\mathcal{U}$ be the total energy of the macrosystem containing $\mathcal{N}$ copies of the system; it is obvious that $\mathcal{U} \le 2\mathcal{N}\epsilon$. The statistical distribution is fixed by giving the number of copies in each microscopic state, i.e. by three non-negative integer numbers N_1, N_2, N_3 constrained by:

$$N_1 + N_2 + N_3 = \mathcal{N}, \tag{3.36}$$

and by:
$$N_2\epsilon + 2N_3\epsilon = \mathcal{U}. \tag{3.37}$$

the multiplicity of the distribution is

$$\mathcal{M}_{(N_i)} \equiv \frac{\mathcal{N}!}{N_1!N_2!N_3!}. \tag{3.38}$$

The simplicity of the model lies in the fact that, due to the constraints, there is actually only one free variable among N_1, N_2 and N_3 on which the distribution is dependent. In particular we choose N_3 and parametrize it as

$$N_3 = x\mathcal{N}.$$

Solving the constraints given above we have:

$$N_2 = \frac{\mathcal{U}}{\epsilon} - 2x\mathcal{N} \equiv (u - 2x)\mathcal{N},$$

where the quantity $u \equiv \mathcal{U}/(\epsilon\mathcal{N})$ has been introduced, which is proportional to the mean energy of the copies $U = u\epsilon$, and:

$$N_1 = (1 - u + x)\mathcal{N}.$$

The fact that the occupation numbers N_i are non-negative integers implies that x must be greater than the maximum between 0 and $u - 1$, and less than $u/2$. Notice also that if $u > 1$ then $N_3 > N_1$. We must exclude this possibility since, as it is clear from the expression of the Canonical Distribution in (3.15), at thermodynamic equilibrium occupation numbers must decrease as the energy increases. There are however cases, like for instance those encountered in *laser* physics, in which the distributions are really reversed (i.e. the most populated levels are those having the highest energies), but they correspond to situations which are not at thermal equilibrium.

In the thermodynamic limit, $\mathcal{N} \to \infty$, we can also say, neglecting distributions with small multiplicities, that each occupation number N_i becomes very large, so that we can rewrite factorials by using Stirling formula (3.9), and the expression giving the multiplicity as:

$$
\begin{aligned}
\mathcal{M}(x) &\simeq c\frac{\mathcal{N}^{\mathcal{N}}}{(x\mathcal{N})^{x\mathcal{N}}((u-2x)\mathcal{N})^{(u-2x)\mathcal{N}}((1-u+x)\mathcal{N})^{(1-u+x)\mathcal{N}}} \\
&= c\left(x^{-x}(u-2x)^{-(u-2x)}(1-u+x)^{-(1-u+x)}\right)^{\mathcal{N}},
\end{aligned}
\tag{3.39}
$$

where c is a constant, which will not enter our considerations.

The important point in our analysis is that the expression in brackets in (3.39) is positive in the allowed range $0 \le x \le u/2$ and has a single maximum, which is strictly inside that range. To find its position we can therefore study, in place of $\mathcal{M}$, its logarithm

$$\ln \mathcal{M}(x) \simeq -\mathcal{N}\left(x\ln x + (u - 2x)\ln(u - 2x) + (1 - u + x)\ln(1 - u + x)\right),$$

whose derivative is

$$(\ln \mathcal{M}(x))' = -\mathcal{N}\left(\ln x - 2\ln(u - 2x) + \ln(1 - u + x)\right),$$

and vanishes if

$$x(1 - u + x) = (u - 2x)^2. \tag{3.40}$$

Equation (3.40) shows that, in the most likely distribution, N_2 is the geometric mean of N_1 and N_3. Hence there is a number $z < 1$ such that $N_2 = zN_1$ and $N_3 = z^2 N_1$. According to the Canonical Distribution, $z = e^{-\beta\epsilon}$. Equation (3.40) has real solutions:

$$x = \frac{1 + 3u \pm \sqrt{(1 + 3u)^2 - 12u^2}}{6}.$$

The one contained in the allowed range $0 \le x \le u/2$ is

$$x_M = \frac{1 + 3u - \sqrt{(1 + 3u)^2 - 12u^2}}{6}.$$

We can compute the second derivative of the multiplicity in x_M by using the relation:

$$\mathcal{M}'(x) = -\mathcal{N}\left(\ln x - 2\ln(u - 2x) + \ln(1 - u + x)\right)\mathcal{M}(x),$$

and, obviously, $\mathcal{M}'(x_M) = 0$. Taking into account (3.40) we have then:

$$\mathcal{M}''(x_M) = -\mathcal{N}\left(\frac{1}{x_M} + \frac{4}{u - 2x_M} + \frac{1}{1 + x_M - u}\right)\mathcal{M}(x_M)$$

$$= -\mathcal{N}\frac{1 + 3u - 6x_M}{(u - 2x_M)^2}\mathcal{M}(x_M). \tag{3.41}$$

Replacing the value found for x_M we arrive finally to:

$$\frac{\mathcal{M}''(x_M)}{\mathcal{M}(x_M)} = -\mathcal{N}\frac{9\sqrt{(1 + 3u)^2 - 12u^2}}{\left(\sqrt{(1 + 3u)^2 - 12u^2} - 1\right)^2} \le -18\mathcal{N}. \tag{3.42}$$

This result, and in particular the fact that $\mathcal{M}''(x_M)$ is of the order of $-\mathcal{N}\mathcal{M}(x_M)$ as $\mathcal{N} \to \infty$, so that

$$\frac{\mathcal{M}(x_M) - \mathcal{M}(x)}{\mathcal{M}(x_M)} \sim \mathcal{N}(x - x_M)^2,$$

demonstrates that the multiplicity has a maximum whose width goes to zero as $1/\sqrt{\mathcal{N}}$. That is also clearly illustrated in Fig. 3.2. Therefore, in the limit $\mathcal{N} \to \infty$, the corresponding probability distribution tends to a Dirac delta function

$$P(x) \equiv \frac{M(x)}{\int_0^{u/2} dy M(y)} \to \delta(x - x_M).$$

Fig. 3.2 Two plots in
arbitrary units of the
multiplicity distribution
$\mathcal{M}(x)$ of the $\mathcal{N}$-elements
Gibbs ensemble of the three
level model. The comparison
of the distribution for $\mathcal{N} = 1$
(*left*) and $\mathcal{N} = 1000$ (*right*)
shows the predicted fast
reduction of the fluctuations
for increasing $\mathcal{N}$

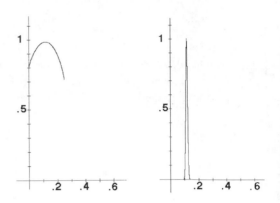

This confirms that the probability is concentrated on a single distribution (a single x
in the present case), which coincides with the most probable one. Even if there are
exceptions to this law, for instance in some systems presenting a critical point (like
the liquid–vapor critical point), that does not regard the systems considered in this
text, so that the equilibrium distribution can be surely identified with the most likely
one.

In Fig. 3.2 the plots of the function given in (3.39) are shown for an arbitrary
choice of the vertical scale. We have set $u = 1/2$ and we show two different cases,
$\mathcal{N} = 1$ (left) and $\mathcal{N} = 1000$ (right).

Making always reference to the three-level system, we notice that the ratios of the
occupation numbers are given by

$$z = e^{-\beta\epsilon} = \frac{x_M}{u - 2x_M}$$
$$= \frac{\sqrt{1 + 6u - 3u^2} + u - 1}{4 - 2u}. \tag{3.43}$$

The plot of z as a function of u is given in the below figure and shows that, in the
range $(0, 1)$, we have $0 \leq z \leq 1$. Hence $\beta \to \infty$ as $u \to 0$ and $\beta \to 0$ as $u \to 1$.

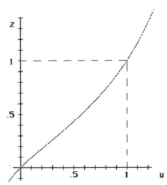

3.4 The Grand Canonical Ensemble and the Perfect Quantum Gas

We shall describe schematically the perfect gas as a system made up of a great number of atoms, molecules or in general particles of the same species, which have negligible interactions among themselves but are subject to external forces. We can consider for instance a gas of particles elastically attracted towards a fixed point, or instead a gas of free particles contained in a box with reflecting walls. We will show detailed computations for the latter case, since it has several interesting applications, but the reader is invited to think about the generalization of the results that will be obtained to the case of different external forces.

The states of every particle in the box can be described as we have done for the single particle in a box, see Sect. 3.1.2. However, classifying the states of many identical particles raises a new problem of quantum nature, which is linked to the *quantum indistinguishability* and to the corresponding statistical choice.

The uncertainty principle is in contrast with the idea of particle trajectory. If a particle is located with a good precision at a given time, then its velocity is highly uncertain and so will be its position at a later time. If two identical particle are located very accurately at a given time t around points r_1 and r_2, their positions at later times will be distributed in a quite random way; if we locate again the particles at time $t + \Delta t$ we could not be able to decide which of the two particles corresponds to the one initially located in r_1 and which to the other.

The fact that the particles cannot be distinguished implies that the probability density for locating the particles in two given points, $\rho(r_1, r_2)$, must necessarily be symmetric under exchange of its arguments:

$$\rho(r_1, r_2) = \rho(r_2, r_1). \tag{3.44}$$

Stated otherwise, indices 1 and 2 refer to the points where the two particles are simultaneously located but in no way identify which particle is located where.

If we consider that also in the case of two (or more) particles the probability density must be linked to the wave functions by the relation:

$$\rho(r_1, r_2) = |\psi(r_1, r_2)|^2,$$

then, taking (3.44) into account, we have that:

$$\psi(r_1, r_2) = e^{i\phi} \psi(r_2, r_1),$$

where ϕ cannot depend on positions since that would change the energy of the corresponding state. A double exchange implies that $e^{2i\phi} = 1$, so that $e^{i\phi} = \pm 1$. Therefore we can state that, in general, the wave function of two identical particles must satisfy the following symmetry relation

$$\psi(r_1, r_2) = \pm \psi(r_2, r_1). \tag{3.45}$$

Since identity among particles is equivalent to the invariance of Schrödinger equation under coordinate exchange, we conclude that Eq. (3.45) is yet another application of the symmetry principle introduced in Sect. 2.6.

Generalizing the same argument to the case of more than two particles, it can be easily noticed that the sign appearing in (3.45) must be the same for all identical particles of the same species. The plus sign applies to photons, to hydrogen and helium atoms, to diatomic molecules made up of identical atoms and to many other particles. There is also a large number of particles for which the minus sign must be used, in particular electrons, protons and neutrons. In general, particles of the first kind are called *bosons*, while particles of the second kind are named *fermions*.

As we have seen at the end of Sect. 2.3, particles may have an internal angular momentum which is called *spin*, whose projection ($\hbar s$) in a given direction, e.g., in the momentum direction, can assume the values $(S - m)\hbar$ where m is an integer such that $0 \leq m \leq 2S$ and S is either integer or half-integer. In the case of particles with non-vanishing mass one has an independent state for each value of m. This is not true for massless particles. Indeed, e.g., for a photon, the spin momentum projection assumes only two possible values ($\pm\hbar$), corresponding to the independent polarizations of light. A general theorem (*spin-statistics theorem*) states that particles carrying half-integer spin are fermions, while those for which S is an integer are bosons.

Going back to the energy levels of a system made up of two particles in a box with reflecting walls, they are given by

$$E = \frac{\pi^2 \hbar^2}{2mL^2} \left[k_{x,1}^2 + k_{y,1}^2 + k_{z,1}^2 + k_{x,2}^2 + k_{y,2}^2 + k_{z,2}^2 \right]. \tag{3.46}$$

The corresponding states are identified by two vectors (*wave vectors*) k_1 and k_2 with positive integer components and, if it applies, by two spin indices s_1 and s_2. Indeed, as we have already said, the generic state of a particle carrying spin is described by a wave function with complex components which can be indicated by $\psi(r, \sigma)$, where σ identifies the single component; in this case $|\psi(r, \sigma)|^2$ represents the probability density of finding the particle around r and in the spin state σ.

Indicating by $\psi_k(r)$ the wave function of a single particle in a box given in (2.109), the total wave function for two particles, which we assume to have well definite spin components s_1 and s_2, should correspond to the product $\psi_{k_1}(r_1)\psi_{k_2}(r_2)\delta_{s_1,\sigma_1}\delta_{s_2,\sigma_2}$, but (3.45) compels us to (anti-)symmetrize the wave function, which can then be written as:

$$\psi(r_1, r_2, \sigma_1, \sigma_2) = N[\psi_{k_1}(r_1)\psi_{k_2}(r_2)\delta_{\sigma_1,s_1}\delta_{\sigma_2,s_2}$$
$$\pm \psi_{k_2}(r_1)\psi_{k_1}(r_2)\delta_{\sigma_1,s_2}\delta_{\sigma_2,s_1}]. \tag{3.47}$$

However that leads to a paradox in case the two wave vectors coincide, $k_1 = k_2$, and the particles are fermions in the same spin state, $s_1 = s_2$. Indeed in this case the minus sign has to be used in (3.47), leading to a vanishing result. The only possible solution to this seeming paradox is *Pauli's Exclusion Principle*, which forbids the presence of two identical fermions having the same quantum numbers (wave vector and spin in the present example).

The identification of the states of two particles which can be obtained by exchanging both wave vectors and spin states suggests that a better way to describe them, alternative to fixing the quantum numbers of the single particles, is that of indicating which combinations (wave vector, spin state) appear in the total wave function, and in case of bosons how many times do they appear: that corresponds to indicating which single particle states (each identified by k and σ) are occupied and by how many particles. In conclusion, the microscopic state of systems made up of many identical particles (quantum gas) can be described in terms of the *occupation numbers* of the quantum states accessible to a single particle: they can be non-negative integers in the case of bosons, while only two possibilities, 0 or 1, are left for fermions. For instance, the wave function in (3.47) can be described in terms of the following occupation numbers:

$$n_{k,\sigma} = \delta_{k,k_1}\delta_{\sigma,s_1} + \delta_{k,k_2}\delta_{\sigma,s_2}.$$

3.4.1 The Perfect Fermionic Gas

According to what stated above, we shall consider a system made up of n identical and non-interacting particles of spin $S = 1/2$, constrained in a cubic box with reflecting walls and an edge of length L. Following Gibbs, the box is supposed to be in thermal contact with $\mathcal{N}$ identical boxes.

The generic microscopic state of the gas, which is indicated with an index i in Gibbs construction, is assigned once the occupation numbers $\{n_{k,s}\}$ are given (with $\{n_{k,s}\} = 0$ or 1) for every value of the wave vector k and of the spin projection $s = \pm 1/2$, with the obvious constraint:

$$\sum_{k,s} n_{k,s} = n. \tag{3.48}$$

The corresponding energy is given by:

$$E_{\{n_{k,s}\}} = \sum_{k,s} \frac{\hbar^2\pi^2}{2mL^2}\left(k_x^2 + k_y^2 + k_z^2\right)n_{k,s} \equiv \sum_{k,s}\frac{\hbar^2\pi^2}{2mL^2}k^2 n_{k,s}. \tag{3.49}$$

Notice that all occupation numbers $n_{k,s}$ refer to the particles present in the specified single particle state: they must not be confused with the numbers describing the distribution of the $\mathcal{N}$ copies of the system in Gibbs method. The partition function of our gas is therefore:

$$Z = \sum_{\{n_{k,s}\}: \sum_{k,s} n_{k,s}=n} e^{-\beta E_{\{n_{k,s}\}}} = \sum_{\{n_{k,s}\}: \sum_{k,s} n_{k,s}=n} e^{-\beta \sum_{k,s} \frac{\hbar^2 \pi^2}{2mL^2} k^2 n_{k,s}}$$

$$= \sum_{\{n_{k,s}\}: \sum_{k,s} n_{k,s}=n} \prod_{k,s} e^{-\beta \frac{\hbar^2 \pi^2}{2mL^2} k^2 n_{k,s}} . \qquad (3.50)$$

The constraint in (3.48) makes the computation of the partition function really difficult. Indeed, without that constraint, last sum in (3.50) would factorize into the product of the different sums over the occupation numbers of the single particle states k, s.

This difficulty can be overcome by relaxing the constraint on the number of particles in each system, keeping only that on the total number n_{tot} of particles in the macrosystem, similarly to what has been done for the energy. Hence the number of particles in the gas, n, will be replaced by the average number $n \equiv n_{tot}/\mathcal{N}$. Also in this case the artifice works well, since the probability of the various possible distributions of the macrosystem is extremely peaked around the most likely distribution, so that the number of particles in each system has negligible fluctuations with respect to the average number $\bar{n}$.

This artifice is equivalent to replacing the Canonical Distribution by the *Grand Canonical* one. In practice, the reflecting walls of our systems are given a small permeability, so that they can exchange not only energy but also particles. In the general case, the Grand Canonical Distribution refers to systems made up of several different particle species, however we shall consider the case of a single species in our computations.

Going along the same lines of the construction given in Sect. 3.1, we notice that the generic state of the system under consideration, identified by the index i, is now characterized by its particle number n_i as well as by its energy E_i. We are therefore looking for the distribution having maximum multiplicity

$$M(\{N_i\}) = \frac{\mathcal{N}!}{\prod_i N_i!},$$

constrained by the total number of considered systems

$$\sum_i N_i = \mathcal{N}, \qquad (3.51)$$

by the total energy of the macrosystem

$$\sum_i N_i E_i = U\mathcal{N} \qquad (3.52)$$

and, as an additional feature of the Grand Canonical Distribution, by the total number of particles of each species. In our case, since we are dealing with a single type of particles, there is only one additional constraint:

$$\sum_i n_i N_i = \bar{n} \, \mathcal{N}.$$ (3.53)

If we apply the method of Lagrange's multipliers we obtain, analogously to what has been done for the Canonical Distribution, see (3.12),

$$\ln \, p_i = -1 + \gamma - \beta \, (E_i - \mu n_i),$$ (3.54)

where we have introduced the new Lagrange multiplier $\beta\mu$, associated with the constraint in (3.53). We arrive finally, in analogy with the canonical case, to the probability in the Grand Canonical Distribution:

$$p_i = \frac{e^{-\beta(E_i-\mu n_i)}}{\sum_j e^{-\beta(E_j-\mu n_j)}} \equiv \frac{e^{-\beta(E_i-\mu n_i)}}{\Xi},$$ (3.55)

where the grand canonical partition function, Ξ, has been implicitly defined.

It can be easily shown that, in the same way as energy exchange (thermal equilibrium) compels the Lagrange multiplier β to be the same for all systems in thermal contact (that has been explicitly shown for the Canonical Distribution), particle exchange forces all systems to have the same *chemical potential* μ for each particle species separately. The chemical potential can be computed through the expression for the average number of particles:

$$\bar{n} = \sum_i n_i \, p_i = \sum_i n_i \, \frac{e^{-\beta(E_i-\mu n_i)}}{\Xi}.$$ (3.56)

Let us now go back to the case of the perfect fermionic gas. The Grand Canonical partition function can be written as:

$$\Xi = \sum_{\{n_{k,s}\}} e^{-\beta\left(E_{\{n_{k,s}\}}-\mu\sum_{k,s} n_{k,s}\right)} = \sum_{\{n_{k,s}\}} e^{-\beta\sum_{k,s}\left(\frac{\hbar^2\pi^2}{2mL^2}k^2-\mu\right)n_{k,s}}$$

$$= \prod_{k,s}\left(\sum_{n_{k,s}=0}^{1} e^{-\beta\left(\frac{\hbar^2\pi^2}{2mL^2}k^2-\mu\right)n_{k,s}}\right) = \prod_{k,s}\left(1+e^{-\beta\left(\frac{\hbar^2\pi^2}{2mL^2}k^2-\mu\right)}\right).$$ (3.57)

Hence, based on (3.55), we can write the probability of the state defined by the occupation numbers $\{n_{k,s}\}$ as

$$p\left(\{n_{k,s}\}\right) = \frac{e^{-\beta\sum_{k,s}\left(\frac{\hbar^2\pi^2k^2}{2mL^2}-\mu\right)n_{k,s}}}{\Xi} = \prod_{k,s}\left(\frac{e^{-\beta n_{k,s}\left(\frac{\hbar^2\pi^2k^2}{2mL^2}-\mu\right)}}{1+e^{-\beta\left(\frac{\hbar^2\pi^2k^2}{2mL^2}-\mu\right)}}\right),$$ (3.58)

which, as can be easily seen, factorizes into the product of the probabilities related to the single particle states: $p\left(\{n_{k,s}\}\right) = \prod_{k,s} p\left(n_{k,s}\right)$, where

$$p\left(n_{k,s}\right) = \frac{e^{-\beta n_{k,s}\left(\frac{\hbar^2 \pi^2 k^2}{2mL^2} - \mu\right)}}{1 + e^{-\beta\left(\frac{\hbar^2 \pi^2 k^2}{2mL^2} - \mu\right)}}. \tag{3.59}$$

Using this result we can easily derive the average occupation number for the single particle state, also known as *Fermi–Dirac distribution*:

$$\bar{n}_{k,s} = \sum_{n_{k,s}=0}^{1} n_{k,s} \, p\left(n_{k,s}\right) = \frac{e^{-\beta\left(\frac{\hbar^2 \pi^2 k^2}{2mL^2} - \mu\right)}}{1 + e^{-\beta\left(\frac{\hbar^2 \pi^2 k^2}{2mL^2} - \mu\right)}} = \frac{1}{1 + e^{\beta\left(\frac{\hbar^2 \pi^2 k^2}{2mL^2} - \mu\right)}}. \tag{3.60}$$

This result can be easily generalized to the case of fermions which are subject to an external force field, leading to single particle energy levels E_α identified by one or more indices α. The average occupation number of the single particle state α is then given by:

$$\bar{n}_\alpha = \frac{1}{1 + e^{\beta(E_\alpha - \mu)}} \tag{3.61}$$

and the chemical potential can be computed making use of (3.56), which according to (3.61) and (3.60) can be written in the following form:

$$\bar{n} = \sum_\alpha \frac{1}{1 + e^{\beta(E_\alpha - \mu)}} = \sum_{k,s} \frac{1}{1 + e^{\beta\left(\frac{\hbar^2 \pi^2 k^2}{2mL^2} - \mu\right)}}, \tag{3.62}$$

where the second equation is valid for free particles.

We have therefore achieved a great simplification in the description of our system by adopting the Grand Canonical construction. This simplification can be easily understood in the following terms. Having relaxed the constraint on the total number of particles in each system, each single particle state can be effectively considered as an independent sub-system making up, together with all other single particle states, the whole system. Each sub-system can be found, for the case of fermions, in only two possible states, with occupation number 0 or 1: its Grand Canonical partition function is therefore trivially given by $1 + \exp(-\beta(E_\alpha - \mu))$, with β and μ being the same for all sub-systems because of thermal and chemical equilibrium. The probability distribution, Eq. (3.59), and the Fermi–Dirac distribution easily follows.

In order to make use of previous formulae, it is convenient to arrange the single particle states k, s according to their energy, thus replacing the sum over state indices by a sum over state energies. With that aim, let us recall that the possible values of k, hence the possible states, correspond to the vertices of a cubic lattice having spacing of length 1. In the below figure we show the lattice for the two-dimensional case. It is clear that, apart from small corrections due to the discontinuity in the distribution of vertices, the number of single particle states having energy less than a given value E is given by

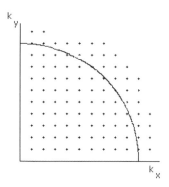

$$n_E = \frac{2}{8} \frac{4\pi \left(\frac{\sqrt{2mE}L}{\pi\hbar}\right)^3}{3} = \frac{\left(\frac{\sqrt{2m}L}{\hbar}\right)^3}{3\pi^2} E^{3/2}, \qquad (3.63)$$

which is equal to the volume of the sphere of radius

$$k = \frac{\sqrt{2mE}L}{\pi\hbar}$$

divided by the number of sectors (which is 8 in three dimensions), since k has only positive components, and multiplied by the number of spin states, e.g., 2 for electrons.

The approximation used above, which improves at fixed particle density $\bar{n}/L^3$ as the volume L^3 increases, consists in considering the single particle states as distributed as a function of their energy in a continuous, instead of discrete, way. On this basis, we can compute the density of single particle states as a function of energy:

$$\frac{dn_E}{dE} = \frac{\sqrt{2m^3}L^3}{\pi^2\hbar^3}\sqrt{E}. \qquad (3.64)$$

Hence we can deduce from (3.60) the distribution of particles as a function of their energy:

$$\frac{d\bar{n}(E)}{dE} = \frac{\sqrt{2m^3}L^3}{\pi^2\hbar^3} \frac{\sqrt{E}}{1 + e^{\beta(E-\mu)}} \qquad (3.65)$$

and replace (3.62) by the following equation:

$$\bar{n} = \int_0^\infty \frac{d\bar{n}(E)}{dE} dE = \int_0^\infty dE \frac{\sqrt{2m^3}L^3}{\pi^2\hbar^3} \frac{\sqrt{E}}{1 + e^{\beta(E-\mu)}}. \qquad (3.66)$$

Equation (3.65) has a simple interpretation in the limit $T \to 0$, i.e. as $\beta \to \infty$. Indeed, in that limit, the exponential in the denominator diverges for all single particle states having energy greater than μ, hence the occupation number vanishes for those states. The exponential instead vanishes for states having energy less that μ, for which the occupation number is one. Therefore the chemical potential in the limit of low temperatures, which is also called *Fermi energy* E_F, can be computed through the equation:

$$n_{E_F} = \frac{\left(\frac{\sqrt{2m}L}{\hbar}\right)^3}{3\pi^2} E_F^{3/2} = \bar{n}. \tag{3.67}$$

Solving for E_F, we obtain:

$$E_F = \mu|_{T=0} = \frac{\hbar^2}{2m} \left(3\pi^2 \rho\right)^{2/3}, \tag{3.68}$$

where $\rho = \bar{n}/L^3$ is the density of particles in the gas.

In order to discuss the opposite limit, in which T is very large ($\beta \to 0$), let us set $z = e^{-\beta\mu}$ and rewrite (3.66) by changing the integration variable ($x = \beta E$):

$$\bar{n} = \frac{\sqrt{2m^3}L^3}{\pi^2 \hbar^3 \beta^{\frac{3}{2}}} \int_0^\infty dx \frac{\sqrt{x}}{1 + ze^x} = \frac{\sqrt{2(mkT)^3}L^3}{\pi^2 \hbar^3} \int_0^\infty dx \frac{\sqrt{x}}{1 + ze^x}. \tag{3.69}$$

We see from this equation that μ must tend to $-\infty$ as $T \to \infty$, i.e. z must diverge, otherwise the right-hand side in (3.69) would diverge like $T^{3/2}$, which is a nonsense since $\bar{n}$ is fixed a priori.

Since at high temperatures z diverges, the exponential in the denominator of (3.60) is much greater than 1, hence (3.65) can be replaced, within a good approximation, by

$$\frac{d\bar{n}(E)}{dE} = \frac{\left(\frac{\sqrt{2m}L}{\hbar}\right)^3}{2\pi^2 z} \sqrt{E} e^{-\beta E} \equiv AL^3 \sqrt{E} e^{-\beta E}. \tag{3.70}$$

The constant A, hence μ, can be computed through

$$\int_0^\infty AL^3 \sqrt{E} e^{-\beta E} dE = 2AL^3 \int_0^\infty x^2 e^{-\beta x^2} dx = -2AL^3 \frac{d}{d\beta} \int_0^\infty e^{-\beta x^2} dx$$

$$= -AL^3 \frac{d}{d\beta} \int_{-\infty}^\infty e^{-\beta x^2} dx = -AL^3 \frac{d}{d\beta} \sqrt{\frac{\pi}{\beta}} = \frac{AL^3}{2} \sqrt{\frac{\pi}{\beta^3}} = \bar{n}. \tag{3.71}$$

We have therefore $A = 2\rho\sqrt{\beta^3/\pi} = e^{\beta\mu}\sqrt{2m^3}/(\pi^2\hbar^3)$, confirming that $\mu \to -\infty$ as $\beta \to 0$ ($\mu \sim \ln\beta/\beta$).

It is remarkable that in the limit under consideration, in which the distribution of particles according to their energy is given by:

$$\frac{d\bar{n}(E)}{dE} = 2\rho\sqrt{\frac{\beta^3}{\pi}}L^3\sqrt{E}e^{-\beta E}, \tag{3.72}$$

Planck's constant has disappeared from formulae. If consequently we adopt the classical formula for the energy of the particles, $E = mv^2/2$, we can find the velocity distribution corresponding to (3.70):

$$\frac{d\bar{n}(v)}{dv} \equiv \frac{d\bar{n}(E)}{dE}\frac{dE}{dv} = \rho\sqrt{\frac{2m^3\beta^3}{\pi}}L^3v^2e^{-\beta mv^2/2}. \tag{3.73}$$

Replacing β by $1/kT$, Eq. (3.73) reproduces the well known Maxwell distribution for velocities, thus confirming the identification $\beta = 1/kT$ made before.

The Fig. 3.3 reproduces the behavior predicted by (3.65) for three different values of kT, and precisely for $kT = 0$, 0.25 and 12.5, measured in the arbitrary energy scale given in the figure, according to which $E_F = 10$. The two curves corresponding to lower temperatures show saturation for states with energy $E < E_F$, in contrast with the third curve which instead reproduces part of the Maxwell distribution and corresponds to small occupation numbers.

Making use of (3.65) we can compute the mean energy U of the gas:

$$U = \int_0^\infty E\frac{d\bar{n}(E)}{dE}dE = \frac{\sqrt{2m^3}L^3}{\pi^2\hbar^3}\int_0^\infty \frac{\sqrt{E^3}}{1+e^{\beta(E-\mu)}}dE, \tag{3.74}$$

obtaining, in the low temperature limit,

$$U = \frac{\sqrt{8m^3}L^3}{5\pi^2\hbar^3}E_F^{5/2} = \frac{(3\bar{n})^{5/3}\pi^{4/3}\hbar^2}{10mL^2} = \frac{(3\bar{n})^{5/3}\pi^{4/3}\hbar^2}{10mV^{2/3}}, \tag{3.75}$$

where $V = L^3$ is the volume occupied by the gas. At high temperatures we have instead:

$$U = 2\rho\sqrt{\frac{\beta^3}{\pi}}L^3\int_0^\infty E^{3/2}e^{-\beta E}dE = 4\rho\sqrt{\frac{\beta^3}{\pi}}L^3\int_0^\infty x^4e^{-\beta x^2}dx$$

$$= \frac{3}{2}\bar{n}\sqrt{\frac{\beta^3}{\pi}}\sqrt{\frac{\pi}{\beta^5}} = \frac{3}{2\beta}\bar{n} \equiv \frac{3}{2}\bar{n}kT. \tag{3.76}$$

This result reproduces what predicted by the classical kinetic theory and in particular the specific heat at constant volume for a gram atom of gas: $C_V = 3/2kN_A \equiv 3/2R$.

In order to compute the specific heat in the low temperature case, we notice first that, for large values of β, Eq.(3.66) leads, after some computations, to $\mu = E_F - O(\beta^{-2})$, hence $\mu = E_F$ also at the first order in T. We can derive (3.74) with respect to T, obtaining:

$$C = k\beta^2 \frac{\sqrt{2m^3}L^3}{\pi^2\hbar^3} \int_0^\infty \frac{\sqrt{E^3}(E-\mu)\,e^{\beta(E-\mu)}}{(1+e^{\beta(E-\mu)})^2}\,dE. \qquad (3.77)$$

For large values of β the exponential factor in the numerator makes the contributions to the integral corresponding to $E \ll \mu$ negligible, while the exponential in the denominator makes negligible contributions from $E \gg \mu$. This permits to make a Taylor expansion of the argument of the integral in (3.77). In particular, if we want to evaluate contributions of order T, taking (3.67) into account we obtain:

$$C \simeq k\beta^2 \frac{3\bar{n}}{8E_F^{\frac{3}{2}}} \int_{-\infty}^\infty \frac{\sqrt{E^3}(E-E_F)}{\left(\cosh\frac{\beta(E-E_F)}{2}\right)^2}\,dE \simeq k\beta^2 \frac{3\bar{n}}{8} \int_{-\infty}^\infty \frac{(E-E_F)}{\left(\cosh\frac{\beta(E-E_F)}{2}\right)^2}\,dE$$

$$+ k\beta^2 \frac{9\bar{n}}{16E_F} \int_{-\infty}^\infty \frac{(E-E_F)^2}{\left(\cosh\frac{\beta(E-E_F)}{2}\right)^2}\,dE = k\bar{n}\frac{9kT}{2E_F} \int_{-\infty}^\infty \left(\frac{x}{\cosh x}\right)^2\,dx$$

$$= k\bar{n}\frac{3\pi^2 kT}{4E_F}, \qquad (3.78)$$

showing that the specific heat has a linear dependence in T at low temperatures.

The linear growth of the specific heat with T for low temperatures can be easily understood in terms of the distribution of particles at low T. In particular we can make reference to (3.65) and to its graphical representation shown in Fig. 3.3: at low

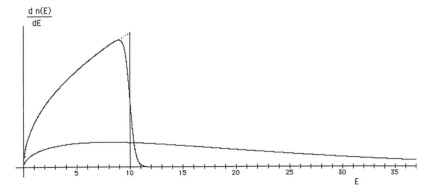

Fig. 3.3 A plot of the Fermi–Dirac energy distribution in arbitrary units for a fermion gas with $E_F = 10$ for $kT = 0$, 0.25 and 12.5. Notice the for the first two values of kT the distribution saturates the Pauli exclusion principle limit, while for $kT = 12.5$ it approaches the Maxwell–Boltzmann distribution

temperatures the particles occupy the lowest possible energy levels, thus saturating the limit imposed by Pauli's principle. In these conditions most of the particles cannot exchange energy with the external environment, since energy exchanges of the order of kT, which are typical at temperature T, would imply transitions of a particle to a different energy level, which however is already completely occupied by other particles. If we refer to the curve corresponding to $kT = 0.25$ in the figure, we see that only particles having energies in a small interval of width kT around the Fermi energy ($E_F = 10$ in the figure), where the occupation number rapidly goes from 1 to 0, can make transitions from one energy level to another, thus exchanging energy with the reservoir and giving a contribution to the specific heat, which is of the order of k for each particle. We expect therefore a specific heat which scales by a factor of the order of kT/E_F with respect to that for the high temperature case: this roughly reproduces the exact result given in (3.78).

Evidently we expect the results we have obtained to apply in particular to the electrons in the conduction band of a metal. It could seem that a gas of electrons be far from being non-interacting, since electrons exchange repulsive Coulomb interactions; however Coulomb forces are largely screened by the other charges present in the metallic lattice, and can therefore be neglected, at least qualitatively, at low energies. We recall that in metals there is one free electron for each atom, therefore, taking iron as an example, which has a mass density $\rho_m \simeq 5 \times 10^3$ kg/m^3 and atomic weight $A \simeq 50$, the electronic density is: $\bar{n}/V = \rho_m(N_A/A)10^3 \sim 6 \times 10^{28}$ particles/m^3. Making use of (3.68) we obtain: $E_F \simeq (10^{-68}/1.8 \times 10^{-30})(3\pi^2 6 \, 10^{28})^{2/3}$ J $\simeq 10^{-18}$ J $\simeq 6$ eV. If we recall that at room temperature $kT \simeq 2.5 \times 10^{-2}$ eV, we see that the order of magnitude of the contribution of electrons to the specific heat is $3 \times 10^{-2}R$ per gram atom, to be compared with $3R$, which is the contribution of atoms according to Dulong and Petit. Had we not taken quantum effects into account, thus applying the equipartition principle, we would have predicted a contribution $3/2R$ coming from electrons. That gives a further confirmation of the relevance of quantum effects for electrons in matter.

Going back to the general case, we can obtain the equation of state for a fermionic gas by using (3.35). From the expression for the energy E_i corresponding to a particular state of the gas given in (3.49) we derive $\partial E_i/\partial V = -2E_i/3V$, hence

$$PV = \frac{2}{3}U , \qquad (3.79)$$

which at high temperatures, where (3.76) is valid, reproduces the well known perfect gas law. Notice that the equation of state in the form given in (3.79) is identical to that obtained in (3.33) and indeed depends only on the dispersion relation (giving energy in terms of momentum) assumed for the free particle states, i.e. the simple form and the factor 2/3 are strictly related to having considered the particles as non-relativistic (see Problem 3.26 for the case of ultrarelativistic particles).

3.4.2 The Perfect Bosonic Gas

To complete our program, let us consider a gas of spinless atoms, hence of bosons; in order to have a phenomenological reference, we shall think in particular of a mono-atomic noble gas like helium. The system can be studied along the same lines followed for the fermionic gas, describing its possible states by assigning the occupation numbers of the single particle states, with the only difference that in this case the wave function must be symmetric in its arguments (the coordinates of the various identical particles), hence there is no limitation on the number of particles occupying a given single particle state. The Grand Canonical partition function for a bosonic gas in a box is therefore

$$
\begin{aligned}
\Xi &= \sum_{\{n_k\}} e^{-\beta\left(E_{\{n_k\}} - \mu \sum_k n_k\right)} = \sum_{\{n_k\}} e^{-\beta \sum_k \left(\frac{\hbar^2 \pi^2}{2mL^2} k^2 - \mu\right) n_k} \\
&= \prod_k \left(\sum_{n_k=0}^{\infty} e^{-\beta\left(\frac{\hbar^2 \pi^2}{2mL^2} k^2 - \mu\right) n_k}\right) \\
&= \prod_k \frac{1}{1 - e^{-\beta\left(\frac{\hbar^2 \pi^2}{2mL^2} k^2 - \mu\right)}},
\end{aligned}
\tag{3.80}
$$

from which the probability of the generic state of the gas follows:

$$
\begin{aligned}
p\left(\{n_k\}\right) &= \frac{e^{-\beta \sum_k \left(\frac{\hbar^2 \pi^2 k^2}{2mL^2} - \mu\right) n_k}}{\Xi} \\
&= \prod_k \left(e^{-\beta n_k \left(\frac{\hbar^2 \pi^2 k^2}{2mL^2} - \mu\right)} \left(1 - e^{-\beta\left(\frac{\hbar^2 \pi^2 k^2}{2mL^2} - \mu\right)}\right)\right),
\end{aligned}
\tag{3.81}
$$

which again can be written as the product of the occupation probabilities relative to each single particle state:

$$
p\left(n_k\right) = e^{-\beta n_k \left(\frac{\hbar^2 \pi^2 k^2}{2mL^2} - \mu\right)} \left(1 - e^{-\beta\left(\frac{\hbar^2 \pi^2 k^2}{2mL^2} - \mu\right)}\right).
\tag{3.82}
$$

The average occupation number of the generic single particle state, k, is thus

$$
\begin{aligned}
\bar{n}_k &= \sum_{n_k=0}^{\infty} n_k \, p\left(n_k\right) = \left(1 - e^{-\beta\left(\frac{\hbar^2 \pi^2 k^2}{2mL^2} - \mu\right)}\right) \sum_{n=0}^{\infty} n \, e^{-\beta n \left(\frac{\hbar^2 \pi^2 k^2}{2mL^2} - \mu\right)} \\
&= \frac{e^{-\beta\left(\frac{\hbar^2 \pi^2 k^2}{2mL^2} - \mu\right)}}{1 - e^{-\beta\left(\frac{\hbar^2 \pi^2 k^2}{2mL^2} - \mu\right)}} = \frac{1}{e^{\beta\left(\frac{\hbar^2 \pi^2 k^2}{2mL^2} - \mu\right)} - 1},
\end{aligned}
\tag{3.83}
$$

which is known as the *Bose–Einstein* distribution. We deduce from last equation that the chemical potential cannot be greater than the energy of the ground state of a single particle, i.e.

$$\mu \leq 3\frac{\hbar^2 \pi^2}{2mL^2},$$

otherwise the average occupation number of that state would be negative. In the limit of large volumes, the ground state of a particle in a box has vanishing energy, hence μ must be negative.

The exact value of the chemical potential is fixed by the relation

$$\bar{n} = \sum_k \bar{n}_k = \sum_k \frac{1}{e^{\beta\left(\frac{\hbar^2 \pi^2 k^2}{2mL^2} - \mu\right)} - 1}. \tag{3.84}$$

For the explicit computation of μ we can make use, as in the fermionic case, of the distribution in energy, considering it approximately as a continuous function:

$$\frac{d\bar{n}(E)}{dE} = \sqrt{\frac{m^3}{2}} \frac{L^3}{\pi^2 \hbar^3} \frac{\sqrt{E}}{e^{\beta(E-\mu)} - 1}. \tag{3.85}$$

Notice that Eq. (3.85) differs from the analogous given in (3.65), which is valid in the fermionic case, both for the sign in the denominator and for a global factor $1/2$ which is due to the absence of the spin degree of freedom.

The continuum approximation for the distribution of the single particle states in energy is quite rough for small energies, where only few levels are present. In the fermionic case, however, that is not a problem, since, due to the Pauli exclusion principle, only a few particles can occupy those levels (2 per level at most in the case of electrons), so that the contribution coming from the low energy region is negligible. The situation is quite different in the bosonic case. If the chemical potential is small, the occupation number of the lowest energy levels can be very large, giving a great contribution to the sum in (3.84). We exclude for the time being this possibility and compute the chemical potential making use of the relation:

$$\bar{n} = \sqrt{\frac{m^3}{2}} \frac{L^3}{\pi^2 \hbar^3} \int_0^\infty dE \frac{\sqrt{E}}{e^{\beta(E-\mu)} - 1} = \sqrt{\frac{(mkT)^3}{2}} \frac{L^3}{\pi^2 \hbar^3} \int_0^\infty dx \frac{\sqrt{x}}{ze^x - 1} \tag{3.86}$$

where again we have set $z = e^{-\beta\mu} \geq 1$. Defining the gas density, $\rho \equiv \bar{n}/L^3$, Eq. (3.86) can be rewritten as:

$$\int_0^\infty dx \frac{\sqrt{x}}{ze^x - 1} = \pi^2 \hbar^3 \rho \sqrt{\frac{2}{(mkT)^3}}. \tag{3.87}$$

On the other hand, recalling that $z \geq 1$, we obtain the following inequality:

$$\int_0^\infty dx \frac{\sqrt{x}}{z e^x - 1} \leq \frac{1}{z} \int_0^\infty dx \frac{\sqrt{x}}{e^x - 1} \leq \int_0^\infty dx \frac{\sqrt{x}}{e^x - 1} \simeq 2.315, \qquad (3.88)$$

which can be replaced in (3.87), giving an upper limit on the ratio $\rho / T^{3/2}$. That limit can be interpreted as follows: for temperatures lower than a certain threshold, the continuum approximation for the energy levels cannot account for the distribution of all particles in the box, so that we must admit a macroscopic contribution coming from the lowest energy states, in particular from the ground state. The limiting temperature can be considered as a critical temperature, and the continuum approximation is valid only if

$$T \geq T_c \simeq 4.38 \frac{\hbar^2 \rho^{2/3}}{m\,k}. \qquad (3.89)$$

As T approaches the critical temperature the chemical potential vanishes and the occupation number of the ground state becomes comparable with $\bar{n}$, hence of macroscopic nature. For temperatures lower than T_c the computation of the total number of particles shown in (3.86) must be rewritten as:

$$\bar{n} = \bar{n}_f + \sqrt{\frac{m^3}{2} \frac{L^3}{\pi^2 \hbar^3}} \int_0^\infty dE \frac{\sqrt{E}}{e^{\beta E} - 1}, \qquad (3.90)$$

where $\bar{n}_f$ refers to the particles occupying the lowest energy states, while the integral over the continuum distribution, in which μ has been neglected, takes into account particles occupying higher energy levels. Changing variables in the integral we obtain:

$$\bar{n} \simeq \bar{n}_f + 2.315 \sqrt{\frac{(mkT)^3}{2} \frac{L^3}{\pi^2 \hbar^3}}, \qquad (3.91)$$

showing that, for $T < T_c$, $\bar{n}_f$ takes macroscopic values, of the order of magnitude of Avogadro's number N_A.

This phenomenon is known as Bose–Einstein condensation. Actually, for the usual densities found in ordinary gases in normal conditions, i.e. $\rho \simeq 10^{25}$ particles/m^3, the critical temperature is of the order of 10^{-2} °K: for this combination of temperature and density, interatomic forces are no more negligible even in the case of helium, so that the perfect gas approximation does not apply. The situation can be completely different at very low densities, indeed Bose–Einstein condensation has been recently observed for temperatures of the order of 10^{-9} °K and densities of the order of 10^{15} particles/m^3.

In the opposite situation, for temperatures much greater than T_c, the exponential clearly dominates in the denominator of the continuum distribution since, analogously to what happens for fermions at high temperatures, one can show that $z \gg 1$.

Hence the -1 term can be neglected, so that the distribution becomes that obtained also in the fermionic case at high temperatures, i.e. the Maxwell distribution.

3.4.3 The Photonic Gas and the Black Body Radiation

We shall consider in brief the case of an electromagnetic field in a box with "almost" completely reflecting walls: we have to give up ideal reflection in order to allow for thermal exchanges with the reservoir. From the classical point of view, the field amplitude can be decomposed in normal oscillation modes corresponding to well defined values of the frequency and to electric and magnetic fields satisfying the well known reflection conditions on the box surface. The modes under consideration, apart from the two possible polarizations of the electric field, are completely analogous to the wave functions of a particle in a box shown in (2.109), that is:

$$\sin\left(\frac{\pi}{L}k_x x\right)\sin\left(\frac{\pi}{L}k_y y\right)\sin\left(\frac{\pi}{L}k_z z\right)\cos\left(\frac{\pi}{L}\sqrt{k_x^2 + k_y^2 + k_z^2}\, ct\right) \qquad (3.92)$$

where, as usual, the integers (k_x, k_y, k_z) define the vector $\boldsymbol{k}$. We have therefore the following frequencies:

$$\nu_k = \frac{c}{2L}|\boldsymbol{k}|. \qquad (3.93)$$

Taking into account the two possible polarizations, the number of modes having frequency less than a given value ν is:

$$n_\nu = \frac{\pi}{3}|\boldsymbol{k}|^3 = \frac{8\pi L^3 \nu^3}{3c^3}, \qquad (3.94)$$

from which the density of modes can be deduced:

$$\frac{dn_\nu}{d\nu} = \frac{8\pi L^3 \nu^2}{c^3}. \qquad (3.95)$$

If the system is placed at thermal equilibrium at a temperature T and we assume equipartition of energy, i.e. that an average energy kT corresponds to each oscillation mode, we arrive to the result found by Rayleigh and Jeans for the energy distribution of the black body[2] radiation as a function of frequency (a quantity which can also be easily measured in the case of an oven):

---

[2]A black body, extending a notion valid for the visible electromagnetic radiation, is defined as an ideal body which is able to emit and absorb electromagnetic radiation of any frequency, so that all oscillation modes interacting with (emitted by) a black body at thermal equilibrium at temperature T can be considered as thermalized at the same temperature.

$$\frac{dU(\nu)}{d\nu} = \frac{8\pi kT}{c^3} L^3 \nu^2. \tag{3.96}$$

This is clearly a paradoxical result, since, integrating over frequencies, we would obtain an infinite internal energy, hence an infinite specific heat. In fact, accurate experimental measurements performed at the end of the 19th century had shown the existence of an exponential suppression of the high energy spectrum, which was later interpreted by Planck based on the hypothesis of quantization of the energy exchanges between atoms and radiation. This hypothesis was then better specified by Einstein who assumed the existence of photons.

Starting from Einstein's hypothesis, Eq. (3.95) can be interpreted as the density of states for a gas of photons, i.e. bosons with energy $E = h\nu$. The density of photons given in (3.64) becomes then:

$$\frac{dn_E}{dE} = \left(\frac{L}{\hbar c}\right)^3 \frac{E^2}{\pi^2}. \tag{3.97}$$

For a gas of photons the collisions with the walls of the box, which thermalize the system, correspond in practice to non-ideal reflection processes in which photons can be absorbed or new photons can be emitted by the walls. Therefore, making always reference to a macrosystem made up of a large number of similar boxes, there is actually no constraint on the total number of particles, hence the chemical potential must vanish.

The distribution law of the photons in energy is then given by:

$$\frac{dn(E)}{dE} = \left(\frac{L}{\hbar c}\right)^3 \frac{E^2}{\pi^2} \frac{1}{e^{\frac{E}{kT}} - 1}. \tag{3.98}$$

From this law we can deduce the distribution in frequency:

$$\frac{dn(\nu)}{d\nu} = \left(\frac{L}{\hbar c}\right)^3 \frac{(h\nu)^2}{\pi^2} \frac{h}{e^{\frac{h\nu}{kT}} - 1} = \frac{8\pi}{c^3} L^3 \frac{\nu^2}{e^{\frac{h\nu}{kT}} - 1} \tag{3.99}$$

and we can finally write the energy distribution of the radiation as a function of frequency by multiplying both sides of (3.99) by the energy carried by each photon:

$$\frac{dU(\nu)}{d\nu} = \frac{8\pi h}{c^3} L^3 \frac{\nu^3}{e^{\frac{h\nu}{kT}} - 1}. \tag{3.100}$$

This distribution was proposed for the first time by Planck and was indeed named after him.

It is evident that at small frequencies Planck distribution is practically equal to that in (3.96). Instead at high frequencies energy quantization leads to an exponential cut in the energy distribution which eliminates the paradox of an infinite internal energy and of an infinite specific heat. We notice that the phenomenon suppressing the high energy modes in the computation of the specific heat is the same leading to a vanishing specific heat for the harmonic oscillator when $kT \ll h\nu$ (the system cannot absorb a quantity of energy less than the minimal quantum $h\nu$) and indeed, as we have already stressed at the end of Sect. 2.7, the radiation field in a box can be considered as an infinite collection of independent harmonic oscillators of frequencies given by the resonant frequencies of the box.

3.5 Gases in the Classical Limit

In the high-temperature limit, when kT becomes much greater than the typical spacing among the energy levels of the system, those can be treated as a continuous spectrum and, in the expression for the canonical partition function given in (3.16), the sum becomes an integral. The measure of the integral corresponds to the density of states. This density has been already considered, e.g., in (3.26) and in (3.64), for ideal free particle gases, that is non-interacting gases, whose energy only depends on momentum. It is apparent from the above mentioned formulae that, for an ideal gas of free, point-like particles, the partition function in the classical limit is written as a momentum integral of the Gibbs factor $\exp(-\beta E)$, where E is the total energy of the system, with the measure

$$d\mu_N(\boldsymbol{p}) = \left(\frac{V}{h^3}\right)^N \prod_{i=1}^{N} d^3 p_i . \tag{3.101}$$

For an ideal gas of N identical particles, the states are still assigned by giving all particles momenta $\boldsymbol{p}_i, i = 1 \ldots, N$, but their density in the N-particle momentum space must be divided by $N!$, because states which are related by permutations of particle momenta must be identified. Thus we have the new measure

$$d\mu_{N,I}(\boldsymbol{p}) = \frac{1}{N!} \left(\frac{V}{h^3}\right)^N \prod_{i=1}^{N} d^3 p_i. \tag{3.102}$$

The question which still needs an answer is how the measure changes in the presence of interaction potentials, i.e. if the energy also depends on the particle positions: in this case one expects an integral over all variables needed to specify the state of the system from a classical point of view, which involve the positions and the momenta (or velocities) of all constituent particles, however we must still define the correct integration measure. Let us start the discussion with a very simple example: the one-dimensional quantum harmonic oscillator defined by the Hamiltonian

$$H = \frac{p^2}{2m} + \frac{m\omega^2 x^2}{2} \tag{3.103}$$

whose energy levels are $\hbar\omega(n + 1/2)$ with n non-negative integer, i.e. they are equally spaced. Therefore, in the limit of high temperatures, $k_B T \gg \hbar\omega$, the partition function corresponds, apart from an irrelevant overall factor, to an integral over a constant density of states

$$Z \sim \int_0^\infty dn e^{-\beta\hbar\omega n} = \frac{1}{\hbar\omega} \int_0^\infty dE e^{-\beta E} = \frac{1}{\beta\hbar\omega}. \tag{3.104}$$

Moreover, it is evident from (3.103) that classical trajectories $H = E$ correspond to ellipses in the x, p plane of semi-axes $\sqrt{2E/(m\omega^2)}$ and $\sqrt{2mE}$, which enclose an area in the x, p plane which is equal to $2\pi E/\omega$. Therefore, from a classical point of view, states which are equally spaced, by $\hbar\omega$, in energy correspond to trajectories which enclose areas in the x, p plane which differ by a fixed amount $2\pi\hbar\omega/\omega = h$. Hence we can make a one-to-one correspondence between physical states and volume elements in x, p space of equal size h, i.e. the classical limit of the partition function can be written as

$$Z = \frac{1}{h} \int_{-\infty}^\infty dx \int_{-\infty}^\infty dp \, e^{-\beta H(x,p)} \tag{3.105}$$

which easily brings to the same result as in (3.104).

The x, p space is usually known as *phase space* and one can show that the result found for the harmonic oscillator is of general nature. At a semi-classical level one can show that, for a generic system described by $3N$ coordinates q_i and $3N$ conjugate momenta p_i, the number of quantum energy levels corresponding to a given portion of phase space $\Gamma = \int \prod_i dq_i dp_i$ is approximately equal to Γ/h^{3N}. We deduce that the correct classical measure for a gas of identical interacting point like particles is

$$d\mu_N(\boldsymbol{p}, \boldsymbol{r}) = \frac{1}{N!} \prod_{i=1}^N \frac{d^3 p_i d^3 r_i}{h^3} , \tag{3.106}$$

as it also emerges from purely dimensional reasons, which suggest to replace the volume V by the space integration measure $d^3 r$.

Actually, phase space emerged as the natural integration space for the foundation of Statistical Mechanics since the early works from Boltzmann and Gibbs in the second half of 19th century, when Quantum Mechanics was far from being discovered; the reason can be explained by recalling a few fundamental properties of Classical Mechanics. If the classical states of particles are identified by points in the phase space, that is, by the conjugate variables P_α and Q_β, with $\alpha, \beta = 1, 2, 3$, the measure $\prod_{i=1}^N dP_{1,i} dP_{2,i} dP_{3,i} dQ_{1,i} dQ_{2,i} dQ_{3,i}$ is invariant under canonical transformations, that is, under redefinitions of the variables keeping unchanged the Hamiltonian form of the dynamical equations. One can show that time evolution itself, defined by the set of Hamilton equations, corresponds to a canonical transformation: in this case the invariance of the phase space integration measure is also known as *Liouville Theorem*. Now, since the whole structure of Statistical Mechanics is based on the principle that all possible microscopic states of the system have equal probabilities a priori, one would like this assumption to be independent of time evolution. By Liouville theorem, this is true only if equal numbers of states are associated with equal portions of phase space: from that it follows that integration over phase space is the correct thing to do. There is an unknown constant, of course, corresponding to the phase space volume associated with a single microscopic state: that remained undefined until Quantum Mechanics fixed it as in (3.106).

In conclusion, disregarding for simplicity the spin of particles, we have that in the classical limit the Canonical partition function of N non-interacting identical particles with mass m in both Fermi and Bose statistics corresponds to

$$
\begin{aligned}
Z_{0,class} &= \frac{1}{N!} \int \prod_{i=1}^{N} \frac{d^3 p_i d^3 r_i}{h^3} e^{-\beta \sum_{i=1}^{N} \frac{p_i^2}{2m}} \\
&\simeq \left(\frac{V}{h^3 N} \int d^3 p \; e^{-\beta \frac{p^2}{2m}} \right)^N = \left(\frac{(2\pi mkT)^{\frac{3}{2}} V}{h^3 N} \right)^N \\
&= \left(\frac{\Phi_0(T) V}{N} \right)^N ,
\end{aligned}
\tag{3.107}
$$

where

$$
\Phi_0(T) = \left(\frac{(2\pi mkT)^{\frac{1}{2}}}{h} \right)^3 ,
\tag{3.108}
$$

and we have replaced $N!$ by its first Stirling approximation N^N. This result, even if written using different symbols, coincides with that given in (3.26).

In the case of interacting gases, if the interaction energy can be expressed by means of a two-body potential, so that the potential energy of the gas is

$$
V(\mathbf{r}_1, \ldots \mathbf{r}_N) = \sum_{i>j=1}^{N} v(|\mathbf{r}_i - \mathbf{r}_j|) ,
\tag{3.109}
$$

the partition function is

$$
\begin{aligned}
Z_{class} &= \frac{1}{N!} \int \prod_{i=1}^{N} \frac{d^3 p_i d^3 r_i}{h^3} e^{-\beta [\sum_{i=1}^{N} \frac{p_i^2}{2m} + \sum_{i>j=1}^{N} v(|\mathbf{r}_i - \mathbf{r}_j|)]} \\
&\simeq \left(\frac{(2\pi mkT)^{\frac{3}{2}}}{h^3 N} \right)^N \int \prod_{i=1}^{N} d^3 r_i \, e^{-\beta \sum_{i>j=1}^{N} v(|\mathbf{r}_i - \mathbf{r}_j|)} .
\end{aligned}
\tag{3.110}
$$

For simplicity we shall assume v bounded from below and vanishing at large inter-particle distances.

In general the calculation of the partition function from Eq. (3.110) is very difficult because it contains an integral with respect to a huge number of variables. Technically one exploits the fact that the function

$$f(r) \equiv e^{-\beta v(r)} - 1, \tag{3.111}$$

is fast vanishing for large r. In terms of $f(r)$ we can write

$$Z_{class} \simeq \left(\frac{(2\pi m k T)^{\frac{3}{2}} V}{h^3 N} \right)^N \left[1 + \frac{N^2}{2V} \int d^3 r f(r) \right.$$
$$\left. + \frac{N^4}{8V^2} \left(\int d^3 r f(r) \right)^2 + \cdots \right]. \tag{3.112}$$

The brackets in the right-hand side contain a power expansion in N/V, where the disregarded terms are increasingly complicated. The expansion is in terms of the particle density $\rho = N/V$ and is called *virial expansion*: for smaller densities interactions become less important, so that the expansion converges faster.

If the two body potential $v(r)$ is the sum of a hard core term[3] and an attractive, negative energy term whose average value is given by C, we can introduce a simple, however rough, approximation of (3.110) setting

$$\int \prod_{i=1}^{N} d^3 r_i e^{-\beta \sum_{i>j=1}^{N} v(|\mathbf{r}_i - \mathbf{r}_j|)} \simeq (V - b)^N e^{\frac{CN^2}{2kTV}}, \tag{3.113}$$

where b is the volume of the hard core of $N - 1$ particles which is excluded from each single particle integral, and C corresponds to the contribution of the attractive part of the potential for each particle pair. We call this approximation the van der Waals approximation because the corresponding equilibrium law, that we shall compute in the next sections, is called the *van der Waals equation*. The approximate partition function is

$$Z_{vdw} \simeq \left(\frac{(2\pi m k T)^{\frac{3}{2}} (V - b)}{h^3 N} e^{\frac{CN}{2kTV}} \right)^N. \tag{3.114}$$

In the following we shall discuss the consequences of Eq. (3.114) and the interpretation of the results. However, before doing this, it is interesting to give a short comparison of the van der Waals approximation to the partition function with its virial expansion. For simplicity we assume the two body potential $v(r)$ constant in the interval $r_h < r < r_a$ and vanishing for r larger than some r_a. If this constant,

---

[3] If the particles were rigid balls of radius r the singular repulsive distance would correspond to the radius $r_h = 2r$.

that we denote by v_a (because, if the potential is constant, its average value coincides with the same constant) is much smaller than kT, We have

$$\int d^3 r f(r) = -\frac{4\pi}{3}(r_h^3 - (e^{\frac{v_a}{kT}} - 1)(r_a^3 - r_h^3)) \sim \frac{C'}{kT} - \frac{2b'}{N} . \qquad (3.115)$$

Therefore, limiting the virial expansion to the first three terms, we get

$$Z_{class} \sim \left(\frac{(2\pi m kT)^{\frac{3}{2}} V}{h^3 N}\right)^N \left[1 + \frac{N^2}{2V}\left(\frac{C'}{kT} - \frac{2b'}{N}\right)\right. \qquad (3.116)$$

$$\left. + \frac{N^4}{8V^2}\left(\frac{C'}{\beta kT} - \frac{2b'}{N}\right)^2 + O\left(\left(N^2/V\right)^3\right)\right] .$$

In the van der Waals approximation, expanding the right-hand side of Eq. (3.114), we have

$$Z_{vdw} \sim \left(\frac{(2\pi m kT)^{\frac{3}{2}} V}{h^3 N}\right)^N \left[1 + \frac{N}{V}\left(\frac{CN}{2kT} - b\right)\right. \qquad (3.117)$$

$$\left. + \frac{N^2}{2V^2}\left(\frac{CN}{2kT} - b\right)^2 + O\left(\left(N^2/V\right)^3\right)\right] .$$

We see that, the first terms of the two expansions coincide if we identify $C = C'$ and $b = b'$. This shows how the virial expansion justifies van der Waals approximation.

A further subject which is worth discussing here is the partition function of a mixture of ideal gases. Let us consider a mixture of gases in a given volume V, let N_i be the number of molecules of the i-th component. If the gases are ideal, we can disregard mutual interactions among the molecules which, however, might have internal degrees of freedom. Computing the partition function of an ideal gas mixture, integration over each different molecule species is completely independent of the others, and a factor $1/N_i!$ must be used for each species separately, in order to take into account the identity among particles belonging to the same component. Therefore, for each species, we can make use of a generalized version of Eq. (3.107) where also the internal energy levels are taken into account; indeed, as it is shown in Problem 3.40, the partition function of an ideal polyatomic molecule is still given by the last expression in (3.107), but with $\Phi_0(T)$ replaced by a different function of the temperature $\Phi_i(T)$ which depends on the nature and the internal degrees of freedom of the component. Finally, it follows that the partition function of a mixture of ideal gases is equal to

$$Z_{mix}(T, V, N_i) = \prod_{i=1}^{n}\left(\Phi_i(T)\frac{V}{N_i}\right)^{N_i} \qquad (3.118)$$

where n is the number of different components.

3.6 Entropy and Thermodynamics

In the preceding sections we have shown by some explicit examples how the equilibrium distribution of a given system can be found once its energy levels are known. We have also computed the mean equilibrium energy by identifying the Lagrange multiplier β with $1/kT$. However, it should be clear that in general the complete analysis of the thermodynamics properties of systems in equilibrium requires some further steps and more information.

In the case of a single harmonic oscillator (which is directly related to Einstein's crystal in the limit of a rigid lattice) the thermodynamic analysis is quite simple. Indeed the model describes a system which exchanges energy with the external environment only through heat transfer. That means that the exchanged heat is a function of the state of the system which does not differ, but for an additive constant, from the mean energy (i.e. from the internal energy $U(T)$). However, also in this simple case we can introduce the concept of entropy S, starting from the differential equation $dS = dU/T$ which, making use of (3.23), gives:

$$dS = \frac{C(T)}{T}dT = \frac{3\hbar^2\omega^2}{kT^3}\frac{e^{\hbar\omega/kT}}{\left(e^{\hbar\omega/kT}-1\right)^2}dT = -3k\hbar^2\omega^2\frac{e^{\beta\hbar\omega}}{\left(e^{\beta\hbar\omega}-1\right)^2}\beta d\beta$$

$$= d\left[3k\left(\frac{\beta\hbar\omega}{e^{\beta\hbar\omega}-1} - \ln\left(1 - e^{-\beta\hbar\omega}\right)\right)\right]. \tag{3.119}$$

If we choose $S(0) = 0$ as the initial condition for $S(T)$, we can easily write the entropy of Einstein's crystal:

$$S(T) = \frac{3\hbar\omega}{T}\frac{1}{e^{\frac{\hbar\omega}{kT}}-1} - 3k\ln\left(1 - e^{-\frac{\hbar\omega}{kT}}\right), \tag{3.120}$$

showing in particular that at high temperatures $S(T)$ grows like $3k\ln T$.

Apart from this result, Eq. (3.120) is particularly interesting since it can be simply interpreted in terms of statistical equilibrium distributions. Indeed, recalling that the probability of the generic state of the system, which is identified with the vector $\boldsymbol{n}$, is given by:

$$p_{\boldsymbol{n}} = e^{-\beta\hbar\omega(n_x+n_y+n_z)}\left(1 - e^{-\beta\hbar\omega}\right)^3,$$

we can compute the following expression:

$$-k\sum_{n_x,n_y,n_z=0}^{\infty} p_{\boldsymbol{n}}\ln p_{\boldsymbol{n}} = k\left(1 - e^{-\beta\hbar\omega}\right)^3\sum_{n_x,n_y,n_z=0}^{\infty} e^{-\beta\hbar\omega(n_x+n_y+n_z)}$$

$$\left[\beta\hbar\omega\left(n_x + n_y + n_z\right) - 3\ln\left(1 - e^{-\beta\hbar\omega}\right)\right] \tag{3.121}$$

$$= -3k \ln \left(1 - e^{-\beta\hbar\omega}\right) \left(\left(1 - e^{-\beta\hbar\omega}\right) \sum_{n=0}^{\infty} e^{-\beta\hbar\omega n} \right)^3$$

$$+3k\beta\hbar\omega \left(1 - e^{-\beta\hbar\omega}\right)^3 \left(\sum_{n=0}^{\infty} e^{-\beta\hbar\omega n} \right)^2 \sum_{m=0}^{\infty} m e^{-\beta\hbar\omega m}.$$

Finally, summing the geometric series by the identity $\sum_{n=0}^{\infty} x^n = 1/(1-x)$, holding true if $|x| < 1$, we get

$$\sum_{n=1}^{\infty} n x^n = x \frac{d}{dx} \frac{1}{1-x} = \frac{x}{(1-x)^2},$$

from which we easily find again the expression in (3.120).

One of the most important consequences suggested by this result is the probabilistic interpretation of entropy, i.e. the general validity of the equation

$$S = -k \sum_{\alpha} p_{\alpha} \ln p_{\alpha}, \qquad (3.122)$$

which is particularly interesting for its simplicity. Indeed, let us consider an isolated system and assume that its accessible states are equally probable, so that p_{α} is constant and equal to the inverse of the number of states. Denoting that number by Ω, it easily follows that $S = k \ln \Omega$, thus entropy measures the number of accessible states. Morover, as one can easily verify by comparing it with (3.10), S coincides, apart from a constant, with the quantity which has been maximized in order to find the equilibrium distribution: that is in agreement with thermodynamics, according to which entropy is maximum at thermal equilibrium.

In order to get further evidence about the generality of Eq. (3.122), let us consider the general case in which the system can exchange work as well as heat with the external environment. For instance, Einstein's model could be made more realistic by assuming that, in the relevant range of pressures, the frequencies of oscillators depend on their density according to $\omega = \alpha (N/V)^{\gamma}$, where typically $\gamma \sim 2$. In these conditions the crystal exchanges also work with the external environment and the pressure can be easily computed by using (3.32):

$$P = \sum_{n_x, n_y, n_z = 0}^{\infty} p_n \frac{\gamma}{V} E_n = \frac{\gamma U}{V}, \qquad (3.123)$$

thus giving the equation of state for the crystal.

Going back to entropy, let us compute, in the most general case of a system whose equilibrium state is determined giving its volume and temperature, the heat exchanged when the parameters β and V undergo infinitesimal variations. From (3.32) we get:

$$dU + PdV = \frac{\partial U}{\partial \beta}d\beta + \frac{\partial U}{\partial V}dV + \frac{1}{\beta}\frac{\partial \ln Z}{\partial V}dV, \tag{3.124}$$

therefore, making use of (3.15) we obtain the following infinitesimal heat transfer:

$$-\frac{\partial^2 \ln Z}{\partial \beta^2}d\beta - \frac{\partial^2 \ln Z}{\partial \beta \partial V}dV + \frac{1}{\beta}\frac{\partial \ln Z}{\partial V}dV. \tag{3.125}$$

Last expression can be put into the definition of entropy, thus giving:

$$dS = k\beta\left[-\frac{\partial^2 \ln Z}{\partial \beta^2}d\beta - \frac{\partial^2 \ln Z}{\partial \beta \partial V}dV + \frac{1}{\beta}\frac{\partial \ln Z}{\partial V}dV\right]$$

$$= k\left[\frac{\partial}{\partial \beta}\left(\ln Z - \beta\frac{\partial \ln Z}{\partial \beta}\right)d\beta + \frac{\partial}{\partial V}\left(\ln Z - \beta\frac{\partial \ln Z}{\partial \beta}\right)dV\right]$$

$$= k\,d\left(\ln Z - \beta\frac{\partial \ln Z}{\partial \beta}\right). \tag{3.126}$$

On the other hand it can be easily verified, using again Eq. (3.15), that:

$$-k\sum_\alpha p_\alpha \ln p_\alpha = k\sum_\alpha p_\alpha(\beta E_\alpha + \ln Z)$$

$$= k(\beta U + \ln Z) = k\left(\ln Z - \beta\frac{\partial \ln Z}{\partial \beta}\right) \equiv S. \tag{3.127}$$

Last equation confirms that the probabilistic interpretation of entropy has a general validity. It implies that, in the quite general situation in which the ground state is non degenerate, the entropy vanishes together with the absolute temperature. This is called the *theorem of Nernst* and has important consequences in the analyses of chemical equilibrium, as it is shown in Problem 3.45. We also have an expression of the partition function in terms of thermodynamic potentials. Indeed, the relation $S = k(\beta U + \ln Z)$ is equivalent to $U = TS - \ln Z/\beta$, hence to $\ln Z = -\beta(U - TS)$. Therefore we can conclude that the logarithm of the partition function equals minus β times the free energy:

$$F = -\frac{\ln Z}{\beta} = U - TS. \tag{3.128}$$

It is well known that the free energy (Helmholtz potential) is the thermodynamic potential suitable for the analysis of the equilibrium at constant volume.

A last remark is in order. In our analysis we have considered only the Canonical Ensemble. If we had considered instead the Grand Canonical distribution, that is Eqs. (3.55) and (3.56) which also give

$$U = \sum_a E_a p_a = \sum_a E_a \frac{e^{-\beta(E_a - \mu n_a)}}{\Xi}, \qquad (3.129)$$

we should have obtained

$$
\begin{aligned}
dU &= \sum_a (E_a dp_a + p_a dE_a) = \sum_a \left(E_a dp_a + p_a \frac{\partial E_a}{\partial V} dV \right) \\
&= -kT \sum_a (\ln p_a + \ln \Xi) dp_a + \mu \sum_a n_a dp_a + P dV \\
&= -kT d \sum_a p_a \ln p_a + \mu dN + P dV, \qquad (3.130)
\end{aligned}
$$

because in the state labeled by a the particle number n_a is also given and hence E_a only depends on V. Here we have denoted by N the average number of particles.

Now, the probabilistic interpretation of entropy leads to the equation

$$T dS = dU + P dV - \mu dN, \qquad (3.131)$$

which, taking into account the properties of the thermodynamic potentials, to be discussed in the next section, confirms that the Grand Canonical multiplier μ is the chemical potential, that is the Gibbs potential divided by the particle number, in other words:

$$G = N\mu. \qquad (3.132)$$

3.7 The Thermodynamic Potentials

Using Eq. (3.128) and the first principle of thermodynamics, written in differential form, we introduce a set of thermodynamic functions called *thermodynamic potentials*. Among these potentials we have the free energy defined by (3.128), the enthalpy

$$H = U + PV, \qquad (3.133)$$

and the free enthalpy (which is also called Gibbs potential or Gibbs free energy):

$$G = F + PV \equiv H - TS, \qquad (3.134)$$

where the pressure is given by Eq. (3.35) ($P = kT \partial \ln Z(T, V)/(\partial V)$). If these functions are continuously differentiable we have

$$dU = T dS - P dV \qquad (3.135)$$
$$dF = -S dT - P dV \qquad (3.136)$$

$$dH = TdS + VdP \tag{3.137}$$

$$dG = -SdT + VdP. \tag{3.138}$$

Such equations involving differentials imply well definite relations among the variations of the various thermodynamic quantities that are obtained while keeping other quantities fixed. For instance, from (3.136) we obtain

$$S = -\left.\frac{\partial F}{\partial T}\right|_V \quad ; \quad P = -\left.\frac{\partial F}{\partial V}\right|_T \quad ; \quad \left.\frac{\partial S}{\partial V}\right|_T = \left.\frac{\partial P}{\partial T}\right|_V \tag{3.139}$$

where last equation, following from the theorem of symmetry for second partial derivatives of differentiable functions, is one of the so-called *Maxwell's relations*. Moreover, while dU is the differential change in the total internal energy of the system, dH is the differential heat exchange at constant pressure, while the differential work made by the system in a isothermal process is given by $-dF$. Actually, the relations existing among the different thermodynamic potentials can be considered as a particular example of what is usually known, in mathematical language, as Legendre transformation, which permits to switch from one independent set of variables to the other. For instance, while F can be viewed as a function of T and V, the Gibbs potential $G = F + PV$ can be viewed as a new function in which, since $dG = dF + d(PV) = -SdT - PdV + PdV + VdP = -SdT + VdP$, the role of V and P as independent variables has been interchanged, so that either F or G should be used depending on whether one considers a thermodynamic process at constant volume or pressure. We will come back to the definition of Legendre transformation when we discuss about phase transitions.

If we consider a system of N identical particles (molecules) both H and G are proportional to N: the thermodynamic potentials and the entropy, at fixed temperature and pressure grow with the volume and hence with the number of molecules, in particular, comparing Eq. (3.131) with the definition of G in (3.134), we have

$$G(T, P) = N\mu(T, P) . \tag{3.140}$$

We now consider the thermodynamic potentials in two important cases. First in the case of a classical ideal monoatomic gas, whose atoms have spin zero. From (3.107) we have

$$\begin{aligned} F(T, V) &= -NkT \left[\ln\left(\frac{V}{N}\right) + \frac{3}{2} \ln\left(\frac{2\pi mkT}{h^2}\right) \right] \\ &= -NkT \left[\ln\left(\frac{V}{N}\right) + \ln \Phi_0(T) \right] , \end{aligned} \tag{3.141}$$

which is apparently a convex function of V. The corresponding entropy[4] is

---

[4]If the atoms have D degenerate ground states (e.g., $2S + 1$ for spin S) from (3.107) we have a further additional contribution equal to $\Delta S = NkT \ln D$, that is a factor D multiplying Φ.

$$S(T, V) = kN \left(\frac{d(T \ln \Phi(T))}{dT} + \ln \frac{V}{N} \right), \qquad (3.142)$$

and the Gibbs potential is

$$G(T, P) = NkT[\ln P - \ln(kT\Phi(T)) + 1]$$
$$\equiv NkT \left[\ln P + \ln \left(\frac{h^3}{(2\pi m)^{\frac{3}{2}}} \right) + 1 - \frac{5}{2} \ln(kT) \right] \qquad (3.143)$$

which is a concave function of P.

Secondly we consider the thermodynamic potentials for a mixture of ideal gases. Considering first of all the entropy, we find just the sum of the contributions of each molecular species, that is

$$S_{mix}(T, V, N_i) = \sum_{i=1}^{n} kN_i \left(\frac{d(T \ln \Phi_i(T))}{dT} + \ln \frac{V}{N_i} \right). \qquad (3.144)$$

For the free energy we get

$$F_{mix}(T, V, N_i) = - \sum_{i=1}^{n} kN_i T \left(\ln \Phi_i(T) + \ln \frac{V}{N_i} \right), \qquad (3.145)$$

while the free enthalpy is

$$G_{mix}(T, V, N_i) = - \sum_{i=1}^{n} kN_i T \left(\ln \Phi_i(T) - 1 + \ln \frac{k \sum_{j=1}^{n} N_j T}{P N_i} \right)$$
$$= - \sum_{i=1}^{n} kN_i T \left(\ln \Phi_i(T) - 1 + \ln(kT) - \ln P_i \right), \qquad (3.146)$$

where $P_i = PN_i/(\sum_{j=1}^{n} N_j)$ is the partial pressure of the species. This formula can also be written as

$$G_{mix}(T, V, N_i) = \sum_{i=1}^{n} N_i \mu_i(T, P_i), \qquad (3.147)$$

introducing the chemical potential of each molecular species

$$\mu_i(T, P) = kT \left(\ln P_i - \ln(kT\Phi_i(T)) + 1 \right). \qquad (3.148)$$

There are many situations in which the physical parameters are not uniformly distributed over the system, even in equilibrium. Typical is the case of a matter system in the presence of time independent electromagnetic fields. In these situations the use of the potentials which have been introduced above, that we call total, is not convenient. We should better make use of potential densities, i.e. potentials per unit volume. These are *intensive* quantities, in contrast with the total potentials which are *extensive* quantities. Other intensive parameters are given by the temperature, the pressure, the particle density ρ (the number of particles per unit volume), while the volume V and the total particle number N are extensive quantities. Studying potential densities we shall only deal with intensive quantities, in particular the relevant state variables will be the temperature, the pressure and the particle densities. We shall denote the potential densities by small letters, hence in particular we shall denote by

$$u(T, \rho) \equiv \frac{U(T, V)}{V} \, , \quad f(T, \rho) \equiv \frac{F(T, V)}{V} \tag{3.149}$$

the internal energy density and the free energy density, respectively: notice that such definitions are only valid in the case of a homogeneous system, while in the presence of inhomogeneities the spatial averages must be taken locally. It is important to note here that a change of the density might correspond to a change of volume or else to a change of number of particles, or both. Therefore, in order to compute the differential of the densities, we must insert into Eqs. (3.135)–(3.138) the differential dN. For the Gibbs potential this follows directly from (3.140). The extension to the other potentials follows from their definitions i.e. Eqs. (3.128) and (3.133). Therefore we have

$$dG = -SdT + VdP + \mu dN$$
$$dF = -SdT - PdV + \mu dN$$
$$dU = TdS - PdV + \mu dN$$
$$dH = TdS + VdP + \mu dN, \tag{3.150}$$

In contrast, considering the potential densities, we have in particular

$$du = Tds + \mu d\rho \, , \quad df = -sdT + \mu d\rho \, . \tag{3.151}$$

Now we consider an interesting example concerning a material in a magnetic field. Let the material be either diamagnetic of paramagnetic, with a linear response, and isotropic. This implies the following linear relation among the magnetic induction[5] $\boldsymbol{B}$, the magnetic field $\boldsymbol{H}$ and the magnetization $\boldsymbol{M}$:

$$\boldsymbol{B} = \mu_0(1 + \chi(T, \rho))\boldsymbol{H} \equiv \mu_0(\boldsymbol{M} + \boldsymbol{H}), \tag{3.152}$$

where χ is the susceptibility of the material and μ_0 is the permeability of the vacuum. In the presence of magnetic fields the energy density of the system receives

---

[5] We adopt here a convention for magnetic fields in materials according to which $\boldsymbol{H}$ is called magnetic field and $\boldsymbol{B}$ is called magnetic induction. An alternative convention is to keep calling $\boldsymbol{B}$ magnetic field and to introduce $\boldsymbol{H}$ as an auxiliary field.

a contribution corresponding to the work per unit volume w done by the external field sources on the system. For a variation of the field $\boldsymbol{B}$ given by $dt\,\dot{\boldsymbol{B}}$, where $\dot{\boldsymbol{B}} = \partial \boldsymbol{B}/\partial t$, on account of the Maxwell equations

$$\nabla \wedge \boldsymbol{E} = -\dot{\boldsymbol{B}}, \quad \nabla \wedge \boldsymbol{H} + \dot{\boldsymbol{D}} = \boldsymbol{j}, \tag{3.153}$$

and disregarding $\dot{\boldsymbol{D}}$, which is infinitesimal, in the second equation, we have

$$dw = -dt\,\boldsymbol{E} \cdot \boldsymbol{j} = -dt\,\boldsymbol{E} \cdot (\nabla \wedge \boldsymbol{H}) = dt\,\boldsymbol{H} \cdot \dot{\boldsymbol{B}} \equiv \boldsymbol{H} \cdot d\boldsymbol{B}, \tag{3.154}$$

where a surface term, proportional to $\nabla \cdot (\boldsymbol{E} \wedge \boldsymbol{H})$, has been disregarded. In other words, if we integrate dw over the whole volume containing the electromagnetic fields, this term does not contribute and the total differential work is equal to the space integral of $\boldsymbol{H} \cdot d\boldsymbol{B}$. Therefore we conclude that, in the presence of magnetic fields, Eq. (3.151) becomes

$$\begin{aligned} du(T, \rho, \boldsymbol{B}) &= T\,ds(T, \rho, \boldsymbol{B}) + \mu(T, \rho, \boldsymbol{B})d\rho + \boldsymbol{H} \cdot d\boldsymbol{B} \\ df(T, \rho, \boldsymbol{B}) &= -s(T, \rho, \boldsymbol{B})dT + \mu(T, \rho, \boldsymbol{B})d\rho + \boldsymbol{H} \cdot d\boldsymbol{B}. \end{aligned} \tag{3.155}$$

Very often, instead of the just presented potential densities for constant $\boldsymbol{B}$, it is convenient to introduce those for constant $\boldsymbol{H}$ that we denote by

$$\begin{aligned} \tilde{u}(T, \rho, \boldsymbol{H}) &= u(T, \rho, \boldsymbol{B}) - \boldsymbol{H} \cdot \boldsymbol{B} \\ &\equiv u(T, \rho, \mu_0(1 + \chi(T, \rho))\boldsymbol{H}) - \mu_0(1 + \chi(T, \rho))|\boldsymbol{H}|^2 \\ \tilde{f}(T, \rho, \boldsymbol{H}) &= f(T, \rho, \boldsymbol{B}) - \boldsymbol{H} \cdot \boldsymbol{B} \\ &\equiv f(T, \rho, \mu_0(1 + \chi(T, \rho))\boldsymbol{H}) - \mu_0(1 + \chi(T, \rho))|\boldsymbol{H}|^2. \end{aligned} \tag{3.156}$$

An example of the use of the new potential densities follows from the expression of the differential

$$\begin{aligned} d\tilde{f}(T, \rho, \boldsymbol{H}) &= -s(T, \rho, \boldsymbol{H})dT + \mu(T, \rho, \boldsymbol{H})d\rho - \boldsymbol{B} \cdot d\boldsymbol{H} \\ &= -s(T, \rho, \boldsymbol{H})dT + \mu(T, \rho, \boldsymbol{H})d\rho - \mu_0(1 + \chi(T, \rho))\boldsymbol{H} \cdot d\boldsymbol{H} \end{aligned} \tag{3.157}$$

from which we get the Maxwell relation

$$\frac{\partial s(T, \rho, \boldsymbol{H})}{\partial \boldsymbol{H}} = \mu_0 \frac{\partial \chi(T, \rho)}{\partial T} \boldsymbol{H}. \tag{3.158}$$

Therefore at constant temperature and density we find

$$s(T, \rho, \boldsymbol{H}) - s(T, \rho, 0) = \mu_0 \frac{\partial \chi(T, \rho)}{\partial T} \frac{|\boldsymbol{H}|^2}{2}. \tag{3.159}$$

In Problem 3.46 we discuss how from this equation it follows the possibility of cooling paramagnetic materials by adiabatic demagnetization.

3.7.1 Phase Transitions

We have seen that, in principle, the statistical calculation of the partition function identifies the Helmholtz potential of a system, even if the equilibrium law deduced from this potential using Eq. (3.35) might not identify the most stable equilibrium state of the system. Discussing in particular the van der Waals example, we shall see how this might happen in a particular situation. In general, we must take into account that, in an equilibrium law at constant temperature, the pressure must be a monotonic decreasing function of the volume, otherwise we would have an unstable situation in which the pressure increases with the volume and, for large volumes, the system behaves like an explosive substance, while for small volumes, if external pressure keeps constant, the system tends to shrink to a point. As a matter of fact, in stable systems the pressure decreases when the volume increases and diverges when the volume tends to zero.[6] Therefore we see that, if $-\ln Z(T, V)$ is a continuously derivable function of V, it is identified with $\beta F(T, V)$ only if it is convex, because its volume derivative must be a non-decreasing function of the volume. However, cases in which $-\ln Z$ is not a convex function may happen, tipically in the framework of some approximation scheme: then we meet the problem of identifying the true Helmholtz potential and the true equilibrium law of the system.

If $F(T, V)$, that for the moment we identify with $-kT \ln Z(T, V)$, is a continuously derivable convex function of V the equilibrium of the system at constant pressure is identified by the minima of the Gibbs potential given in (3.134). As we have already remarked, the relation linking F and G is a particular case of Legendre transformation. A naïve definition of such a transformation is by saying that $G(T, P) = F(T, V) + PV$ where P and V are related by $P = -\partial F/\partial V$: such a definition is well given only if P is a bijective function of V, i.e. if F is convex; moreover, in this case, one also verifies that $G(T, P)$ is the value the function $F(T, V) + PV$ takes at its stationary point, which is unique (and corresponds to a minimum) by convexity of F. When F is non-convex and more solutions exist to the equation $P = -\partial F/\partial V$, the situation is less trivial and the rigorous definition of Legendre transformation (hence of Gibbs potential, in our case), is given by

$$\mathrm{Inf}_V [F(T, V) + PV] \,, \tag{3.160}$$

that is one must take the value of the absolute minimum (with respect to V) of the function $F(T, V) + PV$, for T and P fixed.

---

[6]In the van der Waals case the pressure diverges before the system is shrunk to zero volume.

Now we know that P identifies the slope of the tangent line to $-F$ in the stationary point of $F(T, V) + PV$. Let us denote the volume at this point by V_s and the pressure by P_s, then the equation of the tangent line to $F(T, V)$ is

$$y(x) = F(T, V_s) + (V_s - x)P_s . \tag{3.161}$$

Considering the mentioned situations in which $F(T, V)$ is not a convex function of V, we note that there might be n stationary points of $F(T, V) + PV$ corresponding to the same temperature, T, to the pressure, P_s, and to various different volumes $V_{i,s}$, $(i = 1, \ldots, n)$. The equations of the i-th tangent line would be

$$y_i(x) = F(T, V_{i,s}) + (V_{i,s} - x)P_s. \tag{3.162}$$

According to Eq. (3.160), the Gibbs potential $G(T, P_s)$ corresponds to the minimum value of the $y_i(0)$'s. Let it be $y_j(0)$.

With such a definition, the volume V_s corresponding to a given pressure P_s is identified as follows. One considers the set of all straight lines of given slope $-P_s$ in the $V, F(T, V)$ plane; then, ordering those lines according to their intercept with the y axis, there is one of them which first intersects the function $F(T, V)$. If F admits thermal equilibrium, i.e. if $\partial F/\partial V$ is an increasing function at least in the low and in large volume regime, then one easily realizes that such a first intercepting line is tangent to $F(T, V)$: V_s is the contact point and the whole function $F(T, V)$ lies above it. One also sees, by graphical construction, that as we increase the pressure $P_s \to P_s' > P_s$, i.e. as we decrease the slope, the contact point V_s of the first tangent line moves downwards, i.e. to $V_s' < V_s$, otherwise F would intersect the tangent line in V_s in some point, contrary to what stated above. On the other hand, since one can prove[7] that $V_s = \partial G/\partial P$ computed in P_s, we deduce that $\partial G/\partial P$ is a decreasing function of P, i.e. that $G(T, P)$ is a concave function of P.

Once G is computed we can introduce a new definition of the thermodynamic potential $F(T, V)$, identifying it with the absolute maximum over P of $G(T, P) - PV$, that is

$$\tilde{F}(T, V) = \sup_P[G(T, P) - PV] . \tag{3.163}$$

For the same reasons discussed above, that gives a convex function, which coincides with F only if the original free energy is a convex function. In strictly mathematical terms, it is $-G$ which is the Legendre transform of F, while $\tilde{F}$ is the Legendre transform of $-G$: then, we learn that the Legendre transform is an involution when applied to convex functions.

---

[7] We have indeed

$$\frac{\partial G(T, P_s)}{\partial P_s} = \frac{\partial}{\partial P_s}(F(T, V_s(P_s)) + P_s V_s(P_s)) = \frac{\partial F}{\partial V_s}\frac{\partial V_s}{\partial P_s} + V_s + P_s\frac{\partial V_s}{\partial P_s}$$

$$= -P_s\frac{\partial V_s}{\partial P_s} + V_s + P_s\frac{\partial V_s}{\partial P_s} = V_s.$$

It is possible to understand the consequences of this new definition of F considering that, starting from (3.128), we can write

$$F(T, V) = \int_V^{V_M} P(T, x)dx + F(T, V_M). \qquad (3.164)$$

Suppose that the corresponding equilibrium law of the system has two coinciding solutions

$$P(T, V_1) = P(T, V_2), \qquad (3.165)$$

(this is possible if $F(T, V)$ is not a convex function of V) according to Eq. (3.134) we have different solutions for G, thus we have to choose the solution corresponding to the minimum value of G. But the two choices might coincide, i.e. the two tangent lines given in (3.162) might coincide. In this case in the interval $[V_1, V_2]$ the true potential $\tilde{F}(T, V)$ must be identified with the common tangent line at the extrema of the interval. Let $\tilde{F} = G_0 - P_0 V$ be its equation ($P_0 = P(T, V_1) = P(T, V_2)$). We have

$$G_0 - P_0 V_1 = \int_{V_1}^{V_M} P(T, x)dx + F(T, V_M)$$

$$G_0 - P_0 V_2 = \int_{V_2}^{V_M} P(T, x)dx + F(T, V_M), \qquad (3.166)$$

therefore we have

$$P_0(V_2 - V_1) = \int_{V_1}^{V_2} P(T, x)dx, \qquad (3.167)$$

which means that in the $P - V$ plane at constant T the horizontal line cutting the curve $P(T, V) \equiv -kT \partial \ln Z(T, V)/\partial V$ in the points V_1 and V_2 averages this curve in the considered volume interval.

We understand this equation considering the meaning of P_0. The two volumes V_1 and V_2 corresponding to the same pressure and temperature (P_0, T) are border points of two branches of F and of the equilibrium law. The choice of the most stable branch requires a transition between the two branches at the pressure P_0, which is therefore identified with the transition pressure. Each branch identifies a *phase* of the system and T and P_0 identify an equilibrium point of the two phases. In the van der Waals example the two phases are the gas and the liquid phases. The interval $[V_1, V_2]$ identifies the transition in the sense that, inside the interval, the system appears as a mixture of the two phases, V_1 and V_2 being the volumes of the system in the pure phases. Equation (3.167) allows the identification of the transition pressure P_0 directly from the equilibrium law, it is called the *Maxwell rule*. The value of the Gibbs potential at the transition point is G_0 for both phases. We conclude that a loss of convexity in the curve with equation $\ln Z(T, V)$ at constant T characterizes a *phase transition*.

The possibility of phase transitions clearly appears in the analysis of a real gas in the van der Waals approximation (Eq. (3.114)) whose free energy is

$$-kT \ln Z_{vdw}(T, V) \simeq -NkT \left[\ln \left(\frac{(V - b)}{Nh^3} \right) + \frac{3}{2} \ln(2\pi mkT) + \frac{CN}{2KTV} \right].$$

The pressure is given by

$$P = \frac{NkT}{V - b} - \frac{a}{V^2}, \qquad (3.168)$$

where $a = CN^2/2$ and, of course, $V > b$.

Computing the derivative of the right-hand side of (3.168) we see that the pressure is not a decreasing function of the volume for every value of the temperature. It follows that, for certain values of T and V, the pressure increases with the volume.

We see that for $V \gtrsim b$ the pressure is a decreasing function of V and the same is true for $V \to \infty$. Therefore we see that the instability problem is due to the second term in the right-hand side of (3.168) and unstable points are localized at finite values of the volume. For a deeper analysis of the effects of this term we compute the volume derivative of the pressure looking for stationary points. We have

$$\frac{\partial P}{\partial V} = -\frac{NkT}{(V - b)^2} + 2\frac{a}{V^3}, \qquad (3.169)$$

which vanishes in the solutions of the cubic equation

$$V^3 - \frac{2a}{NkT}(V - b)^2 \equiv V^3 - \alpha(V - b)^2 = 0. \qquad (3.170)$$

Both α and b are positive, therefore Eq. (3.170) has no real solutions for $V < 0$, on the contrary there is an odd number of real solutions for $0 \leq V < b$. Their number must be either one or three, because the cubic equation has, at most, three real solutions. For the same reason, for $V > b$ there are either two real solutions or none. We are interested in the case, i.e. in the ranges of temperature, for which there are two real solutions for $V > b$. Indeed, from (3.168), we see that the pressure diverges if V tends to b from above and vanishes for $V \to \infty$, therefore, either $P(T, V)$ is a monotonic decreasing function of V, $-kT \ln Z_{vdw}(T, V) = F(T, V)$ is a convex function of V and the solutions of the equation of state correspond to stable states, or, for a suitable range of temperature, there is an interval, $b < V_1 < V < V_2$, of instability. For continuity there must exist a temperature for which $V_1 = V_2$, i.e. the pressure has a vanishing second derivative and F has a flex point. The corresponding temperature, volume and pressure are called *critical*.

We perform a quantitative analysis of the solutions of Eq. (3.170) rescaling the volume, i.e. setting $V \equiv \alpha v$ and $b/\alpha \equiv z$ and introducing the new variable $x = v - 1/3$. From the above definitions we see that z is independent of N and proportional to kT. Now Eq. (3.170) assumes the simplified forms

$$v^3 = (v - z)^2, \quad x^3 - (\frac{1}{3} - 2z)x - (\frac{2}{27} - \frac{2}{3}z + z^2) = 0 . \tag{3.171}$$

The last equation is called a depressed cubic equation because it has the form $x^3 + px + q = 0$. Following Cardano's method we see that this equation has three real solutions if and only if $q^2/4 + p^3/(27) \leq 0$. In the case of (3.171) we see that this is equivalent to $z^4/4 = z^3/(27)$. We disregard $z = 0$, which corresponds to vanishing T and for which our approximation does not make sense, and we remain with $z = 4/(27)$ for which $q = -2/3^6$. We have three real solutions. The first one corresponds to $x = -2/9$, that is $v = 1/9 < z$, i.e. $V < b$. This solution, which is unphysical, is expected for any value of the temperature. The other two solutions coincide, they correspond to $x = 1/9$, i.e. $v = 4/9 > z$. Therefore we have a critical point which corresponds to the following temperature, volume and pressure

$$T_c = \frac{8a}{27Nkb} , \quad V_c = 3b , \quad P_c = \frac{a}{27b^2} . \tag{3.172}$$

For temperatures above T_c the pressure increases monotonically when the volume decreases from ∞ to b. For large volumes Eq. (3.168) is not substantially different from the ideal gas law, in contrast when the volume approaches b the pressure diverges and the system becomes incompressible like a liquid. Below T_c we find an unstable regime which corresponds to $-kT \ln Z_{vdw}(T, V)$ loosing its convexity. In this situation we must replace the unstable branch of the free energy for fixed T by its double tangent, i.e. the tangent in the points V_1 and V_2 according to (3.167).

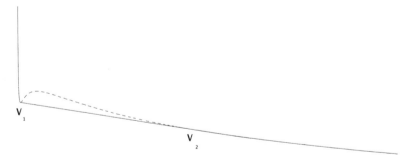

In the figure above it is shown, in arbitrary units, how the free energy at constant $T = 0.4T_c$ is modified replacing the unstable branch (dashed line) by the mentioned double tangent. The branches on the left-hand and on the right-hand side, which are not modified, represent respectively the free energies of the liquid and of the gas phases. The slope of the double tangent gives minus the transition pressure at the chosen temperature. It is apparent that the solid line in the figure is convex. In a plot of the Gibbs potential $G(T, P)$ corresponding to the same temperature, the two branches corresponding to the two phases join continuously, but with a discontinuous derivative, at the pressure P_0.

The study of phase transitions is a very important subject of advanced statistical thermodynamics which is far above the level of the present text.

Suggestions for Supplementary Readings

- F. Reif: *Statistical Physics - Berkeley Physics Course*, volume 5 (Mcgraw-Hill Book Company, New York 1965)
- E. Schrödinger: *Statistical Thermodynamics* (Cambridge University Press, Cambridge 1957)
- T. L. Hill: *An Introduction to Statistical Thermodynamics* (Addison-Wesley Publishing Company Inc., Reading 1960)
- F. Reif: *Fundamentals of Statistical and Thermal Physics* (Mcgraw-Hill Book Company, New York 1965).

Problems

3.1 We have to place four distinct objects into 3 boxes. How many possible different distributions can we choose? What is the multiplicity $\mathcal{M}$ of each distribution? And its probability p?

Answer: We can make 3 different choices for each object, therefore the total number of possible choices is $3^4 = 81$. The total number of possible distributions is instead given by all the possible choices of non-negative integers n_1, n_2, n_3 with $n_1 + n_2 + n_3 = 4$, i.e. $(4 + 1)(4 + 2)/2 = 15$. There are in particular 3 distributions like $(4, 0, 0)$, each with $p = \mathcal{M}/81 = 1/81$; 6 like $(3, 1, 0)$, with $p = \mathcal{M}/81 = 4/81$; 3 like $(2, 2, 0)$, with $p = \mathcal{M}/81 = 6/81$; 3 like $(2, 1, 1)$, with $p = \mathcal{M}/81 = 12/81$.

3.2 The integer number k can take values in the range between 0 and 8 according to the binomial distribution:

$$P(k) = \frac{1}{2^8}\binom{8}{k}.$$

Compute the mean value of k and its mean quadratic deviation.

Answer: $\bar{k} = 4$; $\langle(k - \bar{k})^2\rangle = 2$.

3.3 Let us consider a system which can be found in 4 possible states, enumerated by the index $k = 0, 1, 2, 3$ and with energy $E_k = \epsilon k$, where $\epsilon = 10^{-2}$ eV. The system is at thermal equilibrium at room temperature $T \simeq 300\,°$K. What is the probability for the system being in the highest energy state?

Answer: $Z = \sum_{k=0}^{3} e^{-\beta\epsilon k}$; $U = (1/Z)\sum_{k=0}^{3}\epsilon k e^{-\beta\epsilon k} \simeq 1.035\ 10^{-2}$ eV; $P_{k=3} = (1/Z)e^{-3\beta\epsilon} \simeq 0.127$.

3.4 A diatomic molecule is made up of two particles of equal mass $M = 10^{-27}$ kg which are kept at a fixed distance $L = 4 \times 10^{-10}$ m. A set of $N = 10^9$ such systems, which are not interacting among themselves, is in thermal equilibrium at a temperature $T = 1\,°$K. Estimate the number of systems which have a non-vanishing angular momentum (computed with respect to their center of mass), i.e. the number of rotating molecules, making use of the fact that the number of states with angular momentum $n\hbar$ is equal to $2n + 1$.

Answer: If we quantize rotational energy according to Bohr, then the possible energy levels are $E_n = n^2\hbar^2/(ML^2) \simeq 4.3 \times 10^{-4}n^2$ eV, each corresponding to $2n+1$ different states. These states are occupied according to the Canonical Distribution. The partition function is $Z = \sum_{n=0}^{\infty}(2n+1)e^{-E_n/kT}$. If $T = 1\,°K$, then $kT \simeq 0.862 \times 10^{-4}$ eV, hence $e^{-E_n/kT} \simeq e^{-5.04n^2} \simeq (6.47 \times 10^{-3})^{n^2}$. Therefore the two terms with $n = 0, 1$ give a very good approximation of the partition function, $Z \simeq 1 + 1.94 \times 10^{-2}$. The probability that a molecule has $n = 0$ is $1/Z$, hence the number of rotating molecules is $N_R = N(1-1/Z) \simeq 1.9 \times 10^7$. If we instead make use of Sommerfeld's perfected theory, implying $n^2 \to n(n+1)$ in the expression for E_n, we obtain $Z \simeq 1 + 1.25 \times 10^{-4}$ and $N_R = N(1-1/Z) \simeq 1.25 \times 10^5$. In the present situation, being quantum effects quite relevant, the use of Sommerfeld's correct formula for angular momentum quantization in place of simple Bohr's rule makes a great difference.

3.5 Consider again Problem 3.4 in case the molecules are in equilibrium at room temperature, $T \simeq 300\,°K$. Compute also the average energy of each molecule.

Answer: In this case the partition function is, according to Sommerfeld's theory:

$$Z = \sum_{n=0}^{\infty}(2n+1)e^{-\alpha n(n+1)},$$

with $\alpha \simeq 0.0168$. Since $\alpha \ll 1$, $(2n+1)e^{-\alpha n^2}$ is the product of a linear term times a slowly varying function of n, hence the sum can be replaced by an integral

$$Z \simeq \int_0^{\infty} dn(2n+1)e^{-\alpha n(n+1)} = \frac{1}{\alpha} \simeq 60$$

hence the number of non-rotating molecules is $N_{NR} = N/Z \simeq 1.67 \times 10^7$. From $Z \simeq 1/\alpha = kTML^2/\hbar^2$, we get $U = -\partial/\partial\beta \ln Z = kT$, in agreement with equipartition of energy.

3.6 A system in thermal equilibrium admits 4 possible states: the ground state, having zero energy, plus three degenerate excited states of energy ϵ. Discuss the dependence of its mean energy on the temperature T.

Answer:

$$U = \frac{3\epsilon\, e^{-\epsilon/kT}}{1+3e^{-\epsilon/kT}}\;; \qquad \lim_{T\to 0} U(T) = 0\;; \qquad \lim_{T\to\infty} U(T) = \frac{3}{4}\epsilon\,.$$

3.7 A simple pendulum of length $l = 10$ cm and mass $m = 10$ g is placed on Earth's surface in thermal equilibrium at room temperature, $T = 300\,°K$. What is the mean quadratic displacement of the pendulum from its equilibrium point?

Answer: The potential energy of the pendulum is, for small displacements $s \ll l$, $mgs^2/2l$. From the energy equipartition theorem we infer $mgs^2/2l \simeq kT/2$, hence $\sqrt{\langle s^2 \rangle} \simeq \sqrt{kTl/mg} = 2 \times 10^{-10}$ m.

3.8 What is the length of a pendulum for which quantum effects are important at room temperature?

Answer: $\hbar\omega \sim kT$, so that $l = g/\omega^2 \sim \hbar^2 g/(kT)^2 \sim 6.5 \times 10^{-28}$ m.

3.9 A massless particle is constrained to move along a segment of length L; therefore its wave function vanishes at the ends of the segment. The system is in equilibrium at a temperature T. Compute its mean energy as well as the specific heat at fixed L. What is the force exerted by the particle on the ends of the segment?

Answer: Energy levels are given by $E_n = nc\pi\hbar/L$, hence $Z = \sum_{n=1}^{\infty} e^{-\beta E_n} = 1/(e^{\beta c\pi\hbar/L} - 1)$ from which the mean energy follows

$$U = -\frac{\partial \ln Z}{\partial \beta} = \frac{c\pi\hbar}{L} \frac{1}{(1 - e^{-c\pi\hbar/LkT})} \, ,$$

and the specific heat

$$C_L = k(c\pi\hbar/LkT)^2 \frac{e^{-c\pi\hbar/LkT}}{(1 - e^{-c\pi\hbar/LkT})^2} \, ,$$

which vanishes at low temperatures and approaches k at high temperatures; notice that the equipartition principle does not hold in its usual form in this example, since the energy is not quadratic in the momentum, hence we have k instead of $k/2$. The equation of state can be obtained making use of (3.29) and (3.35), giving for the force $F = (1/\beta)(\partial \ln Z/\partial L) = U/L$. Hence at high temperatures we have $FL = kT$.

3.10 Consider a system made up of N distinguishable and non-interacting particles which can be found each in two possible states of energy 0 and ϵ. The system is in thermal equilibrium at a temperature T. Compute the mean energy and the specific heat of the system.

Answer: The partition function for a single particle is $Z_1 = 1 + e^{-\epsilon/kT}$. That for N independent particles is $Z_N = Z_1^N$. Therefore the average energy is

$$U = \frac{N\epsilon}{1 + e^{\epsilon/kT}}$$

and the specific heat is

$$C = Nk \left(\frac{\epsilon}{kT}\right)^2 \frac{e^{\epsilon/kT}}{(e^{\epsilon/kT} + 1)^2}.$$

3.11 A system consists of a particle of mass m moving in a one-dimensional potential which is harmonic for $x > 0$ ($V = kx^2/2$) and infinite for $x < 0$. If the system is at thermal equilibrium at a temperature T, compute its average energy and its specific heat.

Answer: The wave function must vanish in the origin, hence the possible energy levels are those of the harmonic oscillator having an odd wave function. In particular, setting $\omega = \sqrt{k/m}$, we have $E_n = (2n + 3/2)\hbar\omega$, with $n = 0, 1, \ldots$. The partition function is

$$Z = \frac{e^{-3\beta\hbar\omega/2}}{1 - e^{-2\beta\hbar\omega}},$$

so that the average energy is

$$U = \frac{3\hbar\omega}{2} + \frac{2\hbar\omega}{e^{2\beta\hbar\omega} - 1}$$

and the specific heat is

$$C = k(2\beta\hbar\omega)^2 \frac{e^{2\beta\hbar\omega}}{(e^{2\beta\hbar\omega} - 1)^2} \, .$$

3.12 Compute the average energy of a classical three-dimensional isotropic harmonic oscillator of mass m and oscillation frequency $\nu = 2\pi\omega$ in equilibrium at temperature T.

Answer: The state of the classical system is assigned in terms of the momentum p and the coordinate x of the oscillator, it is therefore represented by a point in phase space corresponding to an energy $E(p, x) = p^2/2m + m\omega^2x^2/2$. The canonical partition function can therefore be written as an integral over phase space

$$Z = \int \frac{d^3p \, d^3x}{\Delta} e^{-\beta p^2/2m} e^{-\beta m\omega^2 x^2/2}$$

where Δ is an arbitrary effective volume in phase space needed to fix how we count states (that is actually not arbitrary according to the quantum theory, which requires $\Delta \sim h^3$). A simple computation of Gaussian integrals gives $Z = \Delta^{-1}(2\pi/\omega\beta)^3$, hence $U = -(\partial/\partial\beta) \ln Z = 3kT$, in agreement with equipartition of energy.

3.13 A particle of mass m moves in the $x - y$ plane under the influence of an anisotropic harmonic potential $V(x, y) = m(\omega_x^2 x^2/2 + \omega_y^2 y^2/2)$, with $\omega_y \ll \omega_x$. Therefore the energy levels coincide with those of a system made up of two distinct particles moving in two different one-dimensional harmonic potentials corresponding respectively to ω_x and ω_y. The system is in thermal equilibrium at a temperature T. Compute the specific heat and discuss its behaviour as a function of T.

Answer: The partition function is the product of the partition functions of the two distinct harmonic oscillators, hence the average energy and the specific heat will be the sum of the respective quantities. In particular

$$C = \frac{(\hbar\omega_x)^2}{kT^2}\frac{e^{\beta\hbar\omega_x}}{(e^{\beta\hbar\omega_x}-1)^2} + \frac{(\hbar\omega_y)^2}{kT^2}\frac{e^{\beta\hbar\omega_y}}{(e^{\beta\hbar\omega_y}-1)^2}.$$

We have three different regimes: $C \sim 0$ if $kT \ll \hbar\omega_y$, $C \sim k$ if $\hbar\omega_y \ll kT \ll \hbar\omega_x$ and finally $C \sim 2k$ if $kT \gg \hbar\omega_x$.

3.14 Compute the mean quadratic velocity for a rarefied and ideal gas of particles of mass $M = 10^{-20}$ kg in equilibrium at room temperature.

Answer: According to Maxwell distribution, $\langle v^2 \rangle = 3kT/M \simeq 1.2 \, \mathrm{m}^2/\mathrm{s}^2$.

3.15 Taking into account that a generic molecule has two rotational degrees of freedom, compute, using the theorem of equipartition of energy, the average square angular momentum J^2 of a molecule whose moment of inertia about the center of mass is $I = 10^{-39}$ kg m^2, independently of the rotation axis, if the temperature is $T = 300 \, ^\circ$K. Discuss the validity of the theorem of equipartition of energy in the given conditions.

Answer: On account of the equipartition of energy, the average rotational kinetic energy of the molecule is $2kT = 8.29 \times 10^{-21}$ J $= \bar{J^2}/2I$, thus $\bar{J^2} = 1.66 \times 10^{-59}$ J s. The theorem of equipartition of energy is based on the assumption that the energy, and hence the square angular momentum, be a continuous variable, while, as a matter of fact (see e.g. Problem 2.1), it is quantized according to the formula $J^2 = n(n+1)\hbar^2$. Therefore the validity of the energy equipartition requires that the difference between two neighboring values of J^2 be much smaller than $\bar{J^2}$, i.e. $(2n+1)/(n(n+1)) \simeq 2/n \ll 1$. In the given conditions $n \simeq J/\hbar \simeq \sqrt{1.49 \times 10^9}$, therefore previous inequality is satisfied.

3.16 Consider a diatomic gas, whose molecules can be described schematically as a pair of pointlike particles of mass $M = 10^{-27}$ kg, which are kept at an equilibrium distance $d = 2 \times 10^{-10}$ m by an elastic force of constant $K = 11.25$ N/m. A quantity equal to 1.66 g atoms of such gas is contained in volume $V = 1 \, \mathrm{m}^3$. Discuss the qualitative behaviour of the specific heat of the system as a function of temperature. Consider the molecules as non-interacting and as if each were contained in a cubic box with reflecting walls of size $L^3 = V/N$, where N is the total number of molecules.

Answer: Three different energy scales must be considered. The effective volume available for each molecule sets an energy scale $E_1 = \hbar^2\pi^2/(4ML^2) \simeq 1.7 \times 10^{-6}$ eV, which is equal to the ground state energy for a particle of mass $2M$ in a cubic box, corresponding to a temperature $T_1 = E_1/k \simeq 0.02 \, ^\circ$K. The minimum rotational energy is instead, according to Sommerfeld, $E_2 = \hbar^2/(Md^2) \simeq 3.5 \times 10^{-3}$ eV, corresponding to a temperature $T_2 = E_2/k \simeq 40 \, ^\circ$K. Finally, the fundamental

888

oscillation energy is $E_3 = \hbar\sqrt{2K/M} \simeq 0.098$ eV, corresponding to a temperature $T_3 = E_2/k \simeq 1140\,°$K. For $T_1 \ll T \ll T_2$ the system can be described as a classical perfect gas of pointlike particles, since rotational and vibrational modes are not yet excited, hence the specific heat per molecule is $C \sim 3k/2$. For $T_2 \ll T \ll T_3$ the system can be described as a classical perfect gas of rigid rotators, hence $C \sim 5k/2$. Finally, for $T \gg T_3$ also the (one-dimensional) vibrational mode is excited and $C \sim 7k/2$. This roughly reproduces, from a qualitative point of view and with an appropriate rescaling of parameters, the observed behavior of real diatomic gases.

3.17 We have a total mass $M = 10^{-6}$ kg of a dust of particles of mass $m = 10^{-17}$ kg. The dust particles can move in the vertical $x - z$ semi-plane defined by $x > 0$, and above a line forming an angle α, with $\tan \alpha = 10^{-3}$, with the positive x axis. The dust is in thermal equilibrium in air at room temperature ($T = 300\,°$K) and hence the particles, which do not interact among themselves, have planar Brownian motion above the mentioned line. What is the distribution of particles along the positive x axis and which their average distance $\bar{x}$ from the vertical z axis?

Answer: We start assuming that, in the mentioned conditions, quantum effects are negligible and hence the particle distribution in the velocity and position plane is given by the Maxwell-Boltzmann law: $d^4n/((d^2v)dxdz) = N \exp(-E/kT) = N \exp(-(mv^2/2 + mgz)/(kT))$ where g is the gravitational acceleration and N is a normalization factor. Then the x-distribution of particles is given by:

$$\frac{dn}{dx} = \int_{-\infty}^{\infty} dv_x \int_{-\infty}^{\infty} dv_y \int_{x\tan\alpha}^{\infty} dz \frac{d^4n}{d^2vdxdz}$$
$$= \frac{2\pi kTN}{m} \int_{x\tan\alpha}^{\infty} dz e^{-mgz/(kT)} = \frac{2\pi(kT)^2N}{m^2g} e^{-mgx\tan\alpha/(kT)}.$$

We can compute N using $\int_0^{\infty}(dn/dx)dx = 2\pi(kT)^3N/(\tan\alpha\, m^3g^2) = M/m = 10^{11}$ and $\bar{x} = kT/(mg\tan\alpha) \simeq 4.22 \times 10^{-2}$ m. Now we discuss the validity of the classical approximation. The average inter-particle distance is of the order of magnitude of $m\bar{x}/M \sim 4 \times 10^{-13}$ m, which should be much larger than the average de Broglie wave length of the particles, which is of the order of magnitude of $h/\sqrt{2mkT} \sim 2 \times 10^{-15}$ m. We conclude that the classical approximation is valid.

3.18 The possible stationary states of a system are distributed in energy as follows:

$$\frac{d\,n}{dE} = \alpha E^3 e^{E/E_0}$$

where E_0 is some given energy scale. Compute the average energy and the specific heat of the system for temperatures $T < E_0/k$, then discuss the possibility of reaching thermal equilibrium at $T = E_0/k$.

Answer: Let us set $\beta_0 = 1/E_0$. The density of states diverges exponentially with energy and the partition function of the system is finite only if the temperature is low enough in order for the Boltzmann factor, which instead decreases exponentially with energy, to be dominant at high energies. For $T < E_0/k$ $(\beta > \beta_0)$ we have:

$$Z = \int_0^\infty dE \alpha E^3 e^{-(\beta-\beta_0)E} = \frac{6\alpha}{(\beta-\beta_0)^4}$$

from which the internal energy U and the specific heat $C = dU/dT$ easily follow:

$$U = \frac{4}{\beta-\beta_0} = \frac{4kTE_0}{E_0-kT} \quad ; \quad C = \frac{4kE_0^2}{(E_0-kT)^2}.$$

The specific heat diverges at $T = E_0/k$: in general that may happen in the presence of a phase transition, but in this specific case also the internal energy diverges as $T \to E_0/k$, meaning that an infinite amount of energy must be spent in order to bring the system at equilibrium at that temperature, i.e. it is not possible to reach thermal equilibrium at that temperature.

3.19 Consider a system made up of two identical fermionic particles which can occupy 4 different states. Enumerate all the possible choices for the occupation numbers of the single particle states. Assuming that the 4 states have the following energies: $E_1 = E_2 = 0$ and $E_3 = E_4 = \epsilon$ and that the system is in thermal equilibrium at a temperature T, compute the mean occupation number of one of the first two states as a function of temperature.

Answer: There are six different possible states for the whole system characterized by the following occupation numbers (n_1, n_2, n_3, n_4) for the single particle states: $(1, 1, 0, 0)$, $(1, 0, 1, 0)$, $(1, 0, 0, 1)$, $(0, 1, 1, 0)$, $(0, 1, 0, 1)$, $(0, 0, 1, 1)$. The corresponding energies are $0, \epsilon, \epsilon, \epsilon, \epsilon, 2\epsilon$. The mean occupation number of the first single particle state (i.e. $\langle n_1 \rangle$) is then given by averaging the value of n_1 over the 6 possible states weighted using the Canonical Distribution, i.e.

$$\langle n_1 \rangle = \frac{(1 + 2e^{-\beta\epsilon})}{(1 + 4e^{-\beta\epsilon} + e^{-2\beta\epsilon})}.$$

3.20 A system, characterized by 3 different single particle states, is filled with 4 identical bosons. Enumerate the possible states of the system specifying the corresponding occupation numbers. Discuss also the case of 4 identical fermions.

Answer: The possible states can be enumerated by indicating all possible choices for the occupation numbers n_1, n_2, n_3 satisfying $n_1 + n_2 + n_3 = 4$. That leads to 15 different states. In the case of fermions, since $n_i = 0, 1$, the constraint on the total number of particles cannot be satisfied and there is actually no possible state for this system.

3.21 A system, characterized by two single particle states of energy $E_1 = 0$ and $E_2 = \epsilon$, is filled with 4 identical bosons. Enumerate all possible choices for the occupation numbers. Assuming that the system is in thermal contact with a reservoir at temperature T and that $e^{-\beta\epsilon} = \frac{1}{2}$, compute the probability of all particles being in the ground state. Compare the answer with that for distinguishable particles.

Answer: Since the occupation numbers must satisfy $N_1 + N_2 = 4$, the possible states are identified by the value of, for instance, N_2, in the case of bosons ($N_2 = 0, 1, 2, 3, 4$), and have energy ϵN_2. In the case of distinguishable particles there are instead $4!/(N_2!(4 - N_2)!)$ different states for each value of N_2. The probability of all the particles being in the ground state is $16/31$ in the first case and $(2/3)^4$ in the second case: notice that this probability is highly enhanced in the case of bosons.

3.22 Consider a gas of electrons at zero temperature. What is the density at which relativistic effects show up? Specify the answer by finding the density for which electrons occupy states corresponding to velocities $v = \sqrt{3}c/2$.

Answer: At $T = 0$ electrons occupy all levels below the Fermi energy E_F, or equivalently below the corresponding Fermi momentum p_F. To answer the question we must impose that

$$p_F = \frac{m_e v}{\sqrt{1 - v^2/c^2}} = \sqrt{3}\, m_e c.$$

On the other hand, the number of states below the Fermi momentum, assuming the gas is contained in a cubic box of size L, is

$$N = \frac{p_F^3 L^3}{3\hbar^3 \pi^2},$$

hence $\rho = p_F^3/(3\pi^2\hbar^3) \simeq 3.04 \times 10^{36}$ particles/m^3.

3.23 The density of states as a function of energy in the case of free electrons is given in (3.64). However in a conduction band the distribution may have a different dependence on energy. Let us consider for instance the simple case in which the density is constant, $dn_E/dE = \gamma V$, where $\gamma = 8 \times 10^{47}$ m^{-3} J^{-1}, the energy varies from zero to $E_0 = 1$ eV and the electronic density is $\rho \equiv \bar{n}/V = 6 \times 10^{28}$ m^{-3}. For T not much greater than room temperature it is possible to assume that the bands above the conduction one are completely free, while those below are completely occupied, hence the thermal properties can be studied solely on the basis of its conduction band. Under these assumptions, compute how the chemical potential μ depends on temperature.

Answer: The average total number of particles comes out to be $\bar{N} = \int_0^{E_0} n(\epsilon)g(\epsilon)d\epsilon$. The density of levels is $g(\epsilon) = dn_E/dE = \gamma V$ and the average occupation number is $n(\epsilon) = 1/(e^{\beta(\epsilon-\mu)} + 1)$. After computing the integral and solving for μ we obtain

$$\mu = kT \ln\left(\frac{e^{\rho/(\gamma kT)} - 1}{1 - e^{\rho/(\gamma kT)}e^{-E_0/kT}}\right).$$

It can be verified that, since by assumption $\rho/\gamma < E_0$, in the limit $T \to 0$ μ is equal to the Fermi energy $E_F = \rho/\gamma$. Instead, in the opposite large temperature limit, $\mu \to -kT \ln(1 - \gamma E_0/\rho)$, hence the distribution of electrons in energy would be constant over the band and simply given by $n(\epsilon)g(\epsilon) = \rho V/E_0$, but of course in this limit we cannot neglect the presence of other bands. Notice also that in this case, due to the different distribution of levels in energy, we have $\mu \to +\infty$, instead of $\mu \to -\infty$, as $T \to \infty$.

3.24 The modern theory of cosmogenesis suggests that cosmic space contains about 10^8 neutrinos per cubic meter and for each species of these particles. Neutrinos can be considered, in a first approximation, as massless fermions having a single spin state instead of two, as for electrons; they belong to 6 different species. Assuming that each species be independent of the others, compute the corresponding Fermi energy.

Answer: Considering a gas of neutrinos placed in a cubic box of size L, the number of single particle states with energy below the Fermi energy E_F is given, for massless particles, by $N_{E_F} = (\pi/6)E_F^3 L^3/(\pi\hbar c)^3$. Putting that equal to the average number of particles in the box, $\bar{N} = \rho L^3$, we have $E_F = \hbar c \, (6\pi^2 \rho)^{1/3} \simeq 3.38 \times 10^{-4}$ eV.

3.25 Suppose now that neutrinos must be described as particles of mass $m_\nu \neq 0$. Consider again Problem 3.24 and give the exact relativistic formula expressing the Fermi energy in terms of the gas density ρ.

Answer: The formula expressing the total number of particles $N = L^3\rho$ in terms of the Fermi momentum p_F is:

$$N_{E_F} = (\pi/6)p_F^3 L^3/(\pi\hbar)^3,$$

hence

$$E_F = \sqrt{m_\nu^2 c^4 + p_F^2 c^2} = \sqrt{m_\nu^2 c^4 + (6\pi^2\rho)^{2/3}(\hbar c)^2}.$$

3.26 Compute the internal energy and the pressure at zero temperature for the system described in Problem 3.24, i.e. for a gas of massless fermions with a single spin state and a density $\rho \equiv \bar{n}V = 10^8$ m^{-3}.

Answer: The density of internal energy is

$$U/V = (81\pi^2\rho^4/32)^{\frac{1}{3}}\hbar c \simeq 4.29 \times 10^{-15} \text{ J/m}^3,$$

and the pressure

$$P = U/3V \simeq 1.43 \times 10^{-15} \text{ Pa}.$$

Notice that last result is different from what obtained for electrons, Eq. (3.79): the factor 1/3 in place of 2/3 is a direct consequence of the linear dependence of energy on momentum taking place for massless or ultrarelativistic particles, in contrast with the quadratic behavior which is valid for (massive) non-relativistic particles.

3.27 10^3 bosons move in a harmonic potential corresponding to a frequency ν such that $h\nu = 1$ eV. Considering that the mean occupation number of the m-th level of the oscillator is given by the Bose–Einstein distribution: $n_m = (e^{\beta(hm\nu - \mu)} - 1)^{-1}$, compute the chemical potential assuming $T = 300\,°K\cdot$

Answer: The total number of particles, $N = 1000$, can be written as

$$N = \sum_m n_m = \frac{1}{e^{-\beta\mu} - 1} + \frac{1}{Ke^{-\beta\mu} - 1} + \frac{1}{K^2 e^{-\beta\mu} - 1} + \cdots$$

where $K = \exp(h\nu/kT) \simeq e^{40}$. Since K is very large and $\exp(-\beta\mu) > 1$ ($\mu \le 0$ for bosons), it is clear that only the first term is appreciably different from zero. Hence

$$\exp(-\beta\mu) = 1 + \frac{1}{N}$$

and finally $\mu \simeq -2.5 \times 10^{-5}$ eV.

3.28 Consider a system of $\bar{n} \gg 1$ spin-less non-interacting bosons. Each boson has two stationary states, the first state with null energy, the second one with energy ϵ. If μ is the chemical potential of the system, $\exp(\beta\mu) = f$ is its fugacity and one has $f \le 1$. Compute $z = f^{-1}$ as a function of the temperature $T = 1/(\beta k)$ (in fact z is a function of $\beta\epsilon$). In particular identify the range of values of z when T varies from 0 to ∞.

Answer: It is convenient to study $\zeta \equiv z \exp(\beta\epsilon/2)$ instead of z. ζ must diverge in the $T \to 0$ limit since $z \ge 1$. The condition that the sum of state occupation numbers be equal to $\bar{n}$ gives the equation:

$$\zeta^2 - 2\cosh(\beta\epsilon/2)(1 + 1/\bar{n})\zeta + 2/\bar{n} + 1 = 0,$$

whose solutions are

$$\zeta_\pm = \cosh(\beta\epsilon/2)(1 + 1/\bar{n}) \pm \sqrt{\cosh^2(\beta\epsilon/2)(1 + 1/\bar{n})^2 - 2/\bar{n} - 1}.$$

One must choose ζ_+ since ζ_- vanishes in the $T \to 0$ limit. Then one has in the $T \to 0$ limit $\zeta \to 2\cosh(\beta\epsilon/2)(1 + 1/\bar{n}) \to \exp(\beta\epsilon/2)(1 + 1/\bar{n})$ and hence the average occupation number of the zero energy state $\bar{n}_0 = 1/(\zeta \exp(-\beta\epsilon/2) - 1) \to \bar{n}$. In the $T \to \infty$ limit one has $\zeta \to 1 + 2/\bar{n}$ and hence $\bar{n}_0 \to 2/\bar{n}$. Therefore z ranges from $1 + 1/\bar{n}$ to $1 + 2/\bar{n}$ and the chemical potential ranges approximately from $-kT/\bar{n}$, for T small, to $-2kT/\bar{n}$ for T large. There is no Bose condensation.

3.29 A system is made up of N identical bosonic particles of mass m moving in a one-dimensional harmonic potential $V(x) = m\omega^2 x^2/2$. What is the distribution of occupation numbers corresponding to the ground state of the system? And that corresponding to the first excited state? Determine the energy of both states.

If $\hbar\omega = 0.1$ eV and if the system is in thermal equilibrium at room temperature, $T = 300\,°$K, what is the ratio R of the probability of the system being in the first excited state to that of the system being in the ground state? How the last answer changes in the case of distinguishable particles?

Answer: In the ground state all particles occupy the single particle state of lowest energy $\hbar\omega/2$, hence $E = N\hbar\omega/2$, while in the first excited state one of the N particles has energy $3\hbar\omega/2$, hence $E = (N+2)\hbar\omega/2$. The ground state has degeneracy 1 both for identical and distinguishable particles. The first excited state has denegeracy 1 in the case of bosons while the degeneracy is N in the other case, since it makes sense to ask which of the N particles has energy $3\hbar\omega/2$. Therefore $R = e^{-\hbar\omega/kT} \simeq 0.021$ for bosons and $R \simeq N\,0.021$ in the second case. For large N the probability of the system being excited is much suppressed in the case of bosons with respect to the case of distinguishable particles.

3.30 Consider again Problem 3.29 in the case of fermions having a single spin state and for $\hbar\omega = 1$ eV and $T \simeq 1000\,°$K.

Answer: In the ground state of the system the first N levels of the harmonic oscillator are occupied, hence its energy is $E_0 = \sum_{i=0}^{N-1}(n + 1/2)\hbar\omega = (N^2/2)\hbar\omega$. The minimum possible excitation of this state corresponds to moving the fermion of highest energy up to the next free level, hence the energy of the first excited state is $E_1 = E_0 + \hbar\omega$. The ratio R is equal to $e^{-\hbar\omega/kT} = 9.12 \times 10^{-6}$.

3.31 A system is made up of $N = 10^8$ electrons which are free to move along a conducting cable of length $L = 1$ cm, which can be roughly described as a one-dimensional segment with reflecting endpoints. Compute the Fermi energy of the system, taking also into account the spin degree of freedom.

Answer: $E_F = \hbar^2 N^2 \pi^2/(8mL^2) = 1.5 \times 10^{-18}$ J.

3.32 Let us consider a system made up of two non-interacting particles at thermal equilibrium at temperature T. Both particles can be found in a set of single particle energy levels ϵ_n, where n is a non-negative integer. Compute the partition function of the system, expressing it in terms of the partition function $Z_1(T)$ of a single particle occupying the same energy levels, for the following three cases: distinguishable particles, identical bosonic particles, identical fermionic particles.

Answer: If the particles are distinguishable, all states are enumerated by specifying the energy level occupied by each particle, hence we can sum over the energy levels of the two particles independently:

$$Z(T) = \sum_n \sum_m e^{-\beta(\epsilon_n + \epsilon_m)} = \left(\sum_n e^{-\beta\epsilon_n}\right)^2 = (Z_1(T))^2.$$

Instead, in case of two bosons, states corresponding to a particle exchange must be counted only once, hence we must treat separately states where the particles are in the same level or not

$$Z(T) = \frac{1}{2}\sum_{n\neq m} e^{-\beta(\epsilon_n+\epsilon_m)} + \sum_n e^{-\beta(\epsilon_n+\epsilon_n)}$$

$$= \frac{1}{2}\sum_{n,m} e^{-\beta(\epsilon_n+\epsilon_m)} - \frac{1}{2}\sum_n e^{-2\beta\epsilon_n} + \sum_n e^{-2\beta\epsilon_n} = \frac{1}{2}\left((Z_1(T))^2 + Z_1(T/2)\right).$$

If the particles are fermions, we must count only states where the particles are in different energy levels

$$Z(T) = \frac{1}{2}\sum_{n\neq m} e^{-\beta(\epsilon_n+\epsilon_m)}$$

$$= \frac{1}{2}\sum_{n,m} e^{-\beta(\epsilon_n+\epsilon_m)} - \frac{1}{2}\sum_n e^{-2\beta\epsilon_n} = \frac{1}{2}\left((Z_1(T))^2 - Z_1(T/2)\right).$$

3.33 A particle of mass $m = 9 \times 10^{-31}$ is placed at thermal equilibrium, at temperature $T = 10^3$ °K, in a potential which can be described as a distribution of spherical wells. Each spherical well has a negligible radius, a single bound state of energy $E_0 = -1$ eV, and the density of spherical wells is $\rho = 10^{24}$ m^{-1}. Assuming that the spectrum of unbound states is unchanged with respect to the free particle case, determine the probability of finding the particle in a "ionized" (i.e., unbound) state.

Answer: Let us discuss at first the case of a single spherical well at the center of a cubic box of volume $V = L^3$. The bound state of the well is not influenced by the walls of the box if $L \gg \hbar/\sqrt{2m|E_0|} \simeq 1.55 \times 10^{-10}$ m. At the given temperature the particle is non-relativistic. The density of free energy levels in the cubic box is $dn_E/dE = V\sqrt{E}(2m)^{3/2}/(4\pi^2\hbar^3)$. Taking into account also the bound state, the partition function is

$$Z = e^{-\beta E_0} + V\frac{(2m)^{3/2}}{4\pi^2\hbar^3}\int_0^\infty dE\,\sqrt{E}e^{-\beta E} = e^{-\beta E_0} + V\frac{(2mkT)^{3/2}}{4\pi^2\hbar^3}\frac{\sqrt{\pi}}{2} = e^{-\beta E_0} + \frac{V}{\lambda_T^3}$$

where $\lambda_T \equiv \sqrt{2\pi\hbar^2/mkT}$ is the de Broglie thermal wavelength of the particle, i.e., its typical wavelength at thermal equilibrium. The probability for the particle being in the bound state is $P_b = e^{-\beta E_0}/(e^{-\beta E_0} + V/\lambda_T^3)$ and it is apparent that, for any given T, $\lim_{V\to\infty} P_b = 0$, i.e., the particle stays mostly in a "ionized" state if the box is large enough. At very low temperatures that may seem strange, since it may be extremely unlikely to provide enough energy to unbind the particle simply by thermal fluctuations; however, once the particle is free, it escapes with an even smaller probability of getting back to the well, if the box is large enough.

In the present case, however, there is a finite density of spherical wells: that is equivalent to considering a single well in a box of volume $V = 1/\rho$. Therefore $P_b = \left(1 + e^{-\beta|E_0|}/(\rho\lambda_T^3)\right)^{-1}$, while the probability for being in an unbound state is $P_{free} = (1 + e^{\beta|E_0|}\rho\lambda_T^3)^{-1} \simeq e^{-\beta|E_0|}(\rho\lambda_T)^{-3} = 2.75 \times 10^{-2}$.

3.34 A gas of monoatomic hydrogen is placed at thermal equilibrium at temperature $T = 2 \times 10^3$ °K. Assuming that the density is low enough to neglect atom-atom interactions, estimate the rate of atoms which can be found in the first excited energy level (principal quantum number $n = 2$).

Answer: According to Gibbs distribution and to the degeneracy of the energy levels of the hydrogen atom, the ratio of the probabilities for a single atom to be in the level with $n = n_2$ or in that with $n = n_1$ is

$$P_{n_2,n_1} = \frac{n_2^2}{n_1^2} \exp\left(-\frac{E_0}{kT}\left(\frac{1}{n_2^2} - \frac{1}{n_1^2}\right)\right)$$

where E_0 is the energy of the ground state, $E_0 = -me^4/(32\pi^2\epsilon_0^2\hbar^2) \simeq -13.6$ eV. Since $kT \simeq 0.1724$ eV, we get $P_{2,1} \simeq 0.8 \times 10^{-24}$. Since in those conditions most atoms will be found in the ground state with $n = 1$, this number can be taken as a good estimate for the rate of atoms in the level with $n = 2$.

The above answer is correct in practice, but needs some further considerations. Indeed, if we try to compute the exact hydrogen atom partition function we find a divergent behaviour even when summing only over bound states: energy levels E_n become denser and denser towards zero energy, where they have an equal Boltzmann weight and become infinite in number: one would conclude that the probability of finding an atom in the ground state is zero at every temperature. However, as n increases, also the atom radius r_n increases: in this situation the atom, which is called a Rydberg atom, interacts strongly with the surrounding black body radiation. Therefore the infinite number of highly excited states should not be taken into account since these strongly interacting states have very short lifetimes. One should however take into account ionized states, in which the electron is free to move far away from the binding proton. Concerning the statistical weight of these states, we can make reference to Problem (3.33): following a similar argument we realize that ionized states become statistically relevant, with respect to the ground state, when the density of atoms is of the order of $e^{-\beta|E_0|}/\lambda_T^3 = e^{-\beta|E_0|}\left(mkT/(2\pi\hbar^2)\right)^{3/2} \sim e^{-\beta|E_0|}10^{28}$ m$^{-3} \sim 10^{-6}$ m^{-3}.

3.35 Consider a homogeneous gas of non-interacting, non-relativistic bosons, which are constrained to move freely on a plane surface. Compute the relation linking the density ρ, the temperature T and the chemical potential μ of the system. Does Bose-Einstein condensation occur? Does the answer to the last question change if an isotropic harmonic potential acts in the directions parallel to the surface?

Answer: An easy computation shows that the density of states for free particles in two dimensions is independent of the energy and given by

$$\frac{dn_E}{dE} = \frac{mA}{2\pi\hbar^2}$$

where m is the particle mass and A is the total area of the surface. According to Bose-Einstein distribution, the total particle density is then given by:

$$\rho = \frac{\overline{N}}{A} = \frac{m}{2\pi\hbar^2} \int_0^\infty dE \, \frac{1}{e^{(E-\mu)/(kT)} - 1} = -\frac{kTm}{2\pi\hbar^2} \ln\left(1 - e^{\mu/(kT)}\right).$$

The chemical potential turns out to be negative, as expected. It is interesting to consider the limit of low densities, in which the typical inter-particle distance is much greater than the typical thermal de Broglie wavelength ($\rho^{-1/2} \gg h/\sqrt{mkT}$) and

$$\mu \simeq kT \ln\left(\frac{2\pi\hbar^2\rho}{mkT}\right),$$

while in the opposite case of high densities one obtains

$$\mu = -kT \exp\left(-\frac{2\pi\hbar^2\rho}{mkT}\right).$$

At variance with the three-dimensional case, ρ diverges as $\mu \to 0$, meaning that the system can account for an arbitrarily large number of particles, with no need for a macroscopic number of particles in the ground state, therefore Bose-Einstein condensation does not occur in two dimensions, due to the different density of states.

If the system is placed in a two-dimensional isotropic harmonic potential of angular frequency ω, the energy levels are $E_n = \hbar\omega(n+1)$, with degeneracy $n+1$, so that the total number of energy levels found below a given energy E is $\sim E^2/(2\hbar^2\omega^2)$ and the density of states becomes $dn_E/dE \simeq E/(\hbar^2\omega^2)$. The average number of particles (the system is not homogeneous and we cannot define a density) is

$$\overline{N} = \frac{1}{\hbar^2\omega^2} \int_0^\infty dE \frac{E}{e^{(E-\mu)/(kT)} - 1} = \frac{k^2T^2}{\hbar^2\omega^2} \int_{-\mu/(kT)}^\infty dx \, \frac{x + \mu/(kT)}{e^x - 1}.$$

In this case the integral in the last member stays finite even in the limit $\mu \to 0$, meaning that condensation in the ground state is necessary to allow for an arbitrary average number of particles $\overline{N}$. The condensation temperature is $T_c \sim \sqrt{\overline{N}}\hbar\omega/k$ and goes to zero, at fixed $\overline{N}$, as $\omega \to 0$ (i.e., going back to the free case).

3.36 A system of massless particles at thermal equilibrium is characterized by the known equation of state $U/V - 3P = 0$, linking the pressure P to the density of internal energy U/V. For a homogeneous system of spinless, non-interacting and distinguishable particles of density ρ, placed at thermal equilibrium at temperature T, compute the lowest order violation to the above relation due to a non-zero particle mass m.

Answer: For one particle in a cubic box of volume $V = L^3$, energy levels are written as $E = \sqrt{p^2c^2 + m^2c^4} = \sqrt{m^2c^4 + (\pi^2\hbar^2c^2/L^2)(n_x^2 + n_y^2 + n_z^2)}$, with n_x,

n_y and n_z positive integers. The number of energy levels below a given threshold $\bar{E} = \sqrt{\bar{p}^2 c^2 + m^2 c^4}$ is given by $\bar{p}^3 V/(3\pi^2 \hbar^3)$, from which one obtains the density of states $dn_E/dE = VpE/(c^2 \hbar^2 \hbar^3)$. The derivative of one energy level with respect to the volume is $\partial E/\partial V = p^2 c^2/(3VE)$.

The internal energy and the pressure are given, for a single particle, by

$$U = \frac{\int_{mc^2}^{\infty} dE (dn_E/dE) e^{-\beta E} E}{\int_{mc^2}^{\infty} dE (dn_E/dE) e^{-\beta E}} \; ; \qquad P = \frac{\int_{mc^2}^{\infty} dE (dn_E/dE) e^{-\beta E} (\partial E/\partial V)}{\int_{mc^2}^{\infty} dE (dn_E/dE) e^{-\beta E}}$$

so that

$$U - 3PV = \frac{\int_{mc^2}^{\infty} dE (dn_E/dE) e^{-\beta E} (E - p^2 c^2/E)}{\int_{mc^2}^{\infty} dE (dn_E/dE) e^{-\beta E}} = \frac{m^2 c^4 \int_{mc^2}^{\infty} dE (dn_E/dE) e^{-\beta E}/E}{\int_{mc^2}^{\infty} dE (dn_E/dE) e^{-\beta E}}.$$

To keep the lowest order in m it is sufficient to evaluate the integrals in the last expression in the limit $m = 0$. Finally, multiplying for the total number of particles in the box, $N = \rho V$, we get

$$\frac{U}{V} - 3P = \frac{\rho\, m^2 c^4}{2kT}.$$

This is the first term of an expansion in terms of the parameter mc^2/kT.

3.37 Consider a rarefied gas of particles of mass m in equilibrium at temperature T. The probability distribution of particle velocities is given by the Gaussian Maxwell-Boltzmann (MB) formula

$$p(v)\, d^3 v = \left(\frac{m}{2\pi kT}\right)^{3/2} e^{-mv^2/(2kT)} d^3 v.$$

Considering a pair of particles in the gas, labelled by 1 and 2, compute the distribution of the relative velocities $v_R \equiv v_1 - v_2$, and that of the velocity of their center of mass $v_B \equiv (v_1 + v_2)/2$.

Answer: The particles in the chosen pair behave as independent systems, hence the probability density in the six-dimensional (v_1, v_2) space is just the product of the two densities:

$$p(v_1, v_2) d^3 v_1 d^3 v_2 = \left(\frac{m}{2\pi kT}\right)^3 e^{-m(v_1^2 + v_2^2)/(2kT)} d^3 v_1 d^3 v_2\,.$$

If we change variables passing from (v_1, v_2) to (v_R, v_B), the new probability density is given by

$$\tilde{p}(v_R, v_B) d^3 v_R d^3 v_B = p(v_1, v_2) d^3 v_1 d^3 v_2$$

hence $\tilde{p}(v_R, v_B) = J(v_R, v_B) p(v_1, v_2)$ where J is the Jacobian matrix, that is, the absolute value of the determinant of the six-dimensional matrix whose elements are

$\partial v_a^i / \partial v_S^j$ where a is either 1 or 2 and S is either R or B and i, j label the Cartesian components. Furthermore $p(\boldsymbol{v}_1, \boldsymbol{v}_2)$ must be considered a function of $(\boldsymbol{v}_R, \boldsymbol{v}_B)$. It is soon verified that $J = 1$ and that $v_1^2 + v_2^2 = (1/2)v_R^2 + 2v_B^2$. Therefore we have

$$\tilde{p}(\boldsymbol{v}_R, \boldsymbol{v}_B)d^3v_R d^3 v_B = \left(\frac{\mu}{2\pi kT}\right)^{3/2} e^{-\mu v_R^2/(2kT)} \left(\frac{M}{2\pi kT}\right)^{3/2} e^{-M v_B^2/(2kT)}$$

where $\mu = m/2$ is the reduced mass of the pair of particles and $M = 2m$ is their total mass.

Then we see that our result corresponds to the product of the MB distribution of a single particle of mass μ and velocity $\boldsymbol{v}_R$ and that of a single particle of mass M and velocity $\boldsymbol{v}_B$. If we search for the $\boldsymbol{v}_R$ distribution of the pair, we must integrate over $\boldsymbol{v}_B$ and we get the first MB distribution, corresponding to the mass μ. If instead we search for the $\boldsymbol{v}_B$ distribution of the pair, we must integrate over $\boldsymbol{v}_R$ and we get the second MB distribution, corresponding to the mass M. It can be easily verified that the result extends unchanged to the case in which the two particles have different masses m_1 and m_2, defining as usual $M = m_1 + m_2$, $\mu = m_1 m_2/M$, $\boldsymbol{v}_B = (m_1 \boldsymbol{v}_1 + m_2 \boldsymbol{v}_2)/M$, $\boldsymbol{v}_R = \boldsymbol{v}_1 - \boldsymbol{v}_2$.

3.38 We have a rarefied gas of electrons for which $mc^2 = 0.511$ MeV, in thermal equilibrium at $kT = 5 \times 10^3$ eV. Since $kT/mc^2 \ll 1$, the rarefied gas in non-relativistic. Therefore the particle momentum $\boldsymbol{p}$ distribution is given by the Maxwell-Boltzmann formula

$$\frac{d\mathcal{P}}{d\boldsymbol{p}} = \left(\frac{1}{2\pi mkT}\right)^{3/2} e^{-p^2/(2mkT)} .$$

The electron gas is mixed with a rarefied positron gas in thermal equilibrium at the same temperature. Positrons are anti-particles of electrons, therefore a positron-electron collision can produce an annihilation into two photons and hence the two gas mixture is a photon source. We would like to compute the photon energy (frequency) distribution.

Answer: If a non-relativistic positron annihilates with a non-relativistic electron and their relative momentum is $\boldsymbol{p}$ while the center of mass one is $\boldsymbol{P}$, forgetting relativistic corrections, we find that the energy of a produced photon is $E = (mc^2 + p^2/(2m))(1 + P_z/Mc)$ where we have chosen the z-axis parallel to the momentum of the photon and the second factor is the transformation factor from the center of mass to the laboratory frame, that is, it accounts for the non-relativistic Doppler effect. As shown in the preceding problem, considering a generic electron-positron pair, one has the center of mass $\boldsymbol{P}$ and relative momentum $\boldsymbol{p}$ distribution $d^6\mathcal{P}/(d\boldsymbol{P}d\boldsymbol{p}) = 1/(2\pi mkT)^3 \exp(-p^2/(mkT)) \exp(-P^2/(4mkT))$. Changing the P_z variable into E and integrating over P_x and P_y we get (the only non-trivial element of the Jacobian matrix is $\partial P_z/\partial E$):

$$\frac{d^4\mathcal{P}}{d\boldsymbol{p}dE} = \frac{2m^2 c}{(2m^2 c^2 + p^2)(\pi mkT)^2} e^{-p^2/(mkT)} \exp\left(\frac{-4m^3 c^2 (E - mc^2 - p^2/(2m))^2}{(2m^2 c^2 + p^2)^2 kT}\right)$$

which should be integrated over p. The Gaussian factor in p corresponds to a mean square value $\bar{p}^2 = 3mkT/2$ and the standard deviation $\Delta^2_{p^2} = 3(mkT)^2/2$. Since in the above distribution, the Gaussian factor put apart, p^2 appears only in the linear combination $mc^2 + p^2/(2m)$ and since $mc^2 = 0.511$ MeV $\gg 3kT/4$ we can replace the linear combination with $mc^2 + 3kT/4$ getting the final photon energy distribution:

$$\frac{dP}{dE} = \frac{1}{\sqrt{\pi mc^2 kT}(1 + 3kT/(4mc^2))} \exp\left[-\frac{(E - mc^2 - 3kT/4)^2}{mc^2 kT(1 + 3kT/(4mc^2))^2} \right].$$

Notice that, however exponentially decaying, the distribution does not vanish for negative E values, this is due to the fact that for these values the non relativistic Doppler formula does not apply. Thus negative E values should not be considered. Notice also that, while the most probable E value is shifted by a factor $3kT/(4mc^2) \sim 10^{-2}$, the width of the distribution is proportional to $\sqrt{mc^2 kT}$, allowing for fluctuations in E of the order of $\sqrt{kT/mc^2} \sim 10\%$, i.e., much larger than the shift. The physical origin of the shift is mostly in the fact that the annihilating pair has an average energy, in the center of mass, which is larger than $2mc^2$, due to thermal motion, by an amount $\sim p^2/m \sim kT$. The broad distribution of energy is instead mostly due to Doppler effect, due to the transformation from the center of mass to the laboratory frame (see also Problem 1.31), and the broadening is proportional to the typical center of mass velocity, which is of the order of $c\sqrt{kT/mc^2}$.

3.39 Compute the magnetic susceptibility of a salt whose magnetic ions (e.g., F_e^{3+}) have molecular number density ρ, magnetic moment $2\mu_B \boldsymbol{J}/\hbar$ where $J = \hbar/2$, $\mu_B = e\hbar\mu_0/(2m)$ is Bohr's magneton and μ_0 is the magnetic permeability of the vacuum.

Answer: The magnetic interaction energy of a single ion is $V = -\boldsymbol{\mu}\cdot\boldsymbol{H}$. If we choose the z axis parallel to the magnetic field we get two interaction energy levels for each ion, the corresponding energies being $\pm\mu_B H$. Therefore, if $\mu_B H/(kT) \ll 1$, the average total magnetic moment per unit volume of the salt is parallel to the magnetic field, the salt being paramagnetic, and its absolute value is, according to Eq. (3.15)

$$M = \rho\mu_B \frac{e^{\mu_B H/(kT)} - e^{-\mu_B H/(kT)}}{e^{\mu_B H/(kT)} + e^{-\mu_B H/(kT)}} = \rho\mu_B \tanh(\mu_B H/kT) \simeq \rho\mu_B^2 H/(kT).$$

Therefore the magnetic susceptibility is $\chi(T, \rho) = \rho\mu_B^2/(kT)$.

3.40 Compute the thermodynamic potentials of a perfect gas whose molecules are diatomic and rigid, that is, they are made of two spinless atoms with masses m_1 and m_2 whose relative distance is fixed and equal to d, when the temperature T satisfies the condition $kT \gg \hbar^2(m_1 + m_2)/(2d^2 m_1 m_2) \equiv \hbar^2/(2m_r d^2) \equiv \hbar^2/(2I)$.

Answer: From its very definition it turns out that the partition function of an ideal gas whose molecules have a given spectrum of excited states is given by

$$Z_{id} = \left(\frac{(2\pi mkT)^{\frac{3}{2}} V}{h^3 N} \sum_{n=0}^{\infty} \nu_n e^{-\frac{E_n}{kT}} \right)^N \equiv \left(\frac{V}{N} \Phi(T) \right)^N$$

where ν_n is the degeneracy of the level with energy E_n and $m = m_1 + m_2$. Indeed the energy of a molecule with momentum p in the n-th excited state is $p^2/(2m) + E_n$. In the case of the rigid diatomic molecule one has $\nu_n = 2n + 1$ and $E_n = -B + \hbar^2 n(n+1)(m_1 + m_2)/(2d^2 m_1 m_2) \equiv -B + \hbar^2 n(n+1)/(2I)$, where $B > 0$ is the binding energy of the two atoms. It follows that the free energy is given by

$$F(T, V) = -NkT \left[\ln \Phi_d(T) + \ln \frac{V}{N} \right].$$

We must compute Φ_d, which, with the above condition for the temperature, is given by

$$\Phi_d(T) = e^{\frac{B}{kT}} \frac{(2\pi mkT)^{\frac{3}{2}}}{h^3} \sum_{n=0}^{\infty} (2n+1) e^{-\frac{\hbar^2 n(n+1)}{2IkT}} \simeq e^{\frac{B}{kT}} \frac{(2\pi mkT)^{\frac{3}{2}}}{h^3} \int_0^{\infty} dx \, e^{-\frac{\hbar^2 x}{2IkT}}$$

$$= e^{\frac{B}{kT}} \frac{(2\pi)^{\frac{7}{2}} m^{\frac{3}{2}} 2I (kT)^{\frac{5}{2}}}{h^5}$$

Therefore the internal energy of the rigid diatomic gas is

$$U(T) = kT^2 N \frac{d \ln \Phi_d(T)}{dT} = \frac{5}{2} NkT + NB,$$

it only depends on the temperature, as it happens for any perfect gas, and the specific heat is $c_v = \frac{5}{2} Nk$. We also have for the free enthalpy

$$G(T, P) = -NkT \left[\ln \Phi_d(T) + \ln \frac{kT}{P} - 1 \right].$$

A final remark is here in order. In the case of two identical atoms, like, e.g., the molecules H_2, N_2 and O_2, one has that only even/odd values of n are allowed, depending on the atomic statistics. As a consequence, the contribution $\Phi_d(T)$ from the angular momentum sum must be divided by two.

3.41 Compute the entropy of the van der Waals gas.

Answer: Using Eqs. (3.114) and (3.127), that we write as $S = k(\ln Z + T\partial_T \ln Z)$, we have

$$S = kN \left(\ln \frac{(2\pi mkT)^{\frac{3}{2}} (V - b)}{h^3 N} + \frac{CN}{2VkT} + \frac{3}{2} - \frac{CN}{2VkT} \right)$$

$$= kN \left(\ln \frac{(2\pi mkT)^{\frac{3}{2}} (V - b)}{h^3 N} + \frac{3}{2} \right).$$

3.42 Compute the equation of a reversible adiabatic transformation for the van der Waals gas.

Answer: Using the entropy computed in the former exercise and noticing that the entropy is constant along a reversible adiabatic transformation we find that its equation is $T^{\frac{3}{2}}(V-b) = constant$.

3.43 Two rigid vessels of volumes V_1 and V_2 contain respectively N_1 and N_2 molecules of two different ideal gases at the same temperature T. The vessels are connected by a tiny tube and a tap whose volumes can be disregarded. Opening the tap, both gases freely diffuse through the tube, until equilibrium is reached, without any temperature change, when the partial pressure of each gas in the two vessels equalize. There has been no heat exchange with the outside, because the gases are ideal and the molecules have kept their kinetic energy during the diffusion. The described transformation is irreversible. The total entropy of the system must have increased of a certain amount called *mixture entropy*. Compute it.

Answer: The initial entropy is given by the sum of the entropies of the two gases. According to Eq. (3.142) it is given by

$$S_I = kN_1\left(\frac{d(T\ln\Phi_1(T))}{dT} + \ln\frac{V_1}{N_1}\right) + kN_2\left(\frac{d(T\ln\Phi_2(T))}{dT} + \ln\frac{V_2}{N_2}\right).$$

The final state is a mixture of the two ideal gases at temperature T and volume V_1+V_2, then, according to (3.144), we have

$$S_F = kN_1\left(\frac{d(T\ln\Phi_1(T))}{dT} + \ln\frac{V_1+V_2}{N_1}\right) + kN_2\left(\frac{d(T\ln\Phi_2(T))}{dT} + \ln\frac{V_1+V_2}{N_2}\right)$$

and hence the difference is

$$S_F - S_I = kN_1\ln\frac{V_1+V_2}{V_1} + kN_2\ln\frac{V_1+V_2}{V_2},$$

which, of course, is positive.

3.44 Gas and liquid of the same substance are in equilibrium at temperature T and pressure P. The equilibrium condition is given by the identity between the specific free enthalpies G (the molar free enthalpies) of the two phases, that is the identity between the chemical potentials. Indeed, the free enthalpy of the whole system, which is given by $(N-\Delta)\mu_1(T,P) + \Delta\,\mu_2(T,P)$, must be stationary, that is Δ independent. This implies $\mu_1(T,P) = \mu_2(T,P)$. Translate this condition into a differential equation for the equilibrium curve in the $T-P$ plane (Clapeyron equation).

Answer: Equation (3.134), translated in terms of the specific thermodynamic functions, becomes $\mu_{g/l}(T,P) = h_{g/l}(T,P) - T\,s_{g/l}(T,P)$ where the indices g/l distinguish the two phases. The equilibrium equation is

$$h_g(T, P) - h_l(T, P) = T(s_g(T, P) - s_l(T, P)).$$

In differential form, from Eqs. (3.135)–(3.138), we have

$$d\mu_g(T, P) = -s_g(T, P)dT + v_g(T, P)dP = d\mu_l(T, P) = -s_l(T, P)dT + v_l(T, P)dP$$

therefore

$$\frac{dP}{dT} = \frac{s_g(T, P) - s_l(T, P)}{v_g(T, P) - v_l(T, P)} = \frac{h_g(T, P) - h_l(T, P)}{T(v_g(T, P) - v_l(T, P))}.$$

Now the difference of specific enthalpies coincides, by definition, with the *specific latent heat* ($\lambda \equiv h_g(T, P) - h_l(T, P)$) thus we have the Clapeyron equation

$$\frac{dP}{dT} = \frac{\lambda(T, P)}{T(v_g(T, P) - v_l(T, P))}.$$

Whenever the specific volume of the gas phase is much larger than that of the liquid and it can be approximated by $v_g \simeq RT/P$ we have

$$\frac{dP}{dT} = P\frac{\lambda(T, P)}{T^2 R}.$$

3.45 Discuss the equilibrium condition for the chemical dissociation of a rigid diatomic ideal gas into the monoatomic component gases. Consider a case analogous to the chemical reaction $O_2 \rightarrow 2O$, where the atoms correspond to the nuclear isotope with mass 16. Note that the result given in Problem 3.40 must be changed, since both the constituent atoms and the O_2 molecule have electronic spin one, this implies a factor $2S + 1 = 3$ multiplying $\Phi_{O_2}(T)$ and $\Phi_O(T)$. Furthermore $I = md^2/2$, where m is the atomic mass, and since the constituent atoms are identical, $\Phi_{O_2}(T)$ must be divided by two. The temperature T and the pressure P are assumed to be high and, respectively, low enough to justify the use of the ideal gas formulae.

Answer: Consider the Eqs. (3.146) and (3.147). Let the initial conditions correspond to N diatomic molecules, if Δ such molecules dissociate into pair of constituent atoms we have a partial pressure $P_d(\Delta) = P(N - \Delta)/(N + \Delta)$ for the diatomic component of the mixture, and $P_a(\Delta) = 2P\Delta/(N + \Delta)$ for the monoatomic ones, therefore the Gibbs potential of the gas mixture resulting from the dissociation is

$$G_{mix}(T, P, \Delta) = (N - \Delta)\mu_d(T, P_d) + 2\Delta\mu_a(T, P_a).$$

The equilibrium condition is $\partial G_{mix}(T, P, \Delta)/\partial \Delta = 0$. From Eqs. (3.135)–(3.138) and (3.140) we have for a mixture of gases $\partial \mu_i/\partial P_i = V/N_i$, therefore we have the equilibrium condition

$$2\mu_a(T, P_a) - \mu_d(T, P_d) + V\left(\frac{\partial P_d}{\partial \Delta} + \frac{\partial P_a}{\partial \Delta}\right) = 2\mu_a(T, P_a) - \mu_d(T, P_d) = 0.$$

If we introduce the *stoichiometric coefficients* $\nu_i = dN_i/d\Delta$ we have $\sum_i \nu_i \mu_i (T, P_i) = 0$ and hence (Eq. (3.148)) $\sum_i \nu_i (\ln P_i - \ln(kT\Phi_i(T)) + 1)$ from which we have

$$\prod_i P_i^{\nu_i} = \left(\frac{kT}{e}\right)^{\sum_i \nu_i} \prod_i \Phi_i^{\nu_i}(T).$$

Taking into account the results of Problem 3.40 and Eq. (3.143) we have

$$\Phi_a(T) = 3\left(\frac{2\pi mkT}{h^2}\right)^{\frac{3}{2}}, \quad \Phi_d(T) = e^{\frac{B}{kT}}\frac{3\pi d^2}{2}\left(\frac{2\pi mkT}{h^2}\right)^{\frac{5}{2}},$$

and hence we finally get

$$\frac{P_a^2}{P_d} = e^{-1}kT\frac{\Phi_a^2(T)}{\Phi_d(T)} = (kT)^{\frac{3}{2}}\left(\frac{2\pi m}{h^2}\right)^{\frac{1}{2}}\frac{6}{\pi d^2}e^{-(\frac{B}{kT}+1)} \equiv K_p^{-1}(T),$$

where K_p is called the (pressure) equilibrium constant of the dissociation reaction.

3.46 We have a sample of the paramagnetic salt considered in Problem 3.39 in the magnetic field H at temperature T_i. We switch off the magnetic field with a reversible adiabatic transformation. Compute the final temperature T_f of the sample assuming its entropy density for $H = 0$ equal to $s(T, \rho, 0) = 3k\rho\ln(T/T_0)$.
Answer: Using Eq. (3.159) and the magnetic susceptibility computed in Problem 3.39, we compute the initial entropy per unit volume of the salt in the magnetic field H

$$s(T_i, \rho, \boldsymbol{H}) = s(T_i, \rho, 0) + \frac{\partial\chi(T_i, \rho)}{\partial T}\frac{|\boldsymbol{H}|^2}{2} = 3k\rho\ln(T_i/T_0) - \frac{\rho\mu_B^2}{kT_i^2}\frac{|\boldsymbol{H}|^2}{2}.$$

At the end of the reversible adiabatic field switch off we have $s(T_i, \rho, \boldsymbol{H}) = s(T_f, \rho, 0)$, therefore we have

$$T_f = T_i \exp\left(-\frac{\mu_B^2|\boldsymbol{H}|^2}{6(kT_i)^2}\right) \simeq T_i - \frac{\mu_B^2|\boldsymbol{H}|^2}{6k^2T_i}.$$

It is apparent that we find a reduction of the temperature of the sample.

Index

A
Action, 27, 28
    minimum action principle, 27
Adiabatic theorem, 178
Alpha particle, 100
    emission, 100
Angular momentum, 75, 184
    Bohr's quantization rule, 75, 122, 218
    intrinsic, 81, 184, 185, 189, 195, 227
    quantization of, 122
    Sommerfeld's quantization rule, 218
Atom, 68, 169
    Bohr model, 72, 74
    energy levels, 88
    hydrogen, 72, 74
    in gases, 177, 179, 183, 184, 191, 194
    in solids, 77, 115, 167, 170, 174–176
    Rutherford model, 72, 76
    Thomson model, 68
Average, 85, 91, 168, 169
    energy, 168, 173, 197
    ensemble, 168
    number, 186
    operator value, 87
    time, 168

B
Balmer series, 73, 74
Band, 119, 120
    conducting, 176, 193
    Kronig-Penney model, 116, 119
    spectra, 115, 119
Bessel spherical functions, 129

Black body radiation, v, 197
    Planck formula, 198
    Rayleigh-Jeans formula, 197
Bloch waves, 119
Bohr model, 72, 74
    correspondence principle, 74
    radii, 74
Boltzmann, 167, 192
    constant, viii, 76, 88, 175
Bose-Einstein
    condensation, 196
    distribution, 195
Bosons, 184, 185, 194, 195, 198
Bound state, 103, 106
Bragg's law, 77

C
Center of mass
    frame, 32–34
Chemical potential, 170, 187, 207
Classical physics, v, 73
Compton effect, 60, 61
Compton wavelength, 61
Conduction band, 193
Conductor, 119, 175
Conservation law, 29, 30, 79, 170
    of electric charge, 79
    of energy, 30, 33, 167, 174
    of momentum, 29–31, 33
    of particle number, 170
    of probability, 79, 114
Constant
    Boltzmann's, viii, 175

Printed in the United States
By Bookmasters